Digital Electronics

for MSc (Physics) and BE (Electronics)

Digital Electronics

for MSc (Physics) and BE (Electronics)

AK Saxena MSc MTech PhD

Associate Professor
Department of Physics
APS University
Rewa (MP)

CBS Publishers & Distributors Pvt Ltd

New Delhi • Bengaluru • Chennai • Kochi • Kolkata • Mumbai
Bhopal • Bhubaneswar • Hyderabad • Jharkhand • Nagpur • Patna • Pune • Uttarakhand • Dhaka (Bangladesh)

Digital
Electronics

ISBN: 978-81-239-2374-1

First Edition: 2014
Reprint: 2019

Published by Satish Kumar Jain and produced by Varun Jain for

CBS Publishers & Distributors Pvt Ltd
4819/XI Prahlad Street, 24 Ansari Road, Daryaganj, New Delhi 110 002, India.
Ph: 23289259, 23266861, 23266867   Website: www.cbspd.com
Fax: 011-23243014   e-mail: delhi@cbspd.com; cbspubs@airtelmail.in.

Corporate Office: 204 FIE, Industrial Area, Patparganj, Delhi 110 092
Ph: 4934 4934   Fax: 4934 4935   e-mail: publishing@cbspd.com; publicity@cbspd.com

Branches

- **Bengaluru:** Seema House 2975, 17th Cross, K.R. Road,
  Banasankari 2nd Stage, Bengaluru 560 070, Karnataka
  Ph: +91-80-26771678/79   Fax: +91-80-26771680   e-mail: bangalore@cbspd.com
- **Chennai:** 7, Subbaraya Street, Shenoy Nagar, Chennai 600 030, Tamil Nadu
  Ph: +91-44-26680620, 26681266   Fax: +91-44-42032115   e-mail: chennai@cbspd.com
- **Kochi:** 42/1325, 1326, Power House Road, Opp KSEB Power House, Ernakulam 682 018, Kochi, Kerala
  Ph: +91-484-4059061-65   Fax: +91-484-4059065   e-mail: kochi@cbspd.com
- **Kolkata:** 6/B, Ground Floor, Rameswar Shaw Road, Kolkata-700 014, West Bengal
  Ph: +91-33-22891126, 22891127, 22891128   e-mail: kolkata@cbspd.com
- **Mumbai:** 83-C, Dr E Moses Road, Worli, Mumbai-400018, Maharashtra
  Ph: +91-22-24902340/41   Fax: +91-22-24902342   e-mail: mumbai@cbspd.com

Representatives

Bhopal 0-8319310552	**Bhubaneswar** 0-9911037372	**Hyderabad**	0-9885175004
Jharkhand 0-9811541605	**Nagpur** 0-9021734563	**Patna**	0-9334159340
Pune 0-9623451994	**Uttarakhand** 0-9716462459	**Dhaka (Bangladesh)**	01912-003485

Printed at India Binding House, Noida (UP), India

Preface

The present book is aimed to serve as a textbook for MSc (physics) and engineering students of all the Indian colleges and Universities. The book has been divided into twelve chapters. The first chapter begins with introductory concepts of digital electronics. This is followed by number systems in Chapter 2. Chapter 3 describes various useful binary codes. Chapter 4 is on Boolean algebra and logic circuits based on various logic gates. Chapter 5 describes various logic families. Chapter 6 discusses combinational and arithmetic logic circuits such as adders and subtractors, comparator and parity generator and checker. Chapter 7 explains various types of flip-flops. Chapter 8 describes shift registers and their working. Chapter 9 deals with asynchronous and synchronous counters and their applications. Chapter 10 is on data processing circuits which include decoders, demultiplexers, multiplexers, encoders, read only memory (ROM), code converters, various types of ROMs, random access memory (RAM), programmable array logic and parity checker and generator. Chapter 11 explains analog to digital and digital to analog converters. Chapter 12 describes 8085 and 8086 microprocessors. The chronology of the contents has been so arranged as to render the readers an easy grasping of the subject.

I am grateful to Prof ON Srivastava (Emeritus Professor, BHU, Varanasi) and Prof DP Tiwari (Head, Department of Physics, APS University, Rewa) for boosting my morale. I am also thankful to Prof SP Agrawal and Prof SK Nigam (Ex-VCs and Heads, Department of Physics, APS University) Prof AP Mishra and Prof SL Agrawal, Dr PK Rai (Computer Centre), Prof Navita Shrivastav and Prof RK Katare (Computer Science Department), Dr CM Tiwari and Dr VK Mishra (lecturers in physics, guest faculty) for providing moral encouragement.

I am thankful to the authors and publishers of various books consulted by me, including those enlisted in the References.

Further, I wish to thank Mr Dharmendra Kumar Saxena for preparing the typescript and Mr YN Arjuna (CBS) for bringing out the book in a short time.

Although utmost care has been taken to minimise errors, suggestions for further improvement and pointing out errors by the readers would be highly welcome.

AK Saxena

Contents

8. Shift Registers   129

9. Binary Counters (Asynchronous and Synchronous Counters)   145

10. Data-Processing Circuits   167

11. Analog to Digital (A/D) and Digital to Analog (D/A) Converters   214

1

Introductory Concepts

1.1 INTRODUCTION

All of us are familiar with the impact of modern computers, communication systems, calculators, watches, etc. on the society. These are all based on the integrated circuits (ICs) whose advent became possible because of the tremendous progress in semiconductor technology in the recent past. The operation of these systems is based on the principles of digital techniques and digital electronics. Digital electronics involves circuits and systems in which there are only two possible states that are typically represented by (two) voltage levels. In digital systems, two states are used to represent numbers, symbols or characters.

1.2 ANALOG AND DIGITAL SIGNALS

There are basically two types of signals: analog and digital.

(a) Analog signals: Analog signal is defined as a voltage or current whose size is proportional to the quantity it represents. Analog signal is continuous and has infinite set of possible values.

(b) Digital signal: A digital signal is one which changes between two discrete levels of voltage. These changes are very sudden. Figure 1.1 illustrates the comparison between analog signal and digital signal. The most positive fixed voltage represents 1 state. Similarly, most negative voltage represents 0 state. Digital signals represent the real quantities by means of groups of 0 and 1.

Group of 0s and 1s in some orderly format can represent unlimited information.

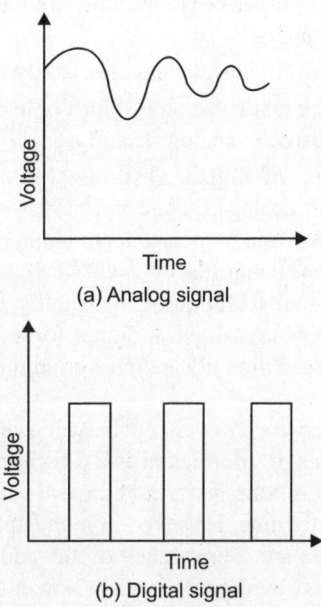

(a) Analog signal

(b) Digital signal

Fig. 1.1: Analog and digital signals

1.3 ANALOG AND DIGITAL SYSTEMS

The different electronic system can be classified as:

(a) Analog systems: An analog system is one which contains devices that manipulate physical quantities which are represented in analog form. In such a system, the quantities can vary over a continuous range of values. Commonly used analog systems are TV receiver, telephone systems and tape recording and play-back equipment.

(b) Digital systems: Digital system consist of devices designed to handle physical quantities that are represented in digital form, i.e. they can take only discrete values. Some examples are digital computers, calculators, digital watches, etc.

Advantages of Digital systems

Now-a-days, most of the applications in electronics use digital methods to perform operations that were performed earlier by using analog methods. The main advantages of digital systems are:

  (i) They are easier to design.
 (ii) Information can be stored very easily.
(iii) Greater accuracy and precision.
 (iv) Operation can be programmed by a set of stored instructions.
  (v) Digital circuits are less affected by noise.
 (vi) Can be fabricated on IC chips of lesser area as compared to analog circuitry.

Limitations of digital systems: Most physical quantities are analog in nature and it is these quantities that are inputs and outputs that are being monitored, processed and controlled by a system. To take advantage of digital techniques, the analog inputs are required to be converted to digital form. These are then processed digitally and converted then back to analog form.

The need for conversion between analog and digital forms of information is a drawback because of the added complexity and expense and it also requires extra time. However, in many applications, these factors are overweighed by the added advantages offered by digital circuits. Now-a-days, both digital and analog techniques are simultaneously of use in some systems.

1.4 DIGITAL SIGNALS

As mentioned in section 1.2, a digital signal has two discrete levels or values. Two different representations of digital signals are shown in Fig. 1.2. In each case there are two discrete levels. These levels can be represented using the terms LOW and HIGH. In Fig. 1.2a, lower of the two levels has been designated as LOW and the higher as high level. On the other hand, in Fig. 1.2b, higher of the two levels has been designated as LOW level and the lower as HIGH level. Digital systems using the representation of signal shown in Fig. 1.2a are said to employ *positive logic*

system and those using representation of signal shown in Fig. 1.2b are said to employ *negative logic* system.

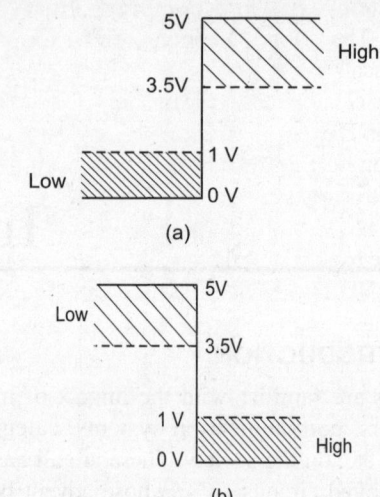

Fig. 1.2: Digital signal representations: (a) positive logic, (b) negative logic

(Unless otherwise specified, we shall be dealing with positive logic system)

The two discrete signal levels HIGH and LOW can also be represented by the binary digits 1 and 0 respectively. A binary digit (0 or 1) is referred to as a *bit*. Since a digital signal can have only one of the two possible levels 1 or 0, the binary number system can be used for the analysis and design of digital systems. The two levels (or states) can also be designated as ON and OFF respectively or TRUE and FALSE. The concept of binary number system was introduced by George Boole to study the mathematical theory of LOGIC which developed later as Boolean algebra.

In digital circuits, two voltage levels represent the two binary digits 1 and 0 and are designed to produce output voltages that fall within the prescribed 0 and 1 voltage ranges such as those defined in Fig. 1.2. The digital circuits are designed to respond to the input voltages within the defined 0 and 1 ranges.

1.5 ELEMENTS OF DIGITAL LOGIC

The term logic refers to something which can be reasoned out. In many situations, the problems and processes that we encounter, can be expressed in the form of logic functions. Since these functions are true/

false or yes/no statements digital circuits with their two state characteristics are extremely useful. Several logic statements when combined, form logic functions. These logic functions can be formulated mathematically using Boolean algebra. There are four basic logic elements using which any digital-system can be built. They are the three basic gates-NOT, AND and OR, and a flip-flop. In fact, a flip-flop can be constructed using gates. So, we can say that any digital circuit can be constructed using only gates. In addition to the three basic gates, there are two universal gates called NAND and NOR. They are called universal gates because any circuit can be constructed using only NAND gates or only NOR gates. There are two more gates called XOR and XNOR.

Using logic gates and flip-flop, more complex logic circuits like counters, shift registers, arithmetic circuits, comparators, encoders, decoders, multiplexers, demultiplexers, memories, etc. can be constructed. More complex logic functions, then, can be combined using these to form complete digital systems to perform specific tasks.

1.6 FUNCTIONS PERFORMED BY DIGITAL LOGIC SYSTEMS

Many operations can be performed by combining logic gates and flip-flops. Some of these are arithmetic operations, comparison, code conversion, encoding, decoding, multiplexing, demultiplexing, shifting, counting and storing. These well be discussed in detail in later chapters. The block diagram operations are given below.

1.6.1 Arithmetic Operations

The basic arithmetic operations are addition, substraction, multiplication and division.

The *addition* operation is performed by a digital logic circuit called *adder*. Its function is to add two numbers *addend* (A) and *augend* (B) with a carry input (CI) and generate a sum term (S) and a carry output term (CO). Figure 1.3a is a block diagram of an adder. It illustrates the addition of the binary equivalents of 8 and 6 with a carry input of 1, which results in a binary sum term 5 and a carry output term 1.

The arithmetic operation of *subtraction* can by performed by a digital logic circuit called the *subtractor*. Its function is to subtract *subtrahend* (A)

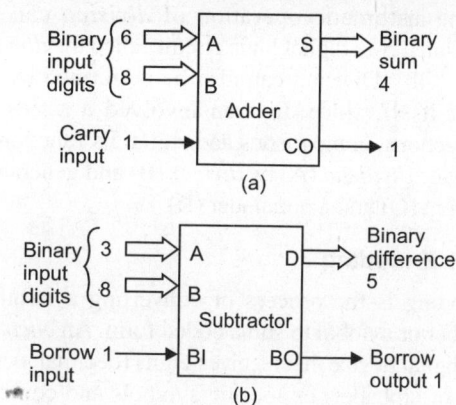

Fig. 1.3 The adder (a) and the subtractor (b)

from *minuend* (B) considering the borrow input (BI) and to generate a difference term (D) and a borrow output term (BO). Since subtraction is equivalent to addition of a negative number, subtraction can be performed by using an adder. Figure 1.3b is a block diagram of a subtractor. It illustrates the subtraction of the binary equivalent of 3 from the binary equivalent of 8 with a borrow input of 1, which results in a binary difference term 5 and a borrow output term 1. The arithmetic operation of *multiplication* can be performed by a digital logic circuit called the *multiplier* Fig. 1.4a. Its function is to multiply *multiplicand* (A) by multiplier (B) and generate the product term (P).

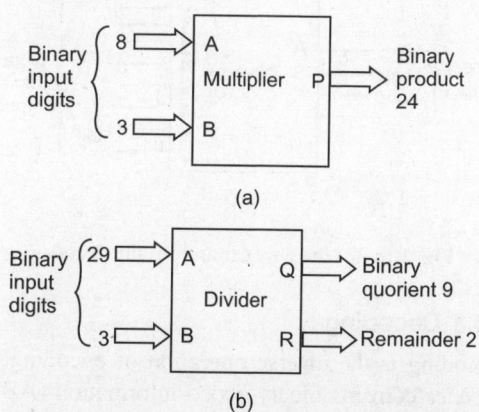

Fig. 1.4: (a) the multiplier and (b) the divider

Since multiplication is simply a series of additions with shifts in the positions of the partial products, it can be performed using an adder.

The arithmetic operation of *division* can be performed by a digital logic circuit called the *divider* (Fig. 1.4b). Division can also be performed by an adder itself, since division involved a series of subtractions, comparisons and shifts. Its function is to divide *dividend* (A) by *divisor* (B) and generate a quotient (Q) and a remainder (R).

1.6.2 Encoding

Encoding is the process of converting a familiar number or symbol to some coded form. An *encoder* is a digital device that receives digits (decimal, octal, etc.) or alphabets or specific symbols and converts them to their respective binary codes. In the octal to binary encoder (Fig. 1.5a), a HIGH level on a given input corresponding to a specific octal digit produces the appropriate 3-bit code (ABC) on the output levels. The figure illustrates encoding of the octal digit 6 to binary 110.

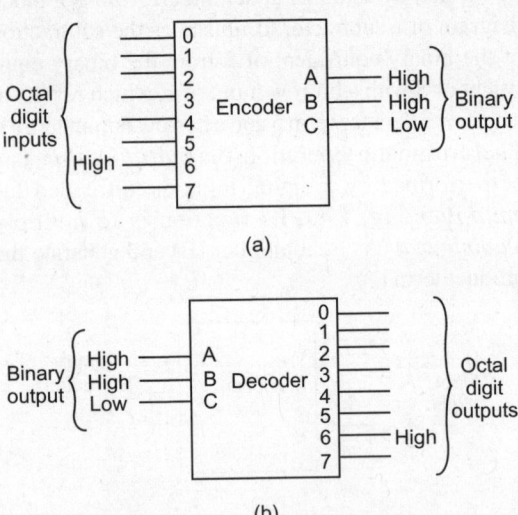

(a)

(b)

Fig. 1.5: (a) the encoder and (b) the decoder

1.6.3 Decoding

Decoding is the inverse operation of encoding. A *decoder* converts binary-coded information (ABC) to unique outputs such as decimal/octal digits, etc. In the binary-to-octal decoder (Fig. 1.5b), a combination of specific levels on the input lines produces a HIGH on the corresponding output line. The figure illustrates decoding of the binary 110 to octal digit 6.

1.6.4 Multiplexing

Multiplexing means sharing. It is the process of switching information from several lines onto a single line in a specified sequence. A multiplexer or data-selector is a logic circuit that accepts several data inputs and allows only one of them to get through to the output. In the multiplexer shown in Fig. 1.6a, if the switch is connected to input A for time t_1, to input B for time t_2, to input C for time t_3 and to input D for time t_4, the output will be as shown in the figure. This figure illustrates 4 to 1 multiplexer.

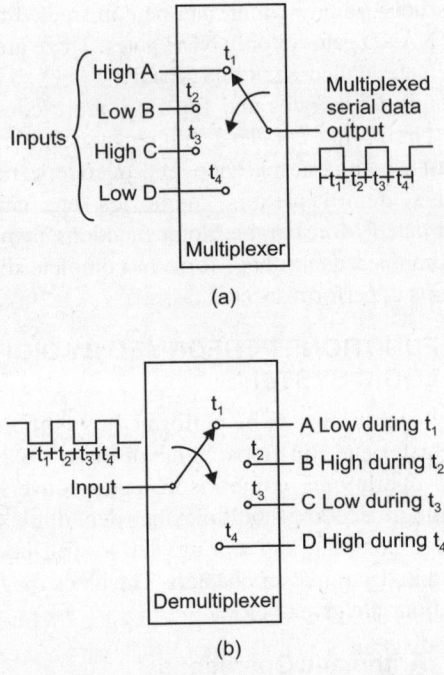

(a)

(b)

Fig. 1.6: (a) the multiplexer and (b) the demultiplexer

1.6.5 Demultiplexing

Demultiplexing operation is the inverse of multiplexing. It is the process of switching information from one input line onto several output lines. A demultiplexer is a digital circuit that takes a single input and distributes it over several outputs. In the demultiplexer shown in Fig. 1.6b, if the switch is connected to output a for time t_1, to output B for time t_2, to output C for time t_3 and to output D for time t_4, the output will be as shown in the figure. This figure illustrates a 1-to-4 demultiplexer.

1.6.6 Comparison

A logic circuit used to compare two quantities and give an output signal indicating whether the two input quantities are equal or not, and if not, which one is

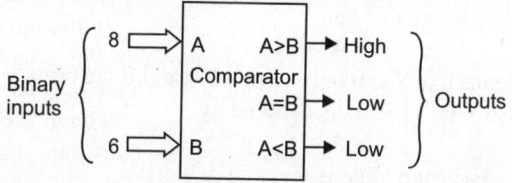

Fig. 1.7: Comparator HIGH level indicates that A is greater than B (8 > 6)

flop is shifted to the flip-flop to its right. Figure 1.8b shows the shifting out of data from the register. The content of the last flip-flop is shifted out and lost.

1.6.8 Counting

A logic circuit used to count the number of pulses inputted to it is called a *counter*. The pulses may represent some events. In order to count, the counter must remember the present number, so that it can go to the next proper number in the sequence when the next pulse comes. So storage elements (i.e. flip-flops) are used to build counters too. Figure 1.9 shows the block diagram of a counter.

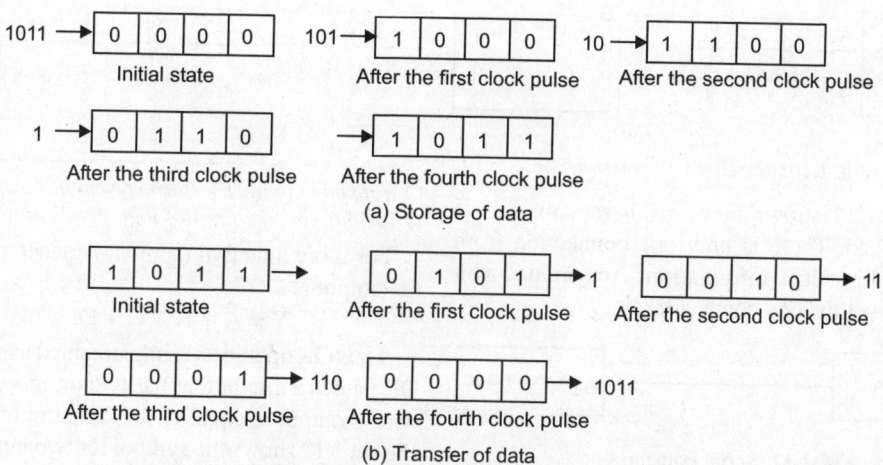

Fig. 1.8: Storage and transfer of data

greater, is called a comparator. Figure 1.7 shows the block diagram of a comparator. The binary represent-ations of the quantities A and B to be compared are applied as inputs to the comparator. One of the outputs A < B, A = B or A > B goes HIGH depending on the magnitudes of the input quantities.

1.6.7 Storage

Storage and shifting of information is very essential in digital systems. Digital circuits used for temporary storage and shifting of information (data), are called *registers*. Registers are made up of storage elements called flip-flops. Figure 1.8a shows the shifting or loading of data into a register made up of four flip-flops. After each clock pulse, the input bit is shifted into the first flip-flop and the content of each flip-

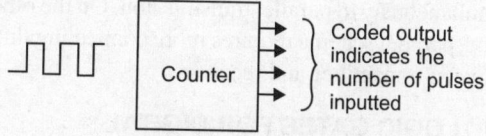

Fig. 1.9: The counter

1.7 DATA TRANSMISSION (Parallel and serial Transmission)

Information (data) is frequently required to be trans-mitted from one place to another in any digital system. The information is in binary form and generally represented as voltages at the outputs of the sending circuit that are connected to the inputs of a receiving circuit. Two basic methods for transmission of digital information are serial and parallel transmissions.

Figure 1.10 shows how the binary number 11111 is transmitted from circuit A to B using parallel transmission. Each bit of the binary number is represented by one of the circuit A outputs, (output A_4 is the most significant bit (MSB) and A_0 is the LSB (least significant bit)). Each of the circuit A outputs are connected to corresponding input of circuit B so that all 5 bits of information are transmitted in parallel. It requires one connecting line per bit as all bits are to be transmitted simultaneously.

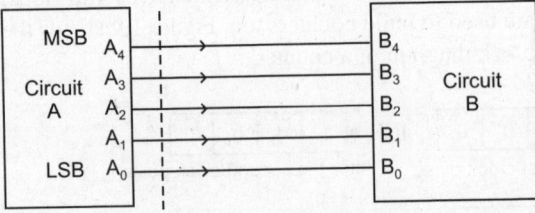

Fig. 1.10: Parallel transmission

Figure 1.11 shows how serial transmission is accomplished. There is only one connection from circuit A to circuit B. Information is transmitted a bit at a time over the one connecting line.

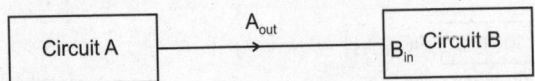

Fig. 1.11 Serial transmission

The transmission using parallel scheme is fast as compared to serial method as all bits are transmitted simultaneously in parallel transmission. On the other hand, parallel scheme requires more connecting lines between transmitter and receiver.

1.8 LOGIC GATES FOR DIGITAL OPERATIONS

A logic gate is most fundamental digital circuit. It is simply a device that has two or more inputs and one output. Its output will be either high or low depending upon the combination of high and low inputs used and the type of gate used. Inputs to the gate are represented by Boolean variables A, B, C, etc. and the output by Boolean variable Y. The function of the gate is represented by Boolean expression and the working or operation of the logic gate is represented by a truth table. There are six types of logic gates:

(i) OR   (ii) AND   (iii) NOT   (iv) NAND
(v) NOR   (vi) XOR   (vii) XNOR

Several logic gates are connected to form a logic network or digital circuit.

(i) OR Gate : If A, B and Y are Boolean variables then, for an OR gate

$$Y = A \text{ or } B$$

means that Y is 0 only if inputs A and B are both '0', else Y is '1'. This is denoted by

$$Y = A + B$$

The truth table is given in Table 1.1

Table 1.1: Truth table for OR operation

Input		Output
A	B	Y
0	0	0
0	1	1
1	0	1
1	1	1

(A truth table shows how the logic circuits output responds to various combinations of logic levels at inputs)

For more than two input variables to an OR gate, the output is

$$Y = A + B + C + D.$$

Thus OR operation is implemented using OR gate. OR gate is a circuit that has two or, more inputs and whose output is equal to the OR sum of the inputs. Figure 1.12 shows the symbol for two input OR gate. The inputs A and B are logic voltage levels and output Y is logic voltage

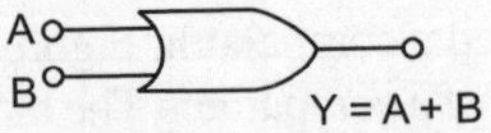

A•⟩⟩ Y = A + B
B•

Fig. 1.12: Logic symbol for an OR gate

level whose value is the result of the OR operation on A and B, i.e. $Y = A + B$.

(ii) AND Gate: If A, B and Y are Boolean variables then

$$Y = A \text{ AND } B$$

means that Y is one (1) only if A and B are both one (1), otherwise Y is zero. This function is also denoted by

$$Y = A.B$$

AND operation can be implemented by using logic circuit known as AND gate. AND gate has two or more inputs and one output which is equal to AND product of the inputs. The truth table and logic symbol for two input AND gate is shown in Fig. 1.13.

Input		Output
A	B	Y
0	0	0
0	1	0
1	0	0
1	1	1

(a)

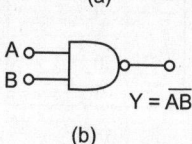

$Y = AB$

(b)

Fig. 1.13: AND gate (a) truth table and (b) logic symbol

It can be seen from the truth table, AND gate output is high (1) only when all its inputs are high (1). For all other cases, the AND gate output is low (0).

(iii) NOT Gate: The NOT operation applied to a Boolean variable A, generates its logical inverse denoted by A, i.e.

$$Y = \overline{A}$$

This operation is implemented using a logic circuit known as inverter. It has one input and one output. The output logic level of inverter's output is always opposite to the logic level of its input. The truth table and logic symbol are given in Fig. 1.14.

Input	Output
A	Y
0	1
1	0

(a)

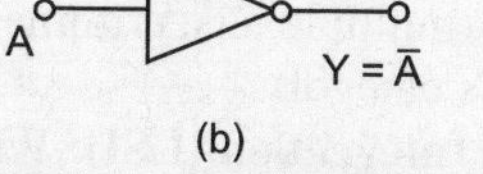

A ————▷o—— $Y = \overline{A}$

(b)

Fig. 1.14: NOT gate (a) truth table and (b) logic symbol

(iv) NAND Gate: If A, B and Y are Boolean variables, then

$$Y = \overline{AB}$$

i.e. first the two variables are ANDed, and then inverted, as indicated by bar over the AND expression.

The truth table for two input NAND gate and its symbol are given in Fig. 1.15.

Input		Output
A	B	Y
0	0	1
0	1	1
1	0	1
1	1	0

(a)

A ———o
B ———o ⊐o— $Y = \overline{AB}$

(b)

Fig. 1.15: NAND gate (a) truth table and (b) logic symbol

From the truth table, it can be seen that NAND gate output is the exact inverse of AND gate for all possible input combinations.

(v) NOR Gate: If A, B and Y are Boolean variables, then

$$Y = \overline{A + B}$$

which means that the two variables are ORed, and then inverted. Thus this operation is equivalent to OR followed by inversion. The truth table and logic symbol are given in Fig. 1.16.

Input		Output
A	B	Y
0	0	1
0	1	0
1	0	0
1	1	0

(a)

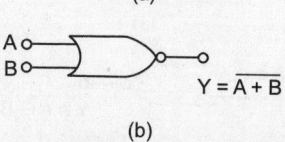

$Y = \overline{A + B}$

(b)

Fig. 1.16: NAND gate (a) truth table and (b) logic symbol

From the truth table, it can be seen that NOR gate output is exact inverse of OR gate output for all possible input conditions.

(vi) Exclusive OR Gate (XOR Gate): The XOR operation gives high output if one of the inputs is high. The Boolean operation for XOR operation can be written as

$$Y = A \oplus B$$
$$= \bar{A}B + A\bar{B}$$

This operation can be implemented using basic AND, OR and invert gates. The symbol and truth table for XOR gate is given in Fig. 1.17.

Input		Output
A	B	Y
0	0	0
0	1	1
1	0	1
1	1	0

(a)

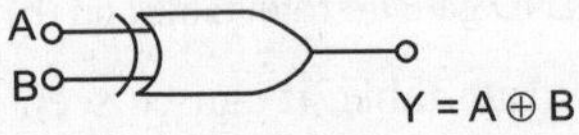

$$Y = A \oplus B$$

(b)

Fig. 1.17: XOR gate (a) truth table and (b) logic symbol

(vii) Exclusive NOR Gate: The Ex-NOR operation gives high output for both inputs low or both inputs high. The Boolean expression is

$$Y = \overline{A \oplus B} = AB + \overline{A}\overline{B}$$

This operation is implemented using basic AND, OR and invert gates. The basic truth table and symbol are shown in Fig. 1.18.

Input		Output
A	B	Y
0	0	1
0	1	0
1	0	0
1	1	1

(a)

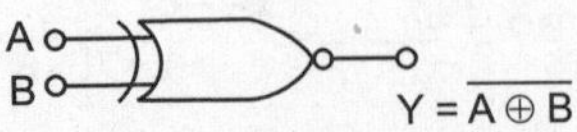

$$Y = \overline{A \oplus B}$$

(b)

Fig. 1.18: Ex-NOR gate (a) truth table and (b) logic symbol

1.9 DIGITAL INTEGRATED CIRCUITS

All the logic functions described (above in section 1.6) and many more are available in the integrated circuit form (IC) form. Modern digital systems utilize ICs in their design. A monolithic IC is an electronic circuit that is constructed entirely on a single piece of semiconductor material (usually silicon) called *substrate* which is commonly referred to as a *chip*.

ICs have the advantages of low cost, low power, smaller size and high reliability over discrete circuitry.

ICs are principally used to perform low power circuit operations such as information processing. ICs cannot handle very large voltages or currents as the heat generated in these tiny devices would result in temperature rise beyond acceptable limits resulting in burning out of ICs.

ICs may be classified as analog and digital. Digital ICs are complete functioning blocks as no additional components are required for their operation. The output may be obtained by applying the input. The output is a logic level 0 or 1.

For analog ICs external components are required. Digital ICs are a collection of resistors, diodes and transistors fabricated on a single chip. The chip is enclosed in a protective plastic or ceramic package from which pins extend for connecting IC to other devices. There are two main types of packages: dual-in-line package (DIP) and the flat package.

1.10 LEVELS OF INTEGRATION

Digital ICs are often categorized according to their circuit complexity as measured by the number of equivalent logic gates on the substrate. There are currently five standard levels of complexity:

Small Scale Integration (SSI): The least complex digital ICs with less than 12 gate circuits on a single chip. Logic gates and flip-flops belong to this category.

Medium scale Integration (MSI): With 12 to 99 gate circuits on a single chip, the more complex logic circuit such as encoders, decoders, counters and registers, multiplexers, arithmetic circuits, etc. belong to this category.

Large Scale Integration (LSI): With 100 to 9999 gate circuits on a single chip, small memories and small microprocessors fall in this category.

Very Large Scale Integration (VLSI): ICs with complexities ranging from 10,000 to 99,999 gate circuits per chip fall in this category. Large memories and large microprocessor systems, etc. belong to this category.

Ultra Large Scale Integration (ULSI): With complexities of over 100,000 gate circuits per chip, very large memories and microprocessor systems and single chip computers come in this category.

Digital ICs can also be categorized according to the principal type of electronic component used in their circuitry. They are-

(a) Bipolar ICs - which use BJT's
(b) Unipolar ICs - which use MOSFET's.

Several integrated circuit fabrication technologies are used to produce digital ICs. Presently, digital ICs are fabricated using TTL, ECL, IIL, MOS and CMOS technologies. Each differs from the other in the type of circuitry used to provide the desired logic operation. While TTL, ECL, and IIL use bipolar transistors as main circuit elements, MOS and CMOS use MOSFETS as main circuit elements. These technologies are also called *logic families*. Several sub-families of these main logic families are also available.

1.11 POPULAR ICS FOR LOGIC GATES

Figures 1.19 to 1.25 show the pin diagrams for ICs employing respectively OR gates, AND gates, NOT gates, NAND gates, NOR gates, XOR gates, XNOR gates.

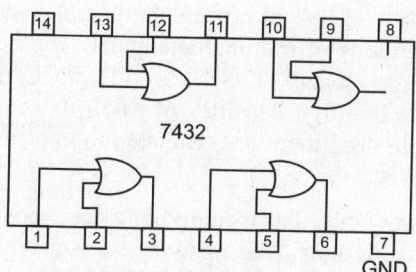

Fig. 1.19: Pin diagram for IC 7432

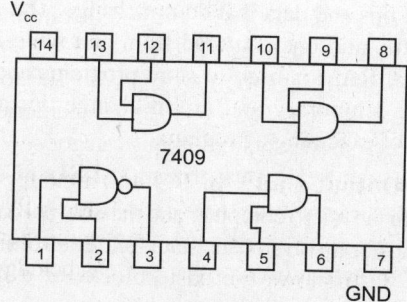

Fig. 1.20: Pin diagram for IC 7409

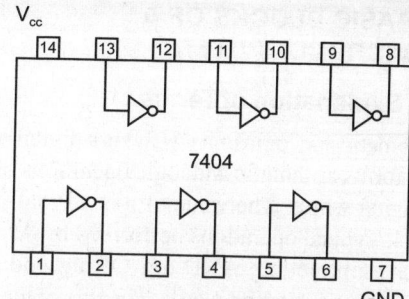

Fig. 1.21: Pin diagram for IC 7404

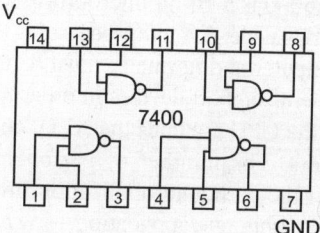

Fig. 1.22: Pin diagram for IC 7400

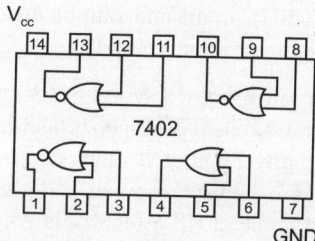

Fig. 1.23: Pin diagram for IC 7402

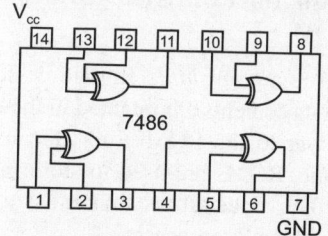

Fig. 1.24: Pin diagram for IC 7486

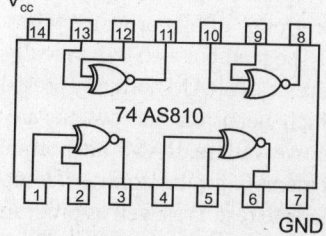

Fig. 1.25: Pin diagram for IC 74AS810

1.12 BASIC BLOCKS OF A MICROCOMPUTER

Some Explanation of Terms

An arithmetic and logic unit (ALU) is a digital circuit that performs arithmetic and logic operations on two n-bit digital words. The value of n is normally 4, 8, 16 or 32. Typical operations performed by ALU are addition, subtraction, ANDing, ORing and comparison of two n-bit digit words. The size of the ALU defines the size of the *microprocessor*, e.g. The Motorola 65000 is a 16-bit microprocessor since its ALU is 16 bits wide. A microprocessor is the CPU of a microcomputer and normally must be augmented with peripheral support devices in order to function. In general the CPU contains the ALU, control units and registers. The number of peripheral devices depends on the particular application involved and even varies within one application.

In general, a *microcomputer* consists of a microprocessor (CPU), input and output means, and a memory to store programs and data.

Read-only memory (ROM) is a storage medium for the groups of bits called *words,* and its contents cannot normally be altered once programmed. A typical ROM is fabricated on an LSI chip and can store, for example, 2048 8-bit words which can be individually accessed by presenting one of 2048 addresses to it. This ROM is referred to as a 2K-word by 8-bit ROM.

A ROM is a *nonvolatile* storage device, which means that its contents are retained in the event of a loss of power to the ROM chip. Because of this characteristic, ROMs are used to store instructions (programs) or data tables that must always be available to the microprocessor.

Random-Access Memory (RAM) is also a storage medium for groups of bits or words whose contents can not only be read but also dynamically altered at specific addresses. A RAM normally provides *volatile* storage, which means that its contents are lost in the event of a power failure. RAMs are normally used as scratchpad memory for the storage of temporary data and intermediate results as well as programs that can be reloaded from a backup non-volatile source.

A *register* can then be considered as volatile storage for a number of bits. There bits may be entered into the register simultaneously (*in parallel*) or sequentially (*serially*) from right to left or from left to right.

The term *bus* refers to a number of conductors organized to provide a means of communication among different elements in a micro computer system.

The conductors in the bus can be grouped in terms of their functions. A microprocessor normally has an address bus, a data bus and a control bus. The address bits to memory or to an external device are sent out on the *address bus*. Instructions from memory and data to and from memory or external devices normally travel on the *data bus*. Control signals for the other buses and among system elements are transmitted on the *control bus*.

A microcomputer has three basic blocks: a CPU, a memory unit and an input/output unit.

Central Processing Unit (CPU): The CPU executes all the instructions and performs arithmetic and logic operations on data. The CPU of the microcomputer is called the *microprocessor*.

The MOS microprocessor is typically a single LSI chip that contains all of the control, arithmetic and logic circuits of the microcomputer. The bipolar microprocessors (TTL, Schottky TTL, ECL) do not provide the high densities of MOS devices and therefore need more than one chip to implement a microprocessor.

Memory Unit: The memory unit stores both data and instructions. The memory section typically contains ROM and RAM chips. The ROM can only be read and is nonvolatile and is used to store instructions and data that do not change. The RAM is volatile and one can read from and write into a RAM. A RAM is used to store programs and data that are temporary and might change during the course of executing a program.

Input/Output Unit: An I/O unit transfers data between the micro computer and the external devices. The transfer involved data, status and control signals. Figure 1.26 shows the basic blocks of a microcomputer.

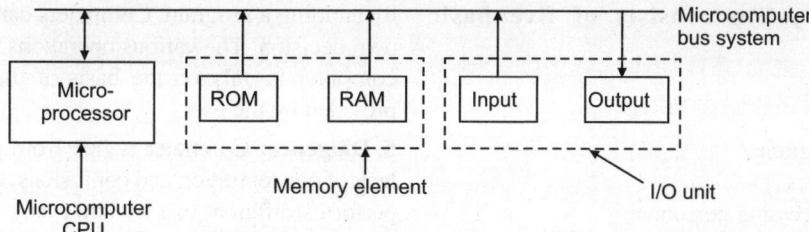

Fig. 1.26: Basic blocks of a microcomputer

1.13 TYPICAL MICROCOMPUTER ARCHITECTURE

Figure 1.27 illustrates the most simplified version of a typical microcomputer. The figure shows basic blocks. The various buses that connect these blocks are also shown. Although this figure looks very simple, it includes all the main elements of a typical microcomputer system.

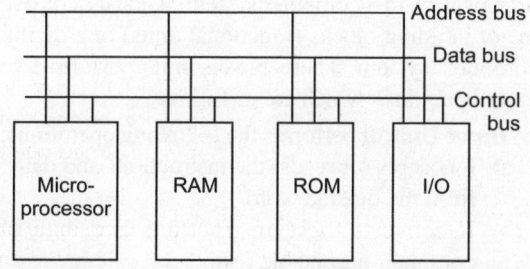

Fig. 1.27: Simplified version of a typical microcomputer structure

1. The Microcomputer Bus: The microcomputer contains three buses, which carry all the address data and control information involved in program execution. These buses connect the microprocessor (CPU) to each of the ROM, RAM and I/O elements so that information transfer between the microprocessor and any of the other elements can take place.

In the microcomputer, most information transfers are carried out with respect to the memory. When the memory is receiving data from another microcomputer element, it is called a WRITE operation and data is written into a selected memory location. When the memory is sending data to another microcomputer element, it is called a READ operation and data is being read from a selected memory location.

Address Bus: In this bus, information transfer takes place in only one direction, from the microprocessor to the memory or I/O elements. Therefore, this is called a *unidirectional bus*.

Data Bus: In this bus, data can flow in both directions, to or from the microprocessor. Therefore, this is a *bidirectional bus*. In some microprocessors, the data pins are used to send other information such as address bits in addition to data. This means that the data pins are time-shared or multiplexed.

Control Bus: This bus consists of a number of signals that are used to synchronize the operation of the individual microcomputer elements. The microprocessor sends some of these control signals to the other elements to indicate the type of operation being performed.

1.14 THE COMPUTER

The word computer has several different levels of meaning:

Level 1: In a very narrow sense, the computer is the part of hardware that performs the data processing, which is done by the central processing unit (CPU).

Level 2: A broader view of the computer which includes all components that are interconnected with each other to perform data processing. The components include not only CPU but also other devices to handle the input data, the storage of data and results. Devices connected to the CPU are sometimes called *peripherals*.

Level 3: A still more comprehensive view is the one that defines the computer as a system which includes the hardware, the software and the people connected to a computer's effective operation.

Computer System consists of five basic components:

1. Hardware
2. Software
3. User programs
4. Procedures
5. Data processing personnel

Hardware is the machine used for data processing. It is a part of computer which we can view. Software consists of the collection of programs. User programs are programs written by the users of the computer systems. Procedures are the rules, policies and guidelines according to which the operation of the computer is performed. Data processing personnel are the people responsible for keeping the data processing department functioning in an effective, convenient and efficient manner.

A modern computer processes the following abilities (i) It can perform complex tasks and repetitive calculations rapidly and accurately (ii) store large amount of data and information for suitable manipulations (iii) able to make decisions (iv) automatically correct or modify data by providing signals.

Characteristics of Computers

1. Speed: Computer is a very fast and accurate device. It can process thousands of instructions within a few seconds, for which a human being can take several days or months.

2. Accuracy: Computer results are accurate because it performs an operation according to given instruction. Errors can occur in computer system, but only when the programmer has made the error or hardware failure. Degree of accuracy is very high in computer systems.

3. Memory: Computers have a large amount of memory to hold a huge amount of data. The information stored in memory is not forgettable by the computer but human beings can forget. Hence memory plays an important role in a computers and stored information can be retrieved later when required for further use.

4. No intelligence: Computers have no intelligence. Intelligence is built in computer by the programmer

by building a program. Computers cannot take their own decision. The various operations performed by computer is only on the basis of the intelligence provided by the user.

5. Diligence: Computer is free from problems like lack of concentration and confusions, etc. Computer performs different tasks without any mixing and it can easily differentiate which type of work is performed by it. It possesses lot of concentration and never gets confused and never gets tired as human beings.

6. Versatility: With the help of computer, we can perform much different tasks. It can be used in any type of application like scientific, commercial, educational or business, etc.

1.15 BASIC ORGANIZATION OF A COMPUTER SYSTEM

Figure 1.28 shows the block diagram of the basic organization of a computer system. It consists of five major building blocks (functional units) of a digital computer system. These blocks/units perform five logic operations which are as follows:

(i) Input Unit: It performs the following operations:
  (a) It accepts (or reads) the instructions and data* from the outside word.
  (b) It converts these instructions and data in computer-acceptable form.
  (c) It supplies the converted instructions and data to the computer system for further processing.[*1]

(ii) Output Unit: The job of an output unit is just the reverse of that of an input unit. The following functions are performed by an output unit:
  (i) It accepts the results produced by the computer which are in coded form and hence, cannot be easily understood by us.
  (ii) It converts these coded results to human acceptable (readable) form.
  (iii) It supplies the converted results to the outside world.

Storage unit: The specific functions of the storage unit are to store:
  (i) The data and instructions required for processing (received from input devices)

[*1] Data means names, numbers, facts, anything needed to work out a problem.

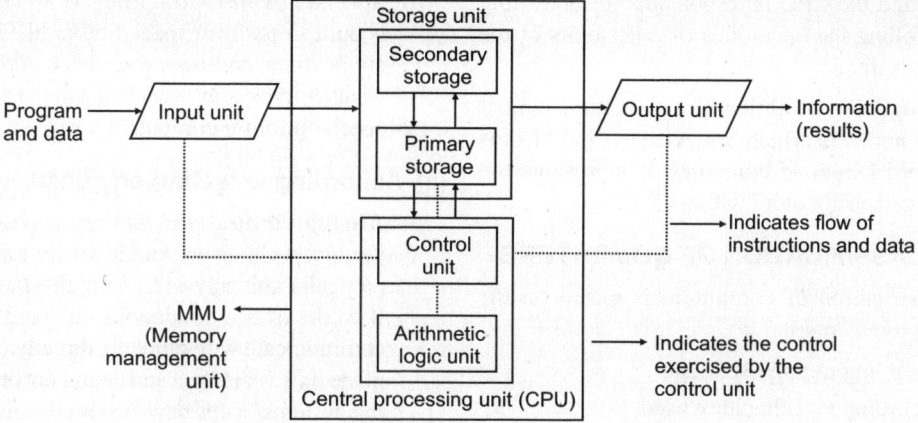

Fig. 1.28: Block diagram of basic organization of a digital computer

(ii) Intermediate results of processing

(iii) Final results of processing, before these results are released to an output device.

The storage unit of all computers is comprised of the following two types of storage:

1. Primary storage: This is also know as *main memory* and is used to hold pieces of program instructions and data, intermediate results of processing of jobs which the computer system is currently working on. While it remains in the memory. The central processing unit (CPU) can access it directly at a very fast speed.

However the information in primary storage can retain as long as the computer system is on. As soon as the computer is switched off or reset the information in primary storage disappears. Moreover, the primary storage normally has limited storage capacity.

2. Secondary storage: This is also known as auxiliary storage, and is used to take care of the limitations of the primary storage. It is much cheaper than primary storage and can retain information even when the computer is switched off or reset. This is normally used to hold program instructions, data and information on which the computer system is not working currently but needs to hold for processing later. The most commonly used secondary storage medium is the magnetic disk.

Arithmetic Logic Unit: The arithmetic logic unit (ALU) of a computer system is the place where actual execution of the instructions takes place. The

calculations are performed and comparisons (decisions) are made in the ALU.

The data and instructions stored in the primary storage before processing are transferred (as and when needed) to the ALU where processing takes place. Intermediate results generated in the ALU are temporarily transferred back to the primary storage until needed later, i.e. data may move from primary storage to ALU and back again to storage many times before the processing is over. The type and number of arithmetic and logic operations which a computer can perform, is determined by the engineering design of the ALU. Generally, all ALUs are designed to perform the four basic arithmetic operations (add, subtract, multiply and divide) and logic operations or comparisons such as less than equal to and greater than.

Control Unit: It does not perform any actual processing on the data, the control unit acts as a central controlling system, for the other components of the computer system. It manages and coordinates the entire computer system. It obtains instructions from the program stored in main memory, interprets the instructions and generates signals, which cause other units of the computer system to execute them.

Central Processing Unit (CPU): The control unit and the ALU of a computer system are jointly known as the central processing unit (CPU). The CPU is the brain of a computer system. In a computer system all major calculations and comparisons are made inside the CPU and the CPU is responsible for activating and controlling the operations of other units of the computer system.

Word Length of Computer: The maximum number of binary numbers which are accepted by ALU is called word length of computer. It represents the processing capacity of all within CPU.

1.16 CLASSIFICATION OF COMPUTERS

The classification of computers is based on the following three criteria:

(i) According to purpose/usage

(ii) According to technology used

(iii) According to size and capacity

Based on this, the classification is shown in Fig. 1.29.

(i) According to purpose

(a) **General Purpose Computers:** Computer that works according to instructions for general requirements such as accounting, invoicing, inventory, etc. are called general purpose computers. Generally, all computers used in offices, for educational, commercial application, etc. are general purpose computers.

(b) **Special Purpose Computer:** Computer which is built to perform special tasks like scientific applications and research, space applications, weather forecasting, medical diagnostic, etc. are special purpose computers.

(ii) According to technology used

(a) **Analog computers:** Analog computers are computers that are made to perform some particular task only and not for all types of work. It works over a continuous data and does not communicate with numbers directly. These are made only for testing and analyzing other existing systems or for new system development.

Features of Analog computers

(i) It performs the job by measuring rather than counting.

(ii) It uses continuous signals rather than discrete (0 or 1)

(iii) Examples are thermometer, speedometer, hydrometer, etc.

(b) **Digital Computers:** Digital Computer is a system that performs various computational tasks. Digital Computers use the binary number system which has two digits; 0 and 1. A binary digit is called

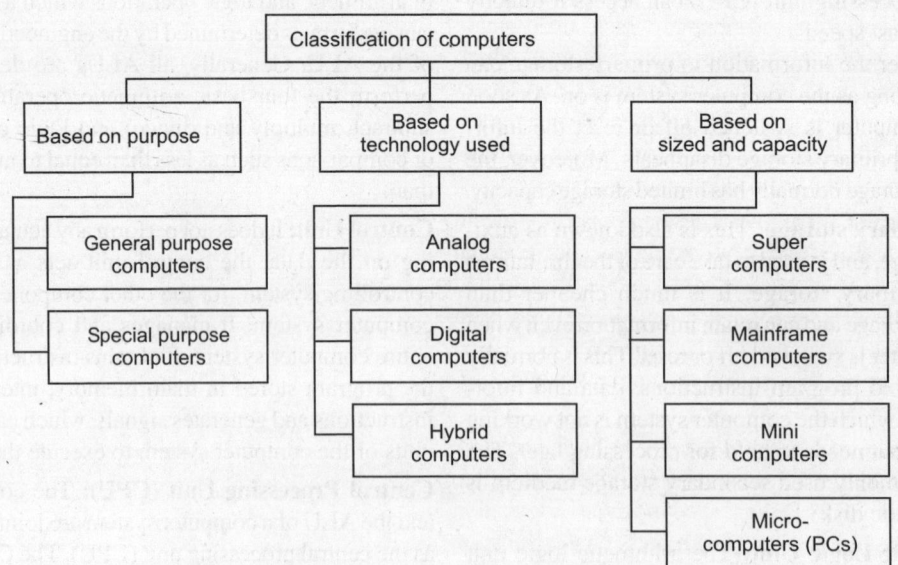

Fig. 1.29 Classification of computers

a *bit*. Information is represented in digital computers in groups of bits. It performs several different tasks and is interactive in nature. If any error has been occurred, we can terminate it due to interactive feature. This feature is not available in hybrid and analog computers.

Features of Digital Computer

(a) Digital computer converts data into digits

(b) It operates essentially on counting instead of measuring

(c) It accepts information in the form of discrete pulses.

(d) It is used for business and scientific applications

(e) It is interactive in nature

(f) These are most popular and widely used computers.

(c) Hybrid Computer: It is a combination of analog and digital computer. These computers are mostly used with process control equipment in continuous production plants like oil refineries, etc. and used at places where signals as well as data are to be entered into computers. Areas of application are nuclear power plants, mines, etc.

(iii) According to size and capacity

(a) Microcomputer: The most common type of computers are microcomputers which are portable personal computers. It is a small computer which mainly consists of single chip. Average data transfer rate of a microcomputer is 5 lac bytes per second. It can hold from 8 to 32 bit word length. Micro-computers can be subdivided into two types:

(i) **Home Computer:** These are basically meant for hobbies and games rather than professional tasks. They consist of a keyboard integrated with CPU in one box and interfaced with ordinary television and multimedia system used for entertainment and training in various computer centers and homes.

(ii) **Personal Computers:** These computers are designed for small business units and office automation. PCs are used in various application areas like: business and professional application and computer learning, word processing, accounting and telecommunication.

(b) Minicomputer: Minicomputers are larger in size than microcomputers and have very fast processing speed. It consists of a multiple processing unit in a single chip. It uses word length of usually 16, 24, 32, or 64 bits. The data transfer rate is about 4 million bytes per second. They can support upto 15 to 25 terminals simultaneously.

(c) Mainframe Computers: These are very large machines with the capability of parallel processing. The data transfer rate of this machine is 8 million bytes per second. It uses the word length of usually 24, 32, 48 and 64 or 128 bits. Mainframe is used for centralized data processing like Train reservation system, Airline reservation. Mainframe computers can support over 500 terminals. Some important mainframe computers are FDM-3090, VAX 8842 and UNIVAC.

Power of 2: Microprocessor design started with 4-bit devices. Then evolved to 8- and 16- bit devices. Thus powers of 2 keep coming up because of the binary nature of computers (Table 1.2). It lists the powers of 2 encountered in microcomputer analysis. As shown the abbreviation K stands for 1024 (approximately 1000). Therefore 1K means 1024, 2K stands for 2048, 4K for 4096 and so on. Some personal microcomputers have 64 K memories that can store upto 65, 536 bytes.

Table 1.2: Power of 2

Powers of 2	Decimal equivalent	Abbreviation
2^0	1	
2^1	2	
2^2	4	
2^3	8	
2^4	16	
2^5	32	
2^6	64	
2^7	128	
2^8	256	
2^9	512	
2^{10}	1024	1 K
2^{11}	2048	2 K
2^{12}	4096	4 K
2^{13}	8192	8 K
2^{14}	16384	16 K
2^{15}	32768	32 K
2^{16}	65536	64 K

(d) Super Computers: Super computers are much faster and more powerful than mainframe computers. Their processing speed like in the range of 400 MIPS-10,000 MIPS, word length 64-96 bit, memory capacity 256 MB and more and machine cycle time 4-6 nanoseconds. Super computers are specially designed to maximize the number of *floating point instructions per second* (FLOPS). Their FLOPS rating is usually more than 1 gigaflops.

Super computer contains a number of CPUs which operate in parallel to make it faster. They are used for massive data processing and solving very sophisticated problems. They are used for weather-forecasting, weapons, research and development, rocketing, in aerodynamics, seismology, atomic, nuclear and plasma physics. Examples are CRAY3 (developed by Control Data Corporation) and SX-2 (developed by Nippon Electric Corporation, Japan), etc. We will discuss more about computers and micro-processor in Chapter 12.

EXERCISE

1. List three examples of analog quantities.
2. What is the difference between analog and digital quantities?
3. What are the advantages of digital techniques over analog?
4. What are the main limitations to the use of digital techniques?
5. What is the difference between analog and digital system? Discuss.
6. Describe the advantages of parallel communication over serial communication.
7. Name different functional units of a computer.
8. For each of the following statements, indicate the logic gate(s) AND, OR, NAND, NOR for which it is true:
   (a) All LOW inputs produce a HIGH output
   (b) Output is HIGH if and only if all inputs are HIGH.
   (c) Output is LOW if and only if all inputs are HIGH.
   (d) Output is LOW if and only if all inputs are LOW.
9. Make truth table for a 3-input
   (a) AND gate   (b) OR gate
   (c) NAND gate   (d) NOR gate.
10. The voltage wave forms shown in Fig. 1.30 are applied at the inputs of 2-input AND, OR, NAND, NOR and X-OR gates.

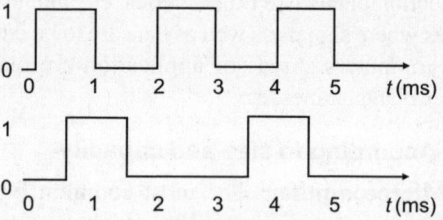

Fig. 1.30

Determine the output wave form in each case.

2

Number Systems

2.1 INTRODUCTION

We are familiar with the number system which is in common use in which an ordered set of ten symbols -0, 1, 2, 3, 4, 5, 6, 7, 8, 9 (known as *digits*) are used to specify any number. This number system is known as the *decimal number system*. The *radix* or *base* of this number system is 10 (number of distinct digits). Any number (e.g. 1986.384) is a collection of these digits. It has an integer part (1986) and a fractional part (0.384) separated from an integer part by a *radix point* (.) also known as *decimal*. There are some other systems also to represent numbers, some of these are: *binary, octal* and *hexadecimal* number systems. These are widely used in digital systems like micro-processors, logic circuits, computers, etc. Therefore, the knowledge of these number systems is very essential for understanding and designing digital systems.

We know that computers and digital circuits use binary signals but are also required to handle data which may be numeric, alphabets or special characters. Therefore the information available needs to be converted into suitable binary form before it can be processed by digital circuits. This means that the information available in the form of numerals, alphabets and special characters or any combination of these must be converted into binary format. To achieve this, a process of coding is employed where each numeral, alphabet or special character is coded in a unique combination of 0s and 1s using a coding scheme known as a *code*. There can be a variety of codes to serve different purposes such as arithmetic operations, data entry, error detection and correction, etc. Selection of a particular code depends on its suitability for the purpose. In any digital systems different codes may be used for different operations and it may be necessary to convert data from one code to another. In this chapter, we will discuss how to make conversion from one system to another and study various codes commonly used in computers.

2.2 NUMBER SYSTEMS

In general, in any number system, there is an ordered set of symbols known as digits. A collection of these digits makes a number which in general has two parts— integer and fractional separated by a radix point (decimal), i.e.

$$(N)_b = \underbrace{d_{n-1} \, d_{n-2} \, ... \, d_i \, ... \, d_1}_{\text{Integer part}} \, d_0 \quad \overset{\bullet}{\uparrow} \atop \text{Radix point}$$

$$\underbrace{d_{-1} \, d_{-2} \, ... \, d_{-f} \, ... \, d_{-m}}_{\text{Fractional part}}$$

Here, N = a number

b = radix or base of the number system

n = number of digits in integer part

m = number of digits in fractional part

d_{n-1} = most significant digit

d_{-m} = least significant digit

and $0 \leq (d_i \text{ or } d_{-f}) \leq b - 1$.

The digits in a number are written side by side (to represent the number) and each position in the number is assigned a *weight* or *index* of importance by some predesigned rule. Table 2.1 gives the details of commonly used number system.

Table 2.1: Characteristics of commonly used number systems

Number system	Base or radix (b)	Symbols used (d_i or d_{-f})	weight assigned to position	
			i	-f
Binary	2	0, 1	2^i	2^{-f}
Octal	8	0, 1, 3, 4, 5, 6, 7,	8^i	8^{-f}
Decimal	10	0, 1, 2, 3, 4, 5, 6, 7, 8, 9	10^i	10^{-f}
Hexadecimal	16	0, 1, 2, 3, 4, 5, 6, 7, 8, 9, A, B, C, D, E, F	16^i	16^{-f}

2.3 DECIMAL NUMBER SYSTEM

The number system which we commonly use in our daily life is called the decimal number system. In this system, the base is equal to 10 (There are altogether ten digits- 0, 1, 2, 3, 4, 5, 6, 7, 8, 9). In this system, for any number, the successive positions to the left of the decimal point represent units, tens, hundreds, thousands, i.e. each position represents a specific power of base 10. For example, the decimal number 3489 (written as 3489_{10}) can be written as

$$(3 \times 1000) + (4 \times 100) + (8 \times 10) + (9 \times 1)$$
$$= 3000 + 400 + 80 + 9 = 3489$$

It may also be noted that the same digit signifies different values depending on the position it occupies in the number. For example

In 3489_{10} the digit 9 signifies $9 \times 10^0 = 9$

In 3498_{10} the digit 9 signifies $9 \times 10^1 = 90$

In 3948_{10} the digit 9 signifies $9 \times 10^2 = 900$

In 9348_{10} the digit 9 signifies $9 \times 10^3 = 9000$

Hence any number can be represented by using the available digits and arranging them in various positions.

The principles which apply to the decimal number system, also apply to any other positional number system. In a positional number system, there are only a few symbols (e.g. 0 to 9 in decimal number system), called digits and these symbols represent different values, depending on the position they occupy in the number, as described above. Thus for any (other) positional number system, it is important to keep track of the base of the number system in which we are working.

Some of the positional number system, which are used in computer design and by computer professionals are describe subsequently.

2.4 BINARY NUMBER SYSTEM

In the binary number system, the base in 2 (instead of 10). We have only two symbols or digits, viz. 0

and 1. The largest single digit is 1 (one less than the base). Each position in a binary number represents a power of the base 2. In this system, the rightmost position is the units (2^0) position, the second position from the right is (2^1), the third position from the right is 2^2's position, then 2^3 position, 2^4 position and so on. For example, the decimal equivalent of the binary number 10101 (written as 10101_2) is

$$(10101_2) = 1 \times 2^4 + (0 \times 2^3) + (1 \times 2^2) + (0 \times 2^1)$$
$$+ (1 \times 2^0)$$
$$= 16 + 0 + 4 + 0 + 1 = 21$$

Thus $\quad 10101_2 = 21_{10}$

Binary digit is often referred to by the common abbreviation "bit". A "bit" in computer terminology means either a 0 or a 1. Table 2.2 lists all the 3-bit numbers along with their decimal equivalents.

Table 2.2: 3-bit numbers with their decimal values

Binary	Decimal equivalents
000	0
001	1
010	2
011	3
100	4
101	5
110	6
111	7

It may be noted that a 3-bit number can have one of the 8 values in the range 0 to 7. In fact it may be seen that any decimal number in the range 0 to 2^{n-1} can be represented in the binary form as an n-bit number.

Binary numbers are after called *binary words* or just *words*. Binary words with certain numbers of bit have also acquired special names. A 4-bit binary word is called a *nibble* and an 8-bit binary word is called a *byte*. A 16-bit binary word is after referred to just a *word*. A 32-bit binary word is referred to as a *double word*. The rightmost or *least significant bit* is usually referred to as the LSB. The leftmost or *most significant bit* of a binary word is usually denoted as the MSB.

2.5 OCTAL NUMBER SYSTEM

In the octal number system, the base is 8. Hence there are eight symbol digits 0, 1, 3, 4, 5, 6, 7, (8 and 9 do not exist in this system). The largest single digit is 7 (one less than the base). Each position in an octal number represents a power of the base (8). Therefore the decimal equivalent of the octal number 3058 is

$$(3 \times 8^3) + (0 \times 8^2) + (5 \times 8^1) + (8 \times 8^0)$$
$$= 1536 + 0 + 40 + 8 = 1584$$
$$3058_8 = 1584_{10}$$

2.6 HEXADECIMAL NUMBER SYSTEM

The hexadecimal system is one with a base of 16, having 16 single character digits or symbols. The first ten digits are the digits of the decimal number system- 0, 1, 2, 3, 4, 5, 6, 7, 8, 9. The remaining six digits are denoted by the symbols A, B, C, D, E and F representing the decimal values 10, 11, 12, 13, 14 and 15 respectively. The largest single digit is F or 15, i.e. one less than the base. Each position in a number of hexadecimal system represents a power of the base 16. Therefore, the decimal equivalent of the hexadecimal number 1AF is

$$(1 \times 16^2) + (A \times 16^1) + (F \times 16^0)$$
$$= (1 \times 256) + (10 \times 16) + (15 \times 1)$$
$$= 256 + 160 + 15$$
$$= 431$$

Hence $1AF_{16} = 431_{10}$

2.7 CONVERTING FROM ONE NUMBER SYSTEM TO ANOTHER

Number expressed in decimal number system are much more meaningful to us than the numbers expressed in any other number system. This is because we have been using decimal numbers in our day-to-day life. However any number in one system can be represented in any other system. Because the input and the final output values are to be in decimal, it is often required to convert numbers in other number system to decimal and vice-versa. We now describe method of converting to base 10 from any other base and a method of converting from base 10 to any other base.

2.7.1 Converting to Decimal from another Base

Step 1: Determine the column (positional) value of each digit

Step 2: Multiply the obtained column values (in step 1) by the digits in the corresponding columns.

Step 3: Sum the products calculated in step 2.

The total is the equivalent value in decimal.

Example 2.1: $11001_2 = ?_{10}$

Solution: Step 1: Determine column values

Column Number (from right)	Column value
1	$2^0 = 1$
2	$2^1 = 2$
3	$2^2 = 4$
4	$2^3 = 4$
5	$2^4 = 16$

Step 2: Multiply column values by corresponding digits.

16	8	4	2	1
×1	×1	×0	×0	×1
16	8	0	0	1

Step 3: Sum the products

$$16 + 8 + 0 + 0 + 1 = 25$$
$$\therefore \quad 11001_2 = 25_{10}.$$

Example 2.2: $4703_8 = ?_{10}$

Solution: Step 1: Determine column values

Column Number (from right)	Column value
1	$8^0 = 1$
2	$8^1 = 8$
3	$8^2 = 64$
4	$8^3 = 512$

Step 2: Multiply column values by corresponding digits.

512	64	8	1
×4	×7	×0	×3
2048	448	0	3

Step 3: Sum the products

$$2048 + 448 + 0 + 3 = 2499$$
$$\therefore \quad 4703_8 = 2499_{10}.$$

Example 2.3: Determine the decimal number represented by the following binary number: 110101.

Solution:

$$(110101)_2 = (1 \times 2^5) + (1 \times 2^4) + (0 \times 2^3)$$
$$+ (1 \times 2^2) + (0 \times 2^1) + (1 \times 2^0)$$
$$= 32 + 16 + 0 + 4 + 0 + 1$$
$$= (53)_{10}$$

Example 2.4: Determine the decimal number represented by the following binary number:

$$101101.10101$$

Solution:

$$(101101.10101)_2 = 1 \times 2^5 + 0 \times 2^4 + 1 \times 2^3 + 1 \times 2^2$$
$$+ 0 \times 2^1 + 1 \times 2^0 + 1 \times 2^{-1}$$
$$+ 0 \times 2^{-2} + 1 \times 2^{-3} + 0 \times 2^{-4}$$
$$+ 1 \times 2^{-5}$$
$$= 32 + 0 + 8 + 4 + 0 + 1$$
$$+ 1/2 + 0 + 1/8 + 0 + 1/32$$
$$= (45.65625)_{10}$$

2.7.2 Decimal to Binary Conversion

Any decimal number can be converted into its equivalent binary number. For integers, the conversion is obtained by continuous division by 2 and keeping track of the remainders, while for fractional parts, the conversion is affected by continuous multiplication by 2 and keeping track of the integers generated. This can be seen by the subsequent examples.

Example 2.5: Convert $(13)_{10}$ to an equivalent base-2 number.

Solution:

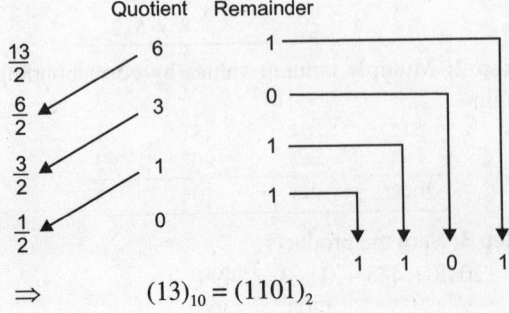

$$\Rightarrow \qquad (13)_{10} = (1101)_2$$

Example 2.6: Convert $(0.65625)_{10}$ to an equivalent base-2 number.

Solution:

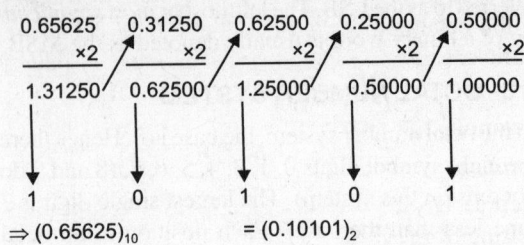

$$\Rightarrow (0.65625)_{10} \qquad = (0.10101)_2$$

Example 2.7: Express the following decimal numbers in the binary form : (a) 10.625 (b) 0.6875.

Solution:

(a) Integer part

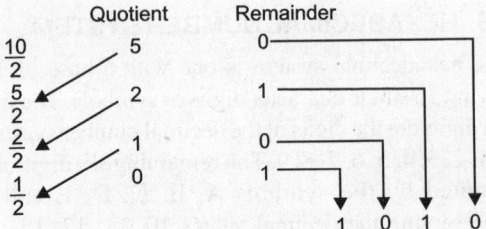

$$\therefore \qquad \text{Integer part}: (10)_{10} = (1010)_2$$

Fractional part:

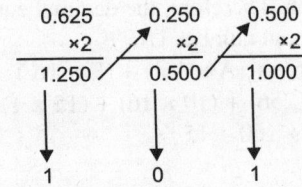

$$\therefore \qquad (0.625)_{10} = (0.101)_2$$
$$\text{and} \quad (10.625)_{10} = (1010.101)_2$$

(b)

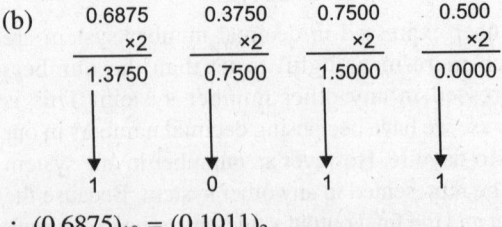

$$\therefore \quad (0.6875)_{10} = (0.1011)_2$$

2.7.3 Binary to Octal Conversion

The following steps are used in this method:

Step 1: Divide the binary digits into groups of three (starting from the right).

Step 2: Convert each group of three binary digits to one octal digit (since there are only 8 digits (0 to 7) in the octal number system (Table 2.2), 3 bits are sufficient to represent any octal number in binary ($2^3 = 8$)

Example 2.8: $(101110)_2 = (?)_8$

Solution:

Step 1: Divide the binary digits into groups of 3 starting from LSD.

$$\underline{101} \quad \underline{110}$$

Step 2: Convert each group into one digits of octal (using binary to decimal conversion)

$$(101)_2 = 1 \times 2^2 + 0 \times 2^1 + 1 \times 2^0,$$
$$= 4 + 0 + 1$$
$$= 5_8$$
$$(110)_2 = 1 \times 2^2 + 1 \times 2^1 + 0 \times 2^0$$
$$= 4 + 2 + 0$$
$$= 6_8$$

Hence $\quad (101110)_2 = (56)_8$

Example 2.9: $(1101010)_2 = (?)_8$

Solution: $(1101010)_2 = \underline{001} \quad \underline{101} \quad \underline{010}$ (groups of 3 digits from right)

$$(001)_2 = 0 \times 2^2 + 0 \times 2^1 + 1 \times 2^0$$
$$= 0 + 0 + 1$$
$$= 1$$
$$(101)_2 = 1 \times 2^2 + 0 \times 2^1 + 1 \times 2^0$$
$$= 4 + 0 + 1$$
$$= 5$$
$$(010)_2 = 0 \times 2^2 + 1 \times 2^1 + 0 \times 2^0$$
$$= 0 + 2 + 0$$
$$= 2$$

Hence $(1101010)_2 = (152)_8$

2.7.4 Octal to Binary Conversion

The following steps are used in this method:

Step 1: Convert each octal digit to a 3 digit binary number (the octal digits may be treated as decimal for this conversion).

Step 2: Combine all the resulting binary groups (of 3 digits each) into a single binary number.

Example 2.10: $(562)_8 = (?)_2$

Solution:

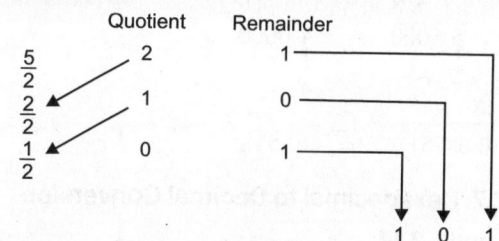

$$\therefore \qquad 5_8 = (101)_2$$
$$\text{Similarly} \quad 6_8 = (110)_2$$
$$2_8 = (010)_2$$

Combining $(562)_8 = \dfrac{101}{5} \quad \dfrac{110}{6} \quad \dfrac{010}{2}$

Hence $\quad (562)_8 = (101110010)_2$

Example 2.11: Find the decimal equivalent of the octal number 127.54

Solution: $(127.54)_8 = 1 \times 8^2 + 2 \times 8^1 + 7 \times 8^0 + 5 \times 8^{-1} + 4 \times 8^{-2}$
$$= 64 + 16 + 7 + \frac{5}{8} + \frac{4}{64}$$
$$= 87 + 0.625 + 0.0625$$
$$= (87.6875)_{10}$$

2.7.5 Octal to Decimal Number Conversion

Example 2.12:
$(6327.4051)_8 = 6 \times 8^3 + 3 \times 8^2 + 2 \times 8^1 + 7 \times 8^0 + 4 \times 8^{-1} + 0 \times 8^{-2} + 5 \times 8^{-3} + 1 \times 8^{-4}$
$$= 3072 + 192 + 16 + 7 + \frac{4}{8} + 0 + \frac{5}{512} + \frac{1}{4096}$$
$$= (3287.5100098)_{10}$$

2.7.6 Decimal to Octal Conversion

Example 2.13: (a) convert $(247)_{10}$ into octal
(b) convert $(0.6875)_{10}$ into octal

Solution: (a)

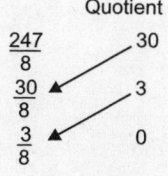 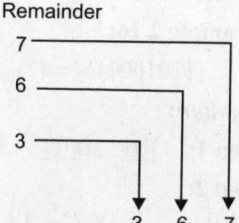

$$\therefore \quad (247)_{10} = (367)_8$$

(b)

0.6875	0.5000
×8	×8
5.5000	4.0000
↓	↓
5	4

$\therefore$ $(0.6875)_{10}$ = $(0.54)_8$

2.7.7 Hexadecimal to Decimal Conversion

Example 2.14:

$(1AC)_{16} = (?)_{10}$

Solution:

$(1AC)_{16} = 1 \times 16^2 + A \times 16^1 + C \times 16^0$

$= 1 \times 256 + 10 \times 16 + 12 \times 1$

$= 256 + 160 + 12$

$= (428)_{10}$

2.7.8 Decimal to Hexadecimal Conversion

Example 2.15: $(428)_{10} = (?)_{16}$

Solution:

16	428	Remainders in Hexadecimal
	26	12 = C
	1	10 = A
	0	1 = 1

Hence $(428)_{10} = (1AC)_{16}$

2.7.9 Binary to Hexadecimal Conversion

The following steps are used in this method:

Step 1: Divide the binary digits into groups of 4, starting from the right (LSD).

Step 2: Convert each group to one hexadecimal digit. Remember that hexadecimal digits 0 to 9 are equal to decimal digits 0 to 9, and hexadecimal digits A to F are equal to decimal values 10 to 15 respectively. Hence, for this step, the binary to decimal conversion can be used but the decimal values 10 to 15 must be represented as hexadecimal A to F. (Table 2.3)

Example 2.16:

$(11010011)_2 = (?)_{16}$

Solution:

Step 1: <u>1101</u> <u>0011</u>

Step 2:

$(1101)_2 = 1 \times 2^3 + 1 \times 2^2 + 0 \times 2^1 + 1 \times 2^0$

$= 8 + 4 + 0 + 1$

$= (13)_{10}$

$= D_{16}$

$(0011)_2 = 0 \times 2^3 + 0 \times 2^2 + 1 \times 2^1 + 1 \times 2^0$

$= 0 + 0 + 2 + 1$

$= 3_{16}$

Hence, $(11010011)_2 = (D3)_{16}$

2.7.10 Hexadecimal to Binary Conversion

Following steps are used in this method:

Step 1: Convert the decimal equivalent of each hexadecimal digit to 4 binary digits.

Step 2: Combine all the resulting binary groups (each of 4 digits) into a single binary number.

Example 2.17:

$(2AB)_{16} = (?)_2$

Solution:

Step 1: $2_{16} = 2_{10} = (0010)_2$

$A_{16} = 10_{10} = (1010)_2$

$B_{16} = 11_{10} = (1011)_2$

Step 2: $(2AB)_{16} = \dfrac{0010}{2} \quad \dfrac{1010}{A} \quad \dfrac{1011}{B}$

Hence $(1101)_2$ $(2AB)_{16} = (001010101011)_2$

Table 2.3: Binary to hexadecimal conversion

Hexadecimal	Binary	Hexadecimal	Binary
0	0000	8	1000
1	0001	9	1001
2	0010	A	1010
3	0011	B	1011
4	0100	C	1100
5	0101	D	1101
6	0110	E	1110
7	0111	F	1111

2.7.11 Octal to Hexadecimal Conversion

To convert an octal number to hexadecimal, first convert it to binary and then the binary to hexadecimal.

Example 2.18: Convert 756.603_8 to hexadecimal

Solution:

(Octal)	7	5	6.	6	0	3
(Binary)	111	101	110.	110	000	011
(Group of 4 bits)	0001	1110	1110.	1100	0001	1000
(Hex)	1	E	E.	C	1	8

Hence $(756.603)_8 = (1EE.C18)_{16}$

2.7.12 Hexadecimal to Octal Conversion

To convert a hexadecimal number to octal, first convert the given hexadecimal number to binary and then the binary to octal

Example 2.19: Convert B9F. AE$_{16}$ to octal

Solution:

(Hex)	B		9	F	.	A	E
(Binary)	1011		1001	1111	.	1010	1110
(Group of 3 bits)	101	110	011	111	.	101	011	100
(Octal bits)	5	6	3	7	.	5	3	4

Hence (B9F.AE)$_{16}$ = (5637.534)$_8$

2.8 SIGNED BINARY NUMBERS

In the decimal number system, a plus (+) sign is used to denote a positive number and a minus (−) sign for denoting a negative number. The plus sign is usually dropped and the absence of any sign means that the number has positive value. This representation of numbers is known as *signed number*. As we know that digital circuits can understand only two symbols 0 and 1, therefore, we must use the same symbols (0 and 1) to indicate the sign of the number also. Normally an additional bit is used as the *sign-bit* and it is placed as the most significant bit. A 0 is used to represent a positive number and a 1 to represent a negative number. For example, an eight bit signed number 01000100 represents a positive number and its value (magnitude) is (01000100) = (68)$_{10}$. The left most 0 (MSB) indicates that the number is positive. On the other hand, in the signed binary form, 11000100 represents a negative number with magnitude (1000100)$_2$ = (68)$_{10}$. The 1 in the left most position (MSB) indicates that the number is negative and the other seven bits give its magnitude. This kind of representation for signed numbers is known as sign-magnitude representation and the user must take care to see the representation used while dealing with these numbers.

Example 2.20: Find the decimal equivalent of the following binary numbers assuming sign magnitude representation of the binary numbers.

  (a) 101100   (b) 001000

Solution:

  (a) Sign bit is 1, which means the number is negative.

Magnitude = 01100 = (12)$_{10}$

∴ (101100)$_2$ = (−12)$_{10}$

  (b) Sign bit is 0, which means the number is positive.

Magnitude = 01000 = 8

∴ (001000)$_2$ = (+8)$_{10}$

2.9 ONE'S COMPLEMENT REPRESENTATION

In a binary number, if each 1 is replaced by 0 and each 0 by 1, the resulting number is known as the one's complement of the first number. In fact, both the numbers are complement of each other. If one of these number is positive, the other will be negative with the same magnitude and vice versa. This method is widely used for representing signed numbers. In this representation, MSB is 0 for positive numbers and 1 for negative numbers.

Example 2.21: Find the one's complement of the following binary numbers:

  (a) 0100111001   (b) 11011010

Solution:

  (a) 1011000110   (b) 00100101

Example 2.22: Represent the following numbers in one's complement form:

  (a) + 7 and −7

  (b) + 8 and −8

Solution:

  (a) $(+7)_{10}$ = (0111)$_2$ and $(−7)_{10}$ = (1000)$_2$

  (b) $(+8)_{10}$ = (01000)$_2$ and $(−8)_{10}$ = (10111)$_2$

2.10 TWO'S COMPLEMENT REPRESENTATION

If 1 is added to 1's complement of a binary number, the resulting number is known as the *two's complement* of the binary number. For example, 2's complement of 0101 is 1011. Since 0101 represents $(+5)_{10}$, Therefore 1011 represents $(−5)_{10}$ in 2's complement representation. In this representation also, if the MSB is 0, the number is positive, whereas if the MSB is 1, the number is negative. It may be observed that the 2's complement of the 2's complement of a number is the number itself.

Example 2.23: Find the 2's complement of the number 01001110.

Solution:

Number   : 01001110
1's complement : 10110001
Add 1   : <u>   +1</u>
                  10110010   (2's complement of
                  the given number)

2.11 FLOATING POINT REPRESENTATION

Consider the decimal number 128.466 which may be written as

(i) 128.466

(ii) 0.128466×10^3

Suppose a register is capable of storing 6 digits and a sign bit and this register is divided into two parts, first part containing the integral portion of the number and second part containing the fractional portion and the decimal point located between the two parts of the register. (Fig. 2.1)

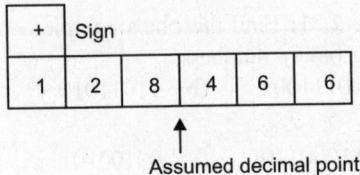

Fig. 2.1

This representation has two drawbacks. The first drawback of this scheme is the need of the user to remember and keep track of the decimal point location. The second drawback is that the range of numbers which can be represented using this scheme is limited to +999.999.

In the second method, called floating point representation, the number is written as a fraction multiplied by a power of 10. The fraction part is known as *mantissa* and the power of 10 (which multiplies the fraction) is known as *exponent*. The register is divided into two parts the first part of 4 digits to contain the mantissa and the second part of 2 digits to hold the exponent. To store both positive and negative exponents, it is desired to split the range of 00 to 99 into two parts. Assuming 0 or origin at 50 all exponents greater than 50, are considered to be positive and all exponents less than 50, as negative. The scheme is thus known as floating point representation with exponent in excess 50 form, the range of exponents will be from −50 to + 49. In this

scheme, the number 1384×10^3 can be stored in a register as in Fig. 2.2

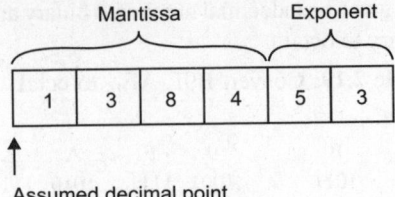

Fig. 2.2

To store the number $-0.0001288 \times 10^{-6}$, the number is first written as -0.1288×10^{-9}, thus keeping all the significant digits in the mantissa, the process is called normalization. This number can then be store in the register as in Fig. 2.3

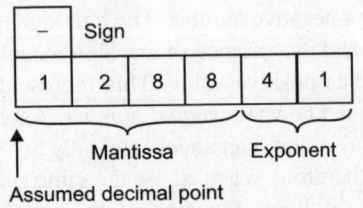

Fig. 2.3

A floating point number is called as in the normalized form if the most significant bit of the mantissa contains a non-zero digit. A floating point binary number is also represented in a similar way except that here, the base or radix is 2.

2.12 COUNTING IN BINARY

Counting in binary is very much similar to decimal counting as shown in Table 2.4

2.13 TWO'S COMPLEMENT ARITHMETIC

The 2's complement system is used to represent negative numbers using modulus arithmetic. The word length of a computer is fixed. That means, if a 4-bit number is added to another 4-bit number, the result will be only of 4-bits. Carry, if any, from the 4th bit will overflow. This is called *modulus arithmetic*. For example 1100 + 1111 = 1011

In the 2's complement subtraction, add the 2's complement of the subtrahend to the minuend. If there is a carry out, ignore it. Look at the sign bit, i.e. MSB of the sum term. If the MSB is a 0, the result is positive

Table 2.4: Counting in binary

Decimal	Binary
0	0000
1	0001
2	0010
3	0011
4	0100
5	0101
6	0110
7	0111
8	1000
9	1001
10	1010
11	1011
12	1100
13	1101
14	1110
15	1111

and is in true binary form. If the MSB is a 1 (whether there is a carry or no carry at all), the result is negative and is in its 2's complement form. Take its 2's complement to find its magnitude in binary.

Example 2.24: Add -75 to + 26, using the 8-bit 2's complement arithmetic.

Solution:

$$
\begin{aligned}
+75 &= 01001011 \\
-75 &= 10110101 \text{ (In 2's complement form)} \\
+26 &= 00011010 \\
-75 &= 10110101 \\
\hline
(-49) &= 11001111 \text{ (No carry)}
\end{aligned}
$$

There is no carry. The MSB is a 1, the result is negative and is in 2's complement form. The required magnitude is 2's complement of 11001111, i.e. 00110001 = 49. The result is therefore - 49.

Example 2.25: Subtract 14 from 46 using the 8-bit 2's complement arithmetic.

Solution:

+14 = 00001110

1's complement of + 14 = 11110001

2's complement of + 14 = 11110001 + 1

                      = 11110010

$$
\begin{aligned}
+46 &\quad 00101110 \\
-14 &\quad +11110010 \\
\hline
+32 &\quad 000100000 \quad \text{(Ignore the carry)}
\end{aligned}
$$

The MSB is 0, so the result is positive and is in normal binary form. Therefore, the result is + 00100000 = + 32

2.14 ONE'S COMPLEMENT ARITHMETIC

The 1's complement of a number is obtained by simply complementing each bit of the number, i.e. by changing all the 0s to 1s and all the 1s to 0s. We can also say that the 1's complement of a number is obtained by subtracting each bit of the number from 1. This complemented value represents the negative of the original number. This is implemented in the hardware by simply feeding all bits through inverters.

One of the difficulties of using 1's complement is its representation of zero. Both 00000000 and its 1's complement 11111111 represent zero. The 00000000 is called *positive zero* and the 11111111 is called *negative zero*.

Example 2.26: Subtract 14 from 25 using the 8 bit 1's complement arithmetic.

Solution:

$$
\begin{aligned}
25 &\quad 00011001 \\
-14 &\quad 11110001 \quad \text{(In 1's complement form)} \\
+11 &\quad ①00001010 \\
&\quad \downarrow \\
&\quad \rightarrow \quad\quad +1 \quad \text{(Add the end around carry)} \\
&\quad 00001011 \quad = + 11_{10}
\end{aligned}
$$

Example 2.27: Add –25 to +14, using the 8-bit 1's complement method.

Solution:

$$
\begin{aligned}
+14 &\quad 00001110 \\
-25 &\quad +11100110 \quad \text{(In 1's complement form)} \\
-11 &\quad 11110100 \quad \text{(No Carry)}
\end{aligned}
$$

There is no carry and MSB is 1. So the result is negative and is in 1's complement form. The 1's complement of 11110100 is 00001011. The result is therefore -11_{10}.

Example 2.28: Add –25 to –14, using the 8-bit 1's complement method.

Solution:

$$
\begin{array}{ll}
-25 & 11100110 \quad \text{(In 1's complement form)} \\
\underline{-14} & \underline{+11110001} \quad \text{(In 1's complement form)} \\
-39 & \text{①}11010111 \\
\end{array}
$$

$$
\begin{array}{ll}
& \downarrow \\
& \rightarrow \qquad +1 \text{ (Add the end around carry)} \\
& \underline{11011000}
\end{array}
$$

The MSB is a 1. So the result is negative and is in its 1's complement form. The 1's complement of 11011000 is 00100111. So, the result is –39.

Example 2.29: Add +25 to + 14, using the 8-bit 1's complement arithmetic.

Solution:

$$
\begin{array}{ll}
+25 & 00011001 \\
\underline{+14} & \underline{00001110} \quad \text{(In 1's complement form)} \\
+39 & 00100111
\end{array}
$$

There is no carry. The MSB is a 0. So the result is positive and is in pure binary. The result is 00100111 = + 39.

Example 2.30: Add +25 to –25, using the 8-bit 1's complement method.

Solution:

$$
\begin{array}{ll}
+25 & 00011001 \\
\underline{-25} & \underline{+11000110} \quad \text{(In 1's complement form)} \\
-00 & 11111111
\end{array}
$$

There is no carry. The MSB is a 1. So the result is negative and is in 1's complement form. The 1's complement of 11111111 is 00000000. Therefore, the result is –0.

EXERCISE

1. Convert the following binary numbers to decimal
   (a) 11   (b) 111   (c) 1111
   (d) 1001   (e) 1011
2. Convert each binary number to decimal
   (a) 110011.11   (b) 1011100.10101
   (c) 1000 001.111   (d) 101110.1010
   (e) 111111.11111
3. Add the binary numbers
   (a) 11 + 1   (b) 1001 + 101
   (c) 101 + 11   (d) 111 + 110
   (e) 1101 + 1011

4. Use direct subtraction on the following binary numbers
   (a) 11–1   (b) 101–100
   (c) 1100–1001   (d) 1110–11
5. Determine the 1's complement of each binary number
   (a) 101   (b) 1010
   (c) 11010111 (d) 00001
6. Solve using 2's complement method:
   (a) $10_2 - 01_2$
   (b) $1101_2 - 1001_2$
   (c) $1111000_2 - 111\ 1111_2$
   (d) $111_2 - 110_2$
   (e) $10101_2 - 10111_2$
7. Determine the binary numbers represented by the following decimal numbers
   (a) 37   (b) 255
   (c) 26.25   (d) 11.75
8. Convert the following numbers from decimal to octal and then to binary
   (a) 375   (b) 249   (c) 27.125
9. Convert the following binary numbers to octal and then to decimal:
   (a) 11011100.10 10 10
   (b) 01010011.010101
   (c) 10110011
10. Convert the following numbers to hexadecimal and then to binary:
    (a) 375   (b) 249   (c) 27.125
11. Convert the following binary numbers to hexadecimal and then to decimal
    (a) 11011100.101010
    (b) 01010011.010101
    (c) 10110011
12. How $(-28.5)_{10}$ will be represented using floating point representation.
13. Solve using 2's complement method:
    (a) $1111000_2 - 1111111_2$
    (b) $10101_2 - 10111_2$
    (c) $1101_2 - 1001_2$
    (d) $111_2 - 110_2$
14. Determine the 1's complement of each binary number
    (a) 101_2
    (b) 1010_2
    (c) 11010111_2
    (d) 00001_2

3

Binary Codes

The digital system works fine for transistors, relays, switches, and integrated circuits. When it is to be used by people for their decimal system it must be custom-designed to fit their system. Here we shall discuss some of the methods used to express both numbers and letters as binary codes.

Though a number of codes are in use, but we shall discuss only a few most commonly used codes:

1. Weighted binary codes
2. Non-weighted binary codes

3.1 WEIGHTED BINARY CODES

Weighted Binary Codes are those which follow the positional weighting principles. Each position of the number denotes a specific weight. The straight binary counting sequence is an example, for each column has a weight 8, 4, 2 or 1, i.e. 2^3, 2^2, 2^1, 2^0.

Several systems of codes are used to express the decimal digits 0 through 9 (Table 3.1). The 8421 and XS3 are both weighted codes, each four-bit group representing one decimal digit, the left three being weighted. The number 761_{10} for example, would be represented in 8421 code as

0111	0110	001
7	6	1

This allows any decimal number to be represented as a series of BCD codes. Using these codes, computer can add in what appears to be decimal and provide decimal answers. One application of an 8421 code appeared at a nuclear rocket test site where BCD lights indicated the time of day for observers two miles away. The time 6:32:40 was represented as

0000	0110	0011	0010	0100	0000
0	6	3	2	4	0

Table 3.1

Decimal	8421	XS3
0	0000	0011
1	0001	0100
2	0010	0101
3	0011	0110
4	0100	0111
5	0101	1000
6	0110	1001
7	0111	1010
8	1000	1011
9	1001	1100

Other codes may be easily processed by employing different hardwares.

3.1.1 Binary Coded Decimal Numbers (BCD)

A code is collection of special group of symbols used to represent numbers, letters, etc. In the BCD code, each decimal digit of the number is represented by its binary equivalent as a nibble, i.e. as a string of 4 bits each. The BCD code is not a number system, but it is a system with each digit encoded in its binary equivalent as a nibble. For example, decimal numbers 5429 and 9637 are expressed in BCD numbers as follows:

5	4	2	9	(Decimal)
0101	0100	0010	1001	(BCD)
9	6	3	7	(Decimal)
1001	0110	0011	0111	(BCD)

The advantage of the BCD code is an easy mode of conversion from decimal to binary and binary to decimal.

The main area of application of BCD numbers is where decimal data is transferred into or out of digital processes. BCD numbers are processed by circuits of calculator, digital clocks, digital voltmeter, etc.

Example 3.1: Convert the following BCD number to its decimal equivalent:

0100 0010 0111 1000

Solution: 0100 0010 0111 1000
          4   2   7   8

Ans: 4278

3.1.2 Converting a given Decimal Number to its BCD Equivalent

Step 1: Write the decimal number
Step 2: Convert each decimal digit to its 4-bit binary equivalent
Step 3: Write the binary number as answer.

Example 3.2: Convert the decimal number 35 into BCD:

Solution: **Step 1:**   3   5
             Step 2:   0011   0101
             Step 3:   00110101

Ans: $35_{10} = 110101$ BCD

3.1.3 Comparison of Number Systems with BCD

Table 3.2 gives the representation of the decimal number 1 through 15 in the binary, octal, hexadecimal and in BCD code for comparison.

3.2 ALPHANUMERIC CHARACTERS IN BCD

Numeric data is not the only form of data to be handled by a computer. It is often required to process alphanumeric data also. An *alphanumeric data* is a string of symbols where a symbol may be one of the letters A, B, C, D.........Z or one of the digits 0, 1, 2, ...9 or a special character such as +, -, *, /, ., ( ), =, space (or blank), etc. An alphabetic data consists of only the letters A, B,....Z and the blank character. Similarly numeric data consists of only numbers 0, 1, 2......9. However, any data must be represented internally by the bits 0 and 1. Hence binary coding schemes are used in computers to represent data internally.

Table 3.2: Comparison of number systems

Decimal	Binary	Octal	Hexadecimal	BCD
0	0	0	0	0000
1	1	1	1	0001
2	10	2	2	0010
3	11	3	3	0011
4	100	4	4	0100
5	101	5	5	0101
6	110	6	6	0110
7	111	7	7	0111
8	1000	10	8	1000
9	1001	11	9	1001
10	1010	12	A	00010000
11	1011	13	B	00010001
12	1100	14	C	00010010
13	1101	15	D	00010011
14	1110	16	E	00010100
15	1111	17	F	00010101

In discussing BCD in the previous article, we have used a group of 4 bits to represent a digit (character) in BCD. 4-bit BCD coding system can be used to represent only decimal numbers and 4 bits are insufficient to represent the various characters used by a computer. Hence, instead of using 4 bits with only 16 possible characters, computer designers commonly use 6 bits to represent characters in BCD codes. In the 6-bit BCD code, the four BCD numeric place positions are retained, but two additional *zero positions* are added (Table 3.3). With 6 bits, it is possible to represent 64 $(=2^6)$ different characters. This is a sufficient number to code the decimal digit (10), alphabetic letters (26) and other special characters (28).

Example 3.3: Show the binary digits used to record the word BASE in BCD.

Solution:

B = 110010 in BCD binary notation
A = 110001 in BCD binary notation
S = 010010 in BCD binary notation
E = 110101 in BCD binary notation

Hence the binary digits

  110010   110001   010010   110101
     B   A   S   E

will record the word BASE in BCD.

Table 3.3: Alphabetic and numeric characters in BCD, along with their octal equivalents

Character	BCD Code		OctalEquivalent
	Zone	Digit	
A	11	0001	61
B	11	0010	62
C	11	0011	63
D	11	0100	64
E	11	0101	65
F	11	0110	66
G	11	0111	67
H	11	1000	70
I	11	1001	71
J	10	0001	41
K	10	0010	42
L	10	0011	43
M	10	0100	44
N	10	0101	45
O	10	0110	46
P	10	0111	47
Q	10	1000	50
R	10	1001	51
S	01	0010	22
T	01	0011	23
U	01	0100	24
V	01	0101	25
W	01	0110	26
X	01	0111	27
Y	01	1000	30
Z	01	1001	31
1	00	0001	01
2	00	0010	02
3	00	0011	03
4	00	0100	04
5	00	0101	05
6	00	0110	06
7	00	0111	07
8	00	1000	10
9	00	1001	11
0	00	1010	12

3.3 REFLECTIVE CODES

A code is said to be reflective when the code for 9 is the inverse for the code for 0, 8 for 1, 7 for 2, 6 for 3 and 5 for 4. Note that the XS3 code is reflective (Table 3.1) whereas the 8421 code is not.

3.4 SEQUENTIAL CODES

A code is said to be sequential when each next code is one binary number greater than its preceding code.

This is mainly used in mathematical manipulation of data. The 8421 and XS3 codes are sequential but the 2421 and 5211 codes are not.

3.5 NON-WEIGHTED CODES

Non weighted codes are codes that are not positionally weighted, i.e. each position within the binary number is not assigned a fixed value. Two such codes are Excess-3 and Gray codes.

3.5.1 Excess-3 Code

Excess-3 (also called XS3) is a non-weighted code used to express decimal numbers. In this code, 3 is added to each of the decimal digits and then each of the resulting digit is converted to equivalent binary number written as a nibble, as is done in BCD code. The code has some very interesting properties when used in addition. To add in XS3, we add the binary numbers. If there is no carry out from the four bit group, subtract 0011. If there is a carry out, we add 0011.

The XS3 code derives its name from the fact that each binary code word is the corresponding 8421 code word plus 0011 (3). It is a sequential code and therefore, can be used for arithmetic operations. It is a self complementing code, i.e. reflective code (Table 3.1).

3.5.2 The Gray Code

The Gray code is a non-weighted code and is not suitable for arithmetic operations. It is not a BCD code. This belongs to a class of codes called *minimum-change codes* in which only one bit in the code-group changes in going from one step to the next. Since successive code words in this code differ in one bit position only, therefore it is a *unit distance* or *cyclic code*. It is also a reflective code: The n least significant bits for 2^n through $2^{n+1}-1$ are the mirror images of those for 0 through 2^n-1, (Table 3.4), i.e. for n = 3 (3 bits after the MSB) the (3) bits for 8 through 15 are the mirror images of those for 0 through 7.

The Gray code is often used in situations where other codes (such as binary) might produce erroneous or ambiguous results, where during successive transitions, more than one bit of the code is changing. For instance, using binary code, going from 0111 to 1000 requires that all four bits change simultaneously.

Table 3.4: Reflection in Gray code

Gray code				Decimal	4-bit binary
1-bit	*2-bit*	*3-bit*	*4-bit*		
0	0 0	0 0 0	0 0 0 0	0	0 0 0 0
1	0 1	0 0 1	0 0 0 1	1	0 0 0 1
	1 1	0 1 1	0 0 1 1	2	0 0 1 0
	1 0	0 1 0	0 0 1 0	3	0 0 1 1
		1 1 0	0 1 1 0	4	0 1 0 0
		1 1 1	0 1 1 1	5	0 1 0 1
		1 0 1	0 1 0 1	6	0 1 1 0
		1 0 0	0 1 0 0	7	0 1 1 1
			1 1 0 0	8	1 0 0 0
			1 1 0 1	9	1 0 0 1
			1 1 1 1	10	1 0 1 0
			1 1 1 0	11	1 0 1 1
			1 0 1 0	12	1 1 0 0
			1 0 1 1	13	1 1 0 1
			1 0 0 1	14	1 1 1 0
			1 0 0 0	15	1 1 1 1

Depending on the device or circuit that is generating the bits, there may be a significant difference in the transition times of the different bits. If so, the transition from 0111 to 1000 could produce one or more intermediate states, e.g. if MSB changes faster than the other bits, the following transition will occur:

0111   decimal 7
↓
1111   (erroneous code)
↓↓↓
1000   decimal 8.

Though, the occurrence of 1111 is only momentary, it could conceivably produce erroneous operation of the elements controlled by the bits. Using the Gray code would eliminate this problem, since only one bit change occurs per transition and no "race" between the bits occurs. However, this code being an unweighted code, it is not suited for arithmetic operations but it finds application in input/output devices and some types of analog-to-digital converters.

3.5.3 (a) Binary to Gray Conversion

To convert binary number into equivalent Gray code number, following rules apply:

(i) The most significant bit (MSB) in the Gray code is the same as corresponding digit in binary number.

(ii) Starting from left to right, add each adjacent pair of binary digits to get the next Gray code digit. Disregard carries if generated.

For example, to convert $(10010)_2$ to Gray code:

(i) The leftmost Gray code digit is same as the leftmost binary digit.

1	0	0	1	0	Binary
↓					
1					Gray

(ii) Add the leftmost binary digit to the adjacent one.

1 +	0	0	1	0	Binary
	↓				
1	1				Gray

(iii) Add next adjacent pair

1	0 +	0	1	0	Binary
		↓			
1	1	1			Gray

(iv) Add the next adjacent pair

1	0	0 +	1	0	Binary
			↓		
1	1	0	1		Gray

(v) Add the last adjacent pair,

1	0	0	1 +	0	Binary
				↓	
1	1	0	1	1	Gray

Hence Gray code for number $(100\ 10)_2$ is 11011.

3.5.3 (b) Gray to Binary Conversion

For converting Gray code to binary following rules apply:

(i) The most significant bit (MSB) in the binary code is the same as the corresponding digit in Gray code.

(ii) Add each binary digit generated to the Gray digit in the next adjacent position. Disregard carries.

For example, the Gray code number 11010 can be converted to binary as follows:

(i) The leftmost digits are the same.

1	1	0	1	0	Gray
↓					
1					Binary

(ii) Add the last binary digit just generated to Gray digit in next position. Neglect carry.

1	1	0	1	0	Gray

1   0   Binary

(iii) Add the last binary digit generated to next Gray digit.

1	1	0	1	0	Gray

1   0   0   Binary .

(iv) Add the last binary digit generated to next Gray digit.

1	1	0	1	0	Gray

1   0   0   1   Binary

(v) Add the last binary digit generated to the next Gray digit

1	1	0	1	0	Gray

1   0   0   1   1   Binary

Hence binary number is $(10011)_2$.

3.6 EXTENDED BINARY-CODED DECIMAL INTERCHANGE CODE (EBCDIP)

The major problem with BCD code is that only 64 $(=2^6)$ different characters can be represented in it. This is not sufficient for providing decimal numbers (10), lower case letters (26), capital letters (26) and a large number of other special characters (28^+). Hence, the BCD code was extended from a 6-bit code to an 8-bit code. The added two bits are used as additional zero-bits, expanding the zone to 4 bits. The resulting code is EBCDIC. In code, it is possible to represent 256 $(=2^8)$ different characters instead of 64 $(=2^6)$. In addition to the various, above character requirements, this also allows a large variety of printable characters and several nonprintable control characters. The control characters are used to control such activities as printer vertical spacing, movement of cursor on the terminal screen, etc.

Since EBCDIC is an 8-bit code, it can be easily divided into two 4-bit groups. Each of these 4-bit groups can be represented by 1 hexadecimal digit. Hence hexadecimal number system is used as short cut notation for memory dump by computers which use EBCDIC for internal representation of characters.

This results in a one-to-four reduction in the volume of memory dump. Table 3.5 shows the alphabetic and numeric characters in EBCDIC along with their hexadecimal equivalents.

Table 3.5: The alphabetic and numeric characters in EBCDIC along with hexadecimal equivalents (*)

Character	EBCDID Code		Hexadecimal
	Zero	Digit	Equivalent
A	1100	0001	C1
B	1100	0010	C2
C	1100	0011	C3
D	1100	0100	C4
E	1100	0101	C5
F	1100	0110	C6
G	1100	0111	C7
H	1100	1000	C8
I	1100	1001	C9
J	1101	0001	D1
K	1101	0010	D2
L	1101	0011	D3
M	1101	0100	D4
N	1101	0101	D5
O	1101	0110	D6
P	1101	0111	D7
Q	1101	1000	D8
R	1101	1001	D9
S	1110	0010	E2
T	1110	0011	E3
U	1110	0100	E4
V	1110	0101	E5
W	1110	0110	E6
X	1110	0111	E7
Y	1110	1000	E8
Z	1110	1001	E9
0	1111	0000	F0
1	1111	0001	F1
2	1111	0010	F2
3	1111	0011	F3
4	1111	0100	F4
5	1111	0101	F5
6	1111	0110	F6
7	1111	0111	F7
8	1111	1000	F8
9	1111	1001	F9

(*) After Pradeep K. Sinha and Priti Sinha, "Computer Fundamentals", BPB Publications New Delhi, India

3.7 ERROR DETECTING CODE

When a binary data is transmitted and processed, it is susceptible to noise that can distort its contents, i.e. the 1s may get changed to 0s and 0s to 1s. Because digital systems must be accurate to the digit, such errors can pose a serious problem. Several schemes have been devised to detect the occurrence of a single-bit error in a binary word and correct the binary word and transmit it.

The simplest technique for detecting errors is that of adding an extra bit, known as the *parity bit* to each word being transmitted. There are two types of parity, odd parity and even-parity. For odd parity, the parity bit is set to a 0 or a1 at the transmitter such that the total number of 1 bits in the word including the parity bit is an odd number. For even parity, the parity bit is set to a 0 or a 1 at the transmitter such that the total number of 1 bits in the word including the parity bit is an even number. Table 3.6 shows the parity bits to be added to transmit decimal digits 0 through 9 in the BCD code.

Table 3.6: Odd and even parity in the BCD code

Decimal	BCD	Odd-parity	Even-parity
0	0000	1	0
1	0001	0	1
2	0010	0	1
3	0011	1	0
4	0100	0	1
5	0101	1	0
6	0110	1	0
7	0111	0	1
8	1000	0	1
9	1001	1	0

When the digital data is received, a parity checking circuit generates an error signal if the total number of 1s is even in an odd-parity system, or odd in an even parity system. The parity check can detect a single bit error. If the code possesses the property by which the occurrence of any single-bit error transforms a valid code into an invalid one, it is said to be an error-detecting code. (A code is an error-detecting code if and only if its minimum distance is two or more). The distance between two words is defined as the number of digits that must change in a word so that the other word results. For example, the distance between 0011 and 1010 is 2 and the distance between 0111 and 1000 is 4.

3.8 ERROR CORRECTING CODE

A code is said to be an error correcting code if the correct code word can always be deduced from an erroneous word. For a code to be a single-bit error-correcting code, the minimum distance of that code must be 3. The minimum distance of a code is the smallest number of bits by which any two code words must differ. A code with minimum distance of three can not only correct single bit errors but also detect (but cannot correct) two-bit errors. The key to error correction is that it must be possible to detect and locate erroneous digits. If the location of an error has been determined, then by complementing the erroneous digit, the message can be corrected. One type error correcting code is Hamming code.

3.9 THE ASCII CODE

To get information into and out of a computer we need to use numbers, letters, and other symbols. This implies some kind of alphanumeric code for the I/O unit of a computer. At one time, every manufacturer had a different code which led to all kinds of confusion. Eventually, industry settled on an input-output code known as the *American Standard Code for Information Inter Change* (ASCII). This code allows manufacturers to standardize I/O hardware such as keyboards, printers, etc.

The ASCII (pronounced *ask'-ee*) code is a 7-bit code whose format (arrangement) is

$X_6 \, X_5 \, X_4 \, X_3 \, X_2 \, X_1 \, X_0$

where each X is a 0 or a 1. For instance, the letter A is coded as 1000001.

Sometimes, a space is inserted for easier reading 100 0001.

Table 3.7 shows the ASCII code. The table is read as follows: e.g. letter A has an $X_6 \, X_5 \, X_4$ of 100 and an $X_3 \, X_2 \, X_1 \, X_0$ of 0001. Therefore, its ASCII code is

100 0001   (A)

Table includes the ASCII code for lowercase letters also. The letter a is coded as

110 0001   (a)

Some more example are

110 0010   (b)

110 0011   (c)

010 0100   ($)

011 1101   (=)

Table 3.7: The ASCII code (*)

X_3X_2			$X_6X_5X_4$			
X_1X_0	*010*	*011*	*100*	*101*	*110*	*111*
0000	SP	0	@	P		p
0001	!	1	A	Q	a	q
0010	"	2	B	R	b	r
0011	#	3	C	S	c	s
0100	$	4	D	T	d	t
0101	%	5	E	U	e	u
0110	&	6	F	V	f	v
0111	'	7	G	W	g	w
1000	(	8	H	X	h	x
1001	)	9	I	Y	i	y
1010	*	:	J	Z	j	z
1011	+	;	K		k	
1100	,	<	L		l	
1101	-	=	M		m	
1110	.	>	N		n	
1111	/	?	O		o	

(*) After Albert Paul Malvino "Digital Computer Electronics"
(TMH)

In Table 3.7, SP stands for space (blank). Hitting the space bar of an ASCII keyboard sends this into a microcomputer010 0000 (for space)

EXERCISE

1. Encode the following decimal numbers in BCD code:

    (a) 20.305   (b) 46   (c) 327.8

2. Encode the decimal number 46 to Gray code.

3. Encode the decimal numbers in problem 1 to Excess-3 code.

4. Write your full name in

    (a) ASCII code   (b) EBCDIC code

5. Convert following to ASCII code

    (a) ABHAY   (b) AKANKSHA

6. Convert each Gray code to binary

    (a) 1011   (b) 10010

7. Convert each binary number to Gray code:

    (a) 11111_2 (b) 1001000_2

8. Convert each Excess-3 code number to decimal

    (a) 1011   (b) 1001

    (c) 0011   (d) 10001110

4

Boolean Algebra and Logic Circuit, Analysis and Design

4.1 INTRODUCTION

In the mid 1800s, an algebra, which simplified the representation and manipulation of propositional logic, was developed by George Boole of UK, (1815-1864). It became known as Boolean algebra, after his name. Later, in 1938, C.E. Shannon proposed the use of Boolean algebra in the design of relay switching circuits. The basic techniques described by Shannon were adopted almost universally, for the design and analysis of switching circuits. Because of the analogous relationship between action of relays and modern electronic circuits, the same techniques, which were developed for the design of relay circuits, are still being used in the design of modern computers.

Any complex logic statement can be expressed by a Boolean function. Boolean algebra differs from both the ordinary algebra and the binary number system. In Boolean algebra A + A = A and A. A = A because the variable A has only a logical value. It does not have any numerical significance. In ordinary algebra A + A = 2A and A . A = A² because the variable A has a numerical value here. In Boolean algebra 1 + 1 = 1 whereas in the binary number system 1 + 1 = 10 and in ordinary algebra 1 + 1 = 2. There is nothing like subtraction or division in Boolean algebra. Also there are no negative and fractional numbers. A variable or function of variables in Boolean algebra can assume only two values, either a 0 or a 1. Logical multiplication is the same as the AND operation and logical addition is the same as the OR operation.

4.2 LOGIC OPERATIONS

The AND, OR and NOT are the three basic operations or functions that are performed in Boolean algebra. In addition, there are some derived operations such as NAND, NOR, XOR and XNOR. These have been described earlier in Chapter 1 in the context of logic gates. However, we will surmise these here very briefly in the context of Boolean algebra.

AND Operation: The AND operation in Boolean algebra is similar to multiplication as performed in ordinary algebra. In fact, it is logical multiplication as performed by the AND gate.

OR Operation: The OR operation in Boolean algebra is similar to addition in ordinary algebra. In fact, it is logical addition as performed by the OR gate.

NOT Operation: The NOT operation in Boolean algebra is nothing but complementation or inversion, i.e. negation as performed by the NOT gate. The NOT operation is indicated by a bar (-) over the variable.

NAND Operation: The NAND operation is equivalent to AND operation plus NOT operation, i.e. it is the AND operation followed by NOT operation for NAND gate.

NOR Operation: The NOR operation in Boolean algebra is equivalent to OR operation followed by NOT operation, i.e. it is the negation of the OR operation as performed by the NOR gate.

X-OR and X-NOR Operations: The X-OR and X-NOR operations on variables A and B in Boolean algebra are denoted by $A \oplus B = A\overline{B} + \overline{A}B$ and $A \odot B = AB + \overline{A}\overline{B}$ respectively.

4.3 BOOLEAN EXPRESSION

Variables and literal-variables are the different letters in a Boolean expression. For example $A.B + \overline{B}\overline{C}D$. Here, there are four variables A, B, C, D.

The complement of variable is not considered as a separate variable.

Each occurrence of variable or its complement is called a *literal*. In the above expression, there are five literals.

4.4 THE PRINCIPLE OF DUALITY

In Boolean algebra, there is a duality between the operations AND and OR, and the digits 0 and 1. For example, consider Table 4.1. We can see that the second

Table 4.1: Illustrating the principle of duality

	Column 1	Column 2	Column 3
Row 1:	1 + 1 = 1	1 + 0 = 0 + 1 = 1	0 + 0 = 0
Row 2:	0 . 0 = 0	0 . 1 = 1 . 0 = 0	1 . 1 = 1

row of the table is obtained from the first row and vice versa, simply by interchanging '+' and '.' and '0' with '1'. This important property is known as the principle of duality in Boolean algebra. Hence, any theorem in Boolean algebra has its dual and if a particular theorem is proved, its dual theorem automatically holds and need not be proved separately.

4.5 DESCRIBING LOGIC CIRCUITS ALGEBRAICALLY

Any logic circuit, however complex, may be completely described using Boolean expressions and operations employing OR, AND and NOT gates. For example, consider the circuit shown in Fig. 4.1

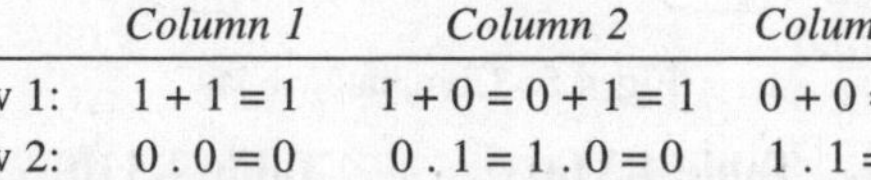

$$A\!\!\!\!\circ\!\!\!\!\!-\!\!\!\!\!\!\begin{array}{c}A.B\\\\B\!\!\!\!\circ\!\!\!\!\!-\end{array}\!\!\!\!\!\!\!\!\!\begin{array}{c}\\\\C\!\!\!\!\circ\!\!\!\!\!-\end{array}\!\!\!\!\!\!\!\!\!\!\!\!\!\!\!\!\!\!\begin{array}{c}\\\\\end{array}Y = A.B + C$$

Fig. 4.1: A logic circuit

The circuit has three inputs A, B and C and a single output. Utilizing the Boolean expression for each gate, we can easily determine the expression for the output. In this circuit, there are two operations (AND and OR) and there may be a confusion as to which operation is to be performed first. To avoid this

confusion, it is to be understood that if an expression contains both AND and OR operations, the AND operations are performed first, unless there are parentheses (( )) in the expression, in which case, the operation inside the parentheses is to be performed first (This rule is similar to that used in ordinary algebra). This is illustrated by example of Fig. 4.2.

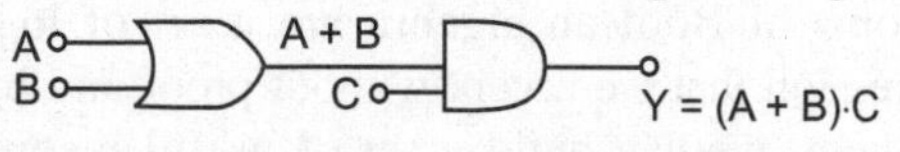

$$A\!\!\!\!\circ\!\!\!\!-\!\!\!\!\!\!\begin{array}{c}A + B\\\\B\!\!\!\!\circ\!\!\!\!-\end{array}\!\!\!\!\!\!\!\!\!\!\!\!\begin{array}{c}\\\\C\!\!\!\!\circ\!\!\!\!-\end{array}\!\!\!\!\!\!\!\!\!\!\!\!\!\!\!\!\!\begin{array}{c}\\\\\end{array}Y = (A + B){\cdot}C$$

Fig. 4.2: Logic circuit whose expression requires parenthesis

4.6 IMPLEMENTING CIRCUITS FROM BOOLEAN EXPRESSIONS

If the operation of a circuit is defined by a Boolean expression, a logic circuit diagram can be implemented directly from the expression for example, if we need a circuit defined by y = A.B.C, we immediately know that we need a three input AND gate. If we need a circuit defined by $y = A + \overline{B}$, we would need a two input OR gate with an inverter on B input. Similar reasoning can be used for more complex circuits.

Now, suppose we want to construct a circuit whose output is $y = AC + B\overline{C} + \overline{A}BC$. This Boolean expression contains three terms $(AC, B\overline{C}, \overline{A}BC)$ which are ORed together. This tells us that a three input OR gate is required with inputs that are equal to AC, $B\overline{C}$ and $\overline{A}BC$. This is illustrated in Fig. 4.3a. Now, each OR gate input is an AND product term which means that an AND gate with appropriate inputs can be used to generate each of these terms. This is shown

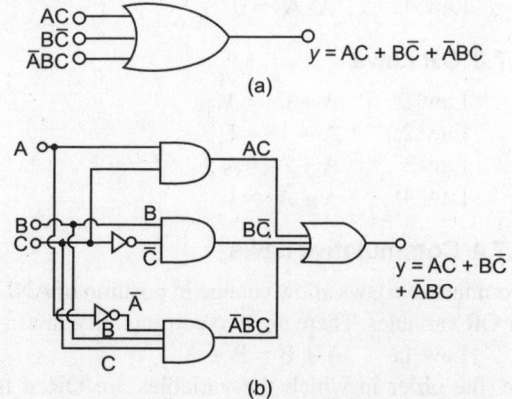

(a)

(b)

Fig. 4.3: Constructing a logic circuit from a Boolean expression

in Fig. 4.3b which is the final circuit diagram. This same general approach can always be followed, although we will later see that there are some more clever and efficient techniques that can be used.

4.7 AXIOMS AND LAWS OF BOOLEAN ALGEBRA

Axioms of Boolean algebra are a set of logical expression that we accept without proof and based on them, we can build a set of useful theorems. Actually axioms are nothing more than the definitions of the three basic logic operations AND, OR and INVERT. Each axiom can be interpreted as the outcome of an operation performed by a logic gate.

Axiom 1:	$0 . 0 = 0$	Axiom 6 : $0 + 1 = 1$
Axiom 2:	$0 . 1 = 0$	Axiom 7 : $1 + 0 = 1$
Axiom 3:	$1 . 0 = 0$	Axiom 8 : $1 + 1 = 1$
Axiom 4:	$1 . 1 = 1$	Axiom 9 : $= 0$
Axiom 5:	$0 + 0 = 0$	Axiom 10 : $= 1$

4.7.1 Complementation Laws: (The term complement means simply, to invert)

Law 1:	$\overline{0} = 1$
Law 2:	$\overline{1} = 0$
Law 3:	If A = 0 then $\overline{A} = 1$
Law 4:	If A = 1 then $\overline{A} = 0$
Law 5:	$\overline{\overline{A}} = A$

4.7.2 AND Laws

Law 1:	$A . 0 = 0$
Law 2:	$A . 1 = A$
Law 3:	$A . A = A$
Law 4:	$A . \overline{A} = 0$

4.7.3 OR Laws

Law 1:	$A + 0 = A$
Law 2:	$A + 1 = 1$
Law 3:	$A + A = A$
Law 4:	$A + \overline{A} = 1$

4.7.4 Commutative Laws

Commutative laws allow change in position of AND or OR variables. There are two commutative laws:

Law 1: $A + B = B + A$

i.e. the order in which the variables are ORed is immaterial. We give below the truth tables illustrating this law.

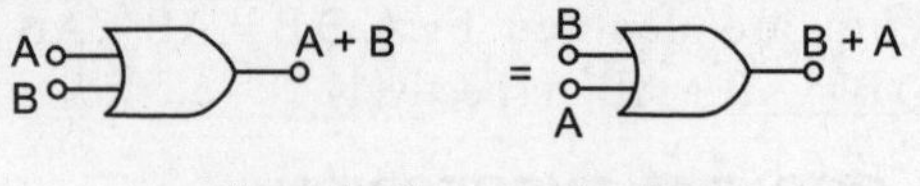

Fig. 4.4: Commutative law

Table 4.2 (a)				Table 4.2 (b)		
A	B	A + B		B	A	B + A
0	0	0		0	0	0
0	1	1	=	0	1	1
1	0	1		1	0	1
1	1	1		1	1	1

This law can be extended to any number of variables:

$A + B + C = B + C + A = C + A + B = B + A + C$

Law 2: $A . B = B . A$

i.e. the order in which the variables are ANDed is immaterial. The truth tables are given below to illustrate this law:

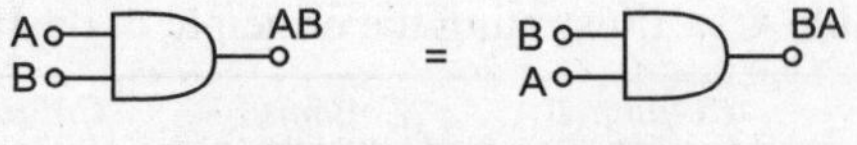

Fig. 4.5: Commutative law

Table 4.3 (a)				Table 4.3 (b)		
A	B	A · B		B	A	B · A
0	0	0		0	0	0
0	1	0	=	0	1	1
1	0	0		1	0	1
1	1	1		1	1	1

This law can be extended to any number of variables:

$A . B . C = B . C . A = C . A . B = B . A . C$

4.7.5 Associative Laws

The associative laws allow grouping of variables. There are two associative laws:

Law 1: $(A + B) + C = A + (B + C)$

This law states that the way the variables are grouped and ORed is immaterial. The following truth table illustrates this law.

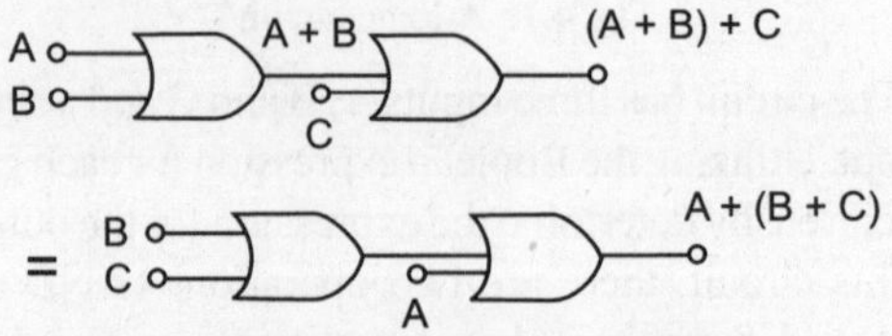

Fig. 4.6: Associative law '1'

Table 4.4 (a)

A	B	C	A+B	(A+B)+C
0	0	0	0	0
0	0	1	0	1
0	1	0	1	1
0	1	1	1	1
1	0	0	1	1
1	0	1	1	1
1	1	0	1	1
1	1	1	1	1

=

Table 4.4 (b)

A	B	C	B+C	A+(B+C)
0	0	0	0	0
0	0	1	1	1
0	1	0	1	1
0	1	1	1	1
1	0	0	0	1
1	0	1	1	1
1	1	0	1	1
1	1	1	1	1

Law 2:   (A.B) C = A (B.C)

This law states that the way variable are grouped and ANDed is immaterial. The following truth table illustrates this.

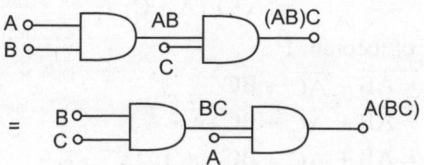

Fig. 4.7: Associative law '2'

Table 4.5 (a)

A	B	C	AB	(AB)C
0	0	0	0	0
0	0	1	0	0
0	1	0	0	0
0	1	1	0	0
1	0	0	0	0
1	0	1	0	0
1	1	0	1	0
1	1	1	1	1

=

Table 4.5 (b)

A	B	C	BC	A(BC)
0	0	0	0	0
0	0	1	0	0
0	1	0	0	0
0	1	1	1	0
1	0	0	0	0
1	0	1	0	0
1	1	0	0	0
1	1	1	1	1

4.7.6 Distributive Law

The distributive laws allow multiplying out of expressions. There are three distributive laws.

Law 1:   A (B + C) = AB + AC

The truth tables given below illustrate this law:

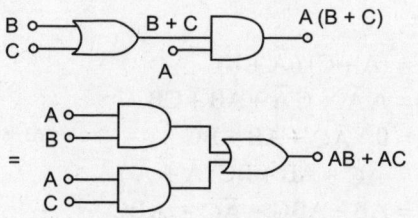

Fig. 4.8: Distributive law

Table 4.6 (a)

A	B	C	AB	(AB)C
0	0	0	0	0
0	0	1	1	0
0	1	0	1	0
0	1	1	1	0
1	0	0	0	0
1	0	1	1	1
1	1	0	1	1
1	1	1	1	1

=

Table 4.6 (b)

A	B	C	AB	AC	AB+AC
0	0	0	0	0	0
0	0	1	0	0	0
0	1	0	0	0	0
0	1	1	1	0	0
1	0	0	0	0	0
1	0	1	0	1	1
1	1	0	1	0	1
1	1	1	1	1	1

This law applies also to a combination of variables, e.g.

$$AB (C + D) = ABC + ABD$$
$$A (BC + DE) = ABC + ADE$$

Law 2:   A + BC = (A + B) (A + C)

This can be proved algebraically as follows:

RHS = (A + B) (A + C)
     = AA + AC + BC + BC
     = A + AC + AB + BC
     = A (1 + C + B) + BC
     = A· 1 + BC   (∴ 1 + C + B = 1 + B = 1)
     = A + BC
     = LHS

Law 3: $A + \overline{A}B = A + B$

This can also be proved algebraically as follows:

$$A + \overline{A}B = (A + \overline{A})(A + B)   \text{(Using law 2)}$$
$$= 1· (A + B)$$
$$= A + B$$

4.7.7 Idempotence Law

Law 1:   A· A = A
Law 2:   A + A = A

(Idempotence means the same value).

4.7.8 Complementation Laws

Law 1: $A· \overline{A} = 0$

Law 2: $A + \overline{A} = 1$

4.7.9 Double Negation Law

$$\overline{\overline{A}} = A$$

4.7.10 Identity Laws

There are two identity laws:

Law 1:   A·1 = A
Law 2:   A + 1 = 1

The first law can be seen as

if $A = 0$ then $A \cdot 1 = 0 \cdot 1 = 0 = A$

if $A = 1$ then $A \cdot 1 = 1 \cdot 1 = 1 = A$

For 2nd law

if $A = 0$ then $A + 0 = 0 + 1 = 1$

if $A = 1$ then $A + 1 = 1 + 1 = 1$

4.7.11 Null Laws

There are two null laws.

Law 1: $A \cdot 0 = 0$

Law 2: $A + 0 = A$

For 1st Law

if $A = 0$ then $A \cdot 0 = 0.0 = 0$

if $A = 1$ then $A \cdot 0 = 1 \cdot 0 = 0$

For 2nd law

if $A = 0$ then $A + 0 = 0 + 0 = 0 = A$

if $A = 1$ then $A + 0 = 1 + 0 = 1 = A$

Figure 4.9 depicts single variable theorems, where, in each case, x is a logic variable that can be either a 0 or a 1. Each theorem is accompanied by a logic circuit diagram that demonstrates its validity.

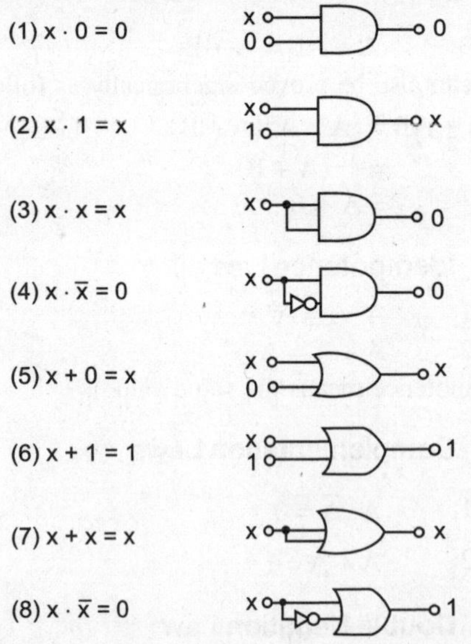

(1) $x \cdot 0 = 0$

(2) $x \cdot 1 = x$

(3) $x \cdot x = x$

(4) $x \cdot \bar{x} = 0$

(5) $x + 0 = x$

(6) $x + 1 = 1$

(7) $x + x = x$

(8) $x \cdot \bar{x} = 0$

Fig. 4.9: Single variable theorems. x is a logic variable that can be either 0 or 1.

4.7.12 Absorption Laws

Law 1: $A + A \cdot B = A$

Law 2: $A(A + B) = A$

Algebraically

   $A + A \cdot B = A(1 + B) = A \cdot 1 = A$

Therefore

   $A + A.$ any term $= A$

Also, we have

   $A(A + B) = A.A + A \cdot B = A + AB$

   $= A(1 + B) = A \cdot 1 = A$

Therefore

   $A(A +$ any term$) = A$

4.7.13 Included Factor Theorem

There are two theorems:

Theorems 1: $AB + \bar{A}C + BC = AB + \bar{A}C$

Theorems 2: $(A + B)(\bar{A} + C)(B + C)$

               $= (A +)(\bar{A} + C)$

Proof of theorem 1:

LHS $= AB + \bar{A}C + BC$

     $= AB + \bar{A}C + BC(A + \bar{A})$

     $= AB + \bar{A}C + BCA + BC\bar{A}$

     $= AB(1 + C) + \bar{A}C(1 + B)$

     $= AB(1) + \bar{A}C(1)$

     $= AB + \bar{A}C = $ RHS

Proof of theorem 2:

LHS $= (A + B)(\bar{A} + C)(B + C)$

     $= (A\bar{A} + AC + B\bar{A} + BC)(B + C)$

     $= (AC + BC + \bar{A}B)(B + C)$

     $= ABC + BC + \bar{A}B + BC + AC + \bar{A}BC$

     $= AC + BC + \bar{A}B$

RHS $= (A + B)(\bar{A} + C)$

     $= A\bar{A} + AC + BC + \bar{A}B$

     $= AC + BC + \bar{A}B$

     $= $ LHS

4.7.14 Transposition Theorem

     $AB + \bar{A}C = (A + C)(\bar{A} + B)$

Proof:

RHS $= (A + C)(\bar{A} + B)$

     $= A\bar{A} + C\bar{A} + AB + CB$

     $= 0 + \bar{A}C + AB + BC$

     $= \bar{A}C + AB + BC(A + \bar{A})$.

     $= AB + ABC + \bar{A}C + \bar{A}BC$

     $= AB + \bar{A}C = $ LHS.

4.8 DE MORGAN'S THEOREMS

De Morgan's theorems represent two of the most powerful laws of Boolean algebra.

Law 1: $\overline{A + B} = \overline{A} \cdot \overline{B}$

Law 2: $\overline{AB} = \overline{A} + \overline{B}$

Table 4.7 proves the law 1 by the method of perfect induction.

Table 4.7: Truth table for proving De Morgan's law 1

A	B	A + B	$\overline{A+B}$	$\overline{A}$	$\overline{B}$	$\overline{A} \cdot \overline{B}$
0	0	0	1	1	1	1
0	1	1	0	1	0	0
1	0	1	0	0	1	0
1	1	1	0	0	1	0

Table 4.8 proves the law 2 by the method of perfect induction.

Table 4.8: Truth table for proving De Morgan's law 2

A	B	$A \cdot B$	$\overline{A \cdot B}$	$\overline{A}$	$\overline{B}$	$\overline{A} + \overline{B}$
0	0	0	1	1	1	1
0	1	0	1	1	0	1
1	0	0	1	0	1	1
1	1	1	0	0	0	0

4.9 IMPLICATIONS OF DE MORGAN'S THEOREMS

Let us examine De Morgan's theorems 1 and 2 from the stand point of logic circuits. The first theorem is

$$\overline{A + B} = \overline{A} \cdot \overline{B}$$

The left hand side of the equation can be viewed as the output of a NOR gate whose inputs are A and B. The right hand side, on the other hand, is the result of first inverting both A and B and then putting them through an AND gate. These two representations are equivalent and are illustrated in Fig. 4.10a

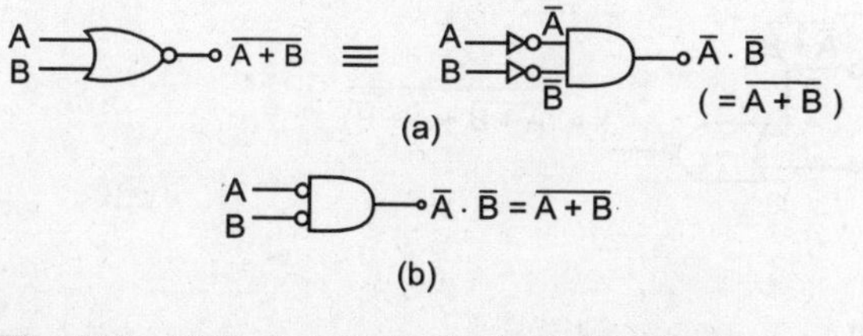

(a)

(b)

Fig. 4.10: (a) Equivalent circuits implied by De Morgan's first theorem, (b) alternative symbol for the NOR function

This means that an AND gate with inverters on each of its inputs is equivalent to a NOR gate (as shown in Fig. 4.10b), where the small circles on the inputs represent the inversion operation. Now, consider De Morgan's 2nd theorem:

$$\overline{A \cdot B} = \overline{A} + \overline{B}$$

The left hand side of the equation can be implemented by a NAND gate with inputs A and B. The right hand side can be implemented by first inverting inputs A and B and then putting them through an OR gate. These two equivalent operations are shown in Fig. 4.10c

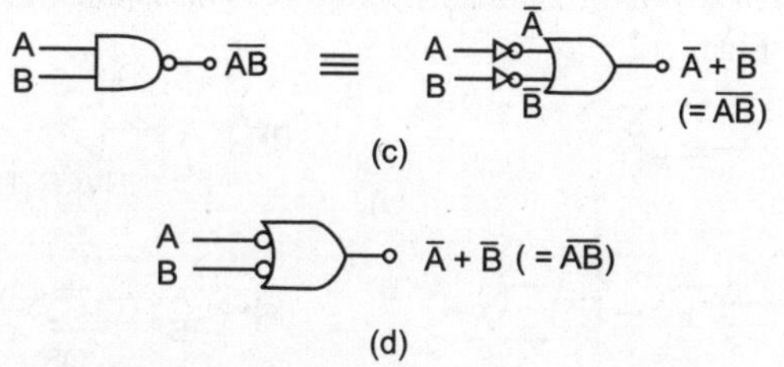

(c)

(d)

Fig. 4.10: (c) Equivalent circuit implied by De Morgan's 2nd theorem (d) alternative symbol for the NAND operation.

The OR gate with inverters on each of the inputs is equivalent to the NAND gate, as shown in Fig. 4.10d, where the circles on the inputs represent the inversion operation.

4.10 UNIVERSALITY OF NAND AND NOR GATES

We have been familiar with Boolean expressions (section 4.3). All Boolean expressions consist of various combinations of the basic operations of OR, AND and INVERT. So, any expression can be

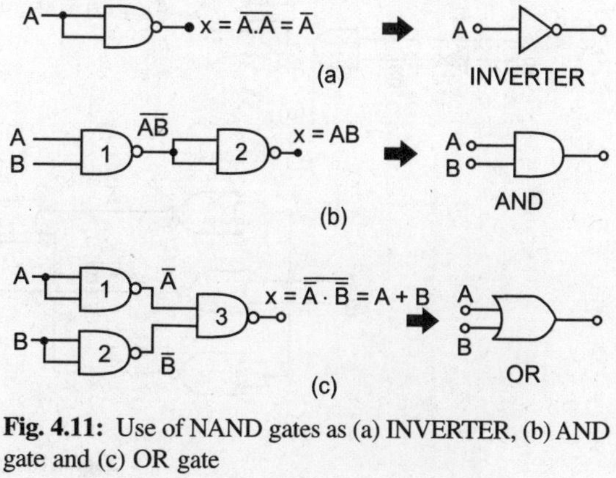

(a)   INVERTER

(b)   AND

(c)   OR

Fig. 4.11: Use of NAND gates as (a) INVERTER, (b) AND gate and (c) OR gate

implemented using combinations of OR, AND and INVERT (i.e. NOT) gates. It is possible however, to implement any logic expression using only NAND gates (and no other type of gate). This is because NAND gates, in the proper combination, can be used to perform each of the Boolean operations OR, AND and INVERT. This is illustrated in Fig. 4.11.

Note that in Fig. 4.11c NAND gates 1 and 2 are used as INVERTERS, so that after NAND gate (3), the final out put is $x = \overline{A.B}$ which can be simplified to $x = A + B$ using De Morgan's theorem.

Similarly, NOR gates can be arranged to implement any of the Boolean operations. This is illustrated in Fig. 4.12.

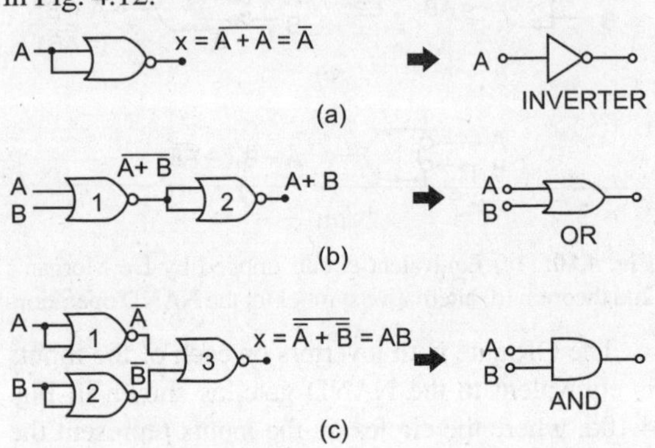

$A \rightarrow\!\!\!>\!\!\circ\; x = \overline{A + A} = \overline{A} \quad\Rightarrow\quad A \circ\!\!-\!\!>\!\!\circ$

INVERTER

(a)

$\begin{array}{c} A \\ B \end{array} \!\!>\!\!\circ^{\;\overline{A+B}}_{1} \!\!-\!\!>\!\!\circ_{2}\; A+B \quad\Rightarrow\quad \begin{array}{c} A \\ B \end{array} \!\!\circ\!\!-\!\!>\!\!\circ$

OR

(b)

$\begin{array}{c} A \\ B \end{array}$ ... $x = \overline{\overline{A} + \overline{B}} = AB \quad\Rightarrow\quad \begin{array}{c} A \\ B \end{array} \!\!\circ\!\!-\!\!>\!\!\circ$

AND

(c)

Fig. 4.12: Use of NOR gates as (a) Inverter (b) OR gate and (c) AND gate

Example 4.1: Realize the XOR function using (a) AND-OR-Invert logic (b) NAND logic and (c) NOR logic.

Solution:

(a) Using AOI logic: $x = \overline{A}B + A\overline{B}$ (Fig. 4.13a)

(b) Using NAND logic:

$x = A\overline{B} + \overline{A}B$

$\quad = A\overline{A} + A\overline{B} + \overline{A}B + B\overline{B}$

$\qquad\qquad$ (adding and $A\overline{A}$ and $B\overline{B}$)

$\quad = A(\overline{A} + \overline{B}) + B(\overline{A} + \overline{B})$

$\quad = A\overline{AB} + B\overline{AB}$

(using De Morgan's theorem $\overline{C + D} = \overline{C} \cdot \overline{D}$)

$\quad = \overline{\overline{A\overline{AB} + B\overline{AB}}}$

$\quad = \overline{A\overline{AB} \cdot B\overline{AB}}$

(Fig. 4.13b shows implementation of this logic)

(c) Using NOR logic:

$x = A\overline{B} + \overline{A}B$

$\quad = A\overline{A} + A\overline{B} + \overline{A}B + B\overline{B}$

$\qquad\qquad$ (adding and $A\overline{A}$ and $B\overline{B}$)

$\quad = A(\overline{A} + \overline{B}) + B(\overline{A} + \overline{B})$

$\quad = (A + B)(\overline{A} + \overline{B})$

$\qquad\qquad$ (Taking $\overline{A} + \overline{B}$ as common)

$\quad = \overline{\overline{(A + B)(\overline{A} + \overline{B})}}$

$\quad = \overline{(\overline{A + B}) (\overline{\overline{A} + \overline{B}})}$

(using De Morgan's theorem $\overline{CD} = \overline{C} + \overline{D}$)

(Fig. 4.13c shows implementation of this logic)

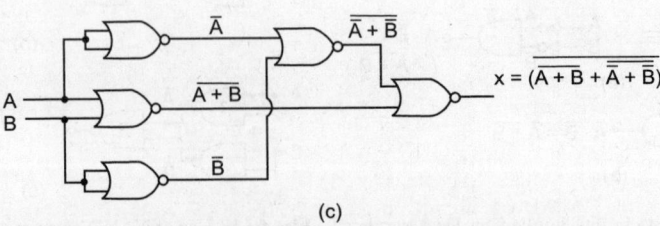

(a)

(b)

(c)

Fig. 4.13: (Example 4.1)

Table 4.9 summarizes all the basic Boolean identities.

Table 4.9: Summary of basic Boolean identities

S. No.	Identities	Dual Identities
1.	$A + 0 = A$	$A \cdot 1 = A$
2.	$A + 1 = 1$	$A \cdot 0 = 0$
3.	$A + A = A$	$A \cdot A = A$
4.	$A + \overline{A} = 1$	$A \cdot \overline{A} = 0$
5.	$\overline{\overline{A}} = A$	
6.	$A + B = B + A$	$A \cdot B = B \cdot A$
7.	$(A + B) + C = A + (B + C)$	$(A \cdot B)C = A \cdot (B \cdot C)$
8.	$A \cdot (B + C) = A \cdot B + A \cdot C$	$A + B \cdot C$ $= (A + B) \cdot (A + C)$
9.	$A + A \cdot B = A$	$A \cdot (A + B) = A$
10.	$A + \overline{A} \cdot B = A + B$	$A \cdot (\overline{A} + B) = A \cdot B$
11.	$\overline{A + B} = \overline{A} \cdot \overline{B}$	$\overline{AB} = \overline{A} + \overline{B}$

Example 4.2: Prove
$$(A + C)(\overline{A} + B) = AB + \overline{A}C$$
Solution:
L.H.S. $= A\overline{A} + AB + C\overline{A} + CB$
$= 0 + AB + C(\overline{A} + B)$
$= AB + C(\overline{A} + AB)$
$\quad (A + \overline{A}B = A + B) \Rightarrow \overline{A} + AB = \overline{A} + B)$
$= AB + \overline{A}C + ABC$
$= AB(C + 1) + \overline{A}C$
$= AB + \overline{A}C = \text{R.H.S.}$ $\quad (\because C + 1 = 1)$

Example 4.3: Prove
$$(A + B)(\overline{A} + C) = B\overline{A} + AC$$
Solution:
L.H.S. $= (A + B)(\overline{A} + C)$
$= A\overline{A} + AC + B\overline{A} + BC$
$= 0 + AC + B(\overline{A} + C),$
$\quad (A + \overline{A}B = A + B \Rightarrow \overline{A} + AB = \overline{A} + B)$
$= AC + B(\overline{A} + AC)$
$= AC + B\overline{A} + BAC$
$= AC(1 + B) + B\overline{A}$
$= B\overline{A} + AC$ $\quad (\because 1 + B = 1)$
$= \text{R.H.S.}$

Example 4.4: Implement $F = AB + \overline{CD}$ with three NAND gates

Solution: $F = AB + C'D' = [(AB)' \, (C'D')']'$
(**Note:** a dash is used instead of a 'bar' to denote a complement or inverse)

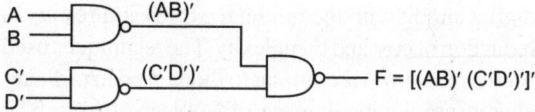

Fig. 4.14 (Example 4.4)

Example 4.5: Prove that $A\overline{B} + \overline{A}B = \overline{(\overline{AB} + \overline{\overline{A}B})}$
Solution:

L.H.S. $= A\overline{B} + \overline{A}B$
$= \left[\overline{(\overline{A\overline{B}}) \cdot (\overline{\overline{A}B})} \right],$
(using $\overline{CD} = \overline{C} + \overline{D}$) where $C = A\overline{B}$, $D = \overline{A}B$
$= \left[\overline{(\overline{A} + B)(A + \overline{B})} \right]$
$\quad (\because \overline{A\overline{B}} = \overline{A} + B \text{ and } \overline{\overline{A}B} = A + \overline{B})$
$= \left[\overline{A}A + \overline{A}\overline{B} + BA + B\overline{B} \right]'$
$= 0 + \overline{A}\overline{B} + BA + 0$
$= \overline{(AB + \overline{A}\overline{B})} = \text{R.H.S.}$

Example 4.6: Prove
$$AB + \overline{A}\overline{B} = (A + \overline{B})(\overline{A} + B) = \overline{(A\overline{B} + \overline{A}B)}$$
Solution:

L.H.S. $= AB + \overline{A}\overline{B} = \left[\overline{(\overline{AB}) + \overline{\overline{A}\overline{B}}} \right]'$
(using $\overline{(CD)} = \overline{C} + \overline{D}$ with $C = AB$, $D = \overline{A}\overline{B}$)
$= [(\overline{A} + \overline{B})(A + B)]'$ (using $\overline{AB} = \overline{A} + \overline{B}$)
$= \overline{(\overline{A}A + \overline{A}B + \overline{B}A + \overline{B}B)}$
$= \overline{(0 + \overline{A}B + \overline{A}B + 0)}$
$= \overline{(A\overline{B} + \overline{A}B)}$ (**Ans.:** ii)
$= (\overline{A\overline{B}}) \cdot (\overline{\overline{A}B}),$ using $\overline{C + D} = \overline{C}\overline{D}$
$= (\overline{A} + B)(A + \overline{B}),$ (**Ans.:** i)
using $\overline{CD} = \overline{C} + \overline{D}$

Example 4.7 Diagram the logic circuit for
$F = A\overline{C} + B\overline{C}D + ACD$, using NAND gates.
$F = [(AC)' \, (BC'D)' \, (ACD)]'$

The logic circuit can be drawn similar to example 4.4.

4.11 REDUCING BOOLEAN EXPRESSIONS

Every Boolean expressions must be reduced to as simple form as possible, because realization of a digital circuit with the minimal expression results in reduction of cost and complexity. The techniques used for the reductions are similar to those used in ordinary algebra employing the laws of Boolean algebra. The procedure is as follows:

1. Multiply all variables, removing the parentheses.

2. Look for identical terms. Only one of the identical terms be retained and remaining dropped, e.g.
$$AB + AB + AB = AB$$

3. Look for a variable and its negation in the same term. This term can be dropped, e.g.
$$A \cdot B\overline{B} = A \cdot 0 = 0$$
$$ABC \cdot \overline{C} = AB \times 0 = 0 \quad \text{(using } C \cdot \overline{C} = 0)$$

4. Look for pairs of terms that are identical except for one variable which may be missing in one of the terms. The larger term can be dropped:
$$AB\overline{C}\overline{D} + AB\overline{C} = AB\overline{C}(\overline{D} + 1) = AB\overline{C} \cdot 1$$
$$= AB\overline{C} \quad \text{(using } A + 1 = 1)$$

5. Look for the pairs of terms which have the same variables, with one or more variables complemented. If a variable in one term of such a pair is complemented while in the second term is not, then use
$$A + \overline{A} = 1, \text{ e.g.}$$
$$AB\overline{C}\overline{D} + AB\overline{C}D = AB\overline{C}(\overline{D} + D) = AB\overline{C}$$
$$AB(C + D) + AB(\overline{C + D})$$
$$AB[(C + D) + (\overline{C + D})] = AB \cdot 1 = AB$$

6. $A + AB = A$ or $A + \overline{A}B = A + B$.

Example 4.8: Prove
$$(X + X\overline{Y})\left[XZ + X\overline{Z}(Y + \overline{X})\right](Z + Y) = XY + XZ$$
using algebraic reduction.

Solution:

L.H.S. $= (X + X\overline{Y})[X\overline{Z}T + X\overline{Z}\overline{X}](Z + Y)$
$$= X[XZ + X\overline{Z}Y + 0](Z + Y);$$
$$\quad (\because A + AB = A \text{ and } X\overline{X} = 0)$$
$$= X \cdot X[Z + \overline{Z}Y](Z + Y)$$
$$\quad \text{(taking X common)}.$$
$$= X(Z + \overline{Z}Y)(Z + Y)$$

$$= X(Z + Y)(Z + Y) \quad (\because A + \overline{A}B = A + B)$$
$$= X(Z + Y) \quad\quad (\because A \times A = A)$$
$$= XZ + XY = \text{R.H.S.}$$

Example 4.9: Show that
$$(X + \overline{Y} + XY)(X + \overline{Y})\overline{X}Y = 0$$

Solution:

L.H.S. $= (X + XY + \overline{Y})(X + \overline{Y})\overline{X}Y$
$$= (X + \overline{Y})(X + \overline{Y})\overline{X}Y \quad (\because A + AB = A)$$
$$= (X + \overline{Y})\overline{X}Y \quad\quad (\because AA = A)$$
$$= X\overline{X}Y + \overline{Y}\overline{X}Y$$
$$= 0 = \text{R.H.S.} \quad (\because X\overline{X} = 0; \overline{Y}Y = 0)$$

Example 4.10: Prove that
$$(C + D)(\overline{A} + B)(A + \overline{B})(A + B) = AB(C + D)$$

Solution:

L.H.S. $= (C + D)[\overline{A}A + \overline{A}B + AB + B\overline{B}](A + B)$
$$= (C + D)[0 + \overline{A}B + AB + 0](A + B)$$
$$= (C + D)(ABA + ABB + \overline{A}BA + \overline{A}BB)$$
$$= (C + D)(AB + AB + 0 + 0)$$
$$\quad (\because AA = A, BB = B)$$
$$= (C + D)AB = \text{R.H.S.} \quad (\because A + A = A)$$

Example 4.11: Prove
$$(A + B)(\overline{A} + B) + (C + \overline{D})(C + D)$$
$$+ (A + E)(A + \overline{E}) = A + B + C$$

Solution:

L.H.S. $= (\overline{A}B + AB + B) + (C + C\overline{D} + CD)$
$$+ (A + AE + A\overline{E})$$
$$\quad \text{(using } A\overline{A} = 0, BB = B \text{ etc.)}$$
$$= (\overline{A}B + B) + (C + C\overline{D}) + (A + A\overline{E})$$
$$\quad \text{(repeated use of } A + AB = A)$$
$$= B + C + A = \text{R.H.S.}$$

Example 4.12: Negate the expression:
$$F = A\overline{B}C + A\overline{B}\overline{C} + ABC$$

Solution: $\overline{F} = \overline{(A\overline{B}C + A\overline{B}\overline{C} + ABC)}$
$$= [(A\overline{B}C + A\overline{B}\overline{C}) + ABC]'$$

(**Note:** Here a 'dash' (′) outside the square bracket denotes inversion, to avoid inconvenience in using a 'bar' over the expression we have used dashes).

$$= (AB'C + AB'C')' (ABC)'$$
$$\text{(using } (A + B)' = A'B')$$
$$= (AB'C)' (AB'C')' (ABC)'$$
$$= (A' + B + C') (A' + B + C) (A' + B' + C')$$
$$\text{(repeated use of } (AB)' = A' + B')$$
$$= (A' + A'B + A'C + B + BC + C'A' + C'B)$$
$$(A' + B' + C')$$
$$(\because A \cdot A = A, \; AB + AB = AB)$$
$$= (A' + A'C + B + C'A' + BC') (A' + B' + C')$$
$$\text{(using } A' + A'B = A', \; B + BC = B)$$
$$= [A' + B + A'(C + C') + BC'] (A' + B' + C')$$
$$= (A' + B + BC') (A' + B' + C')$$
$$(\because C + C' = 1)$$
$$= (A' + B) (A' + B' + C') = A' + BA' + A'B'$$
$$+ A'C' + BC'$$
$$= A' + A'B' + A'C' + BC' = A' + BC' \text{ **Ans.**}$$

4.12 SUM OF PRODUCTS METHOD

Figure 4.15 shows the four possible ways to AND two input signals that are in complemented and uncomplemented form:

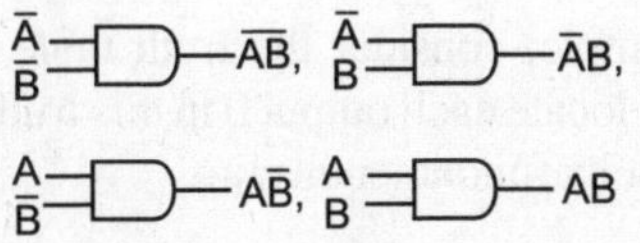

$$\frac{\overline{A}}{B} \;\rightarrow\; \overline{A}B, \quad \frac{\overline{A}}{B} \;\rightarrow\; \overline{A}B,$$

$$\frac{A}{B} \;\rightarrow\; A\overline{B}, \quad \frac{A}{B} \;\rightarrow\; AB$$

Fig. 4.15: ANDing two variables and their complements

These outputs are called fundamental products (Table 4.10)

Table 4.10: Fundamental products for two inputs

A	B	Fundamental produc
0	0	$\overline{A}\overline{B}$
0	1	$\overline{A}B$
1	0	$A\overline{B}$
1	1	AB

The idea of fundamental products also applies to three or more input variables. Assume three input variables A, B, C and their complements. There are 8 ways of ANDing these variables and their complements resulting in following fundamental products

$$\overline{A}\overline{B}\overline{C}, \; \overline{A}\overline{B}C, \; \overline{A}B\overline{C}, \; \overline{A}BC$$
$$A\overline{B}\overline{C}, \; A\overline{B}C, \; AB\overline{C}, \; \overline{A}BC.$$

The each of these 8 fundamental products can be represented by 8 AND gates similar to Fig. 4.15 and

Table 4.11 summarizes these fundamental products by listing each one next to the input condition that results in a high output, e.g. when A = 1, B = 0, C = 0, the

Table 4.11: Fundamental products for three inputs

A	B	C	Fundamental product
0	0	0	$\overline{A}\overline{B}\overline{C}$
0	0	1	$\overline{A}\overline{B}C$
0	1	0	$\overline{A}B\overline{C}$
0	1	1	$\overline{A}BC$
1	0	0	$A\overline{B}\overline{C}$
1	0	1	$A\overline{B}C$
1	1	0	$AB\overline{C}$
1	1	1	ABC

Fundamental product results in an output of

$$Y = A\overline{B}\overline{C} = 1 \cdot \overline{0}\,\overline{0} = 1 \cdot 1 = 1.$$

4.12.1 Sum of products equation for a given truth table

Given a truth table like Table 4.12, we can get sum of products solution as follows:

Table 4.12: Design truth table

A	B	C	Y
0	0	0	0
0	0	1	0
0	1	0	0
0	1	1	1
1	0	0	0
1	0	1	1
1	1	0	1
1	1	1	1

For this, we have to locate each output 1 in the truth table and write down the corresponding fundamental product. For example, the first output 1 appears for an input of A = 0, B = 1 and C = 1. The corresponding fundamental product is $\overline{A}BC$ fundamental product is $A\overline{B}C$. Continuing. The next output 1 is for A = 1, B = 0 and C = 1. The corresponding fundamental product is $A\overline{B}C$. Continuing similarly, we can identify all the fundamental products as shown in Table 4.13

Table 4.13: Identifying fundamental products

A	B	C	Y	
0	0	0	0	
0	0	1	0	
0	1	0	0	
0	1	1	1	$\to \overline{A}BC$
1	0	0	0	
1	0	1	1	$\to A\overline{B}C$
1	1	0	1	$\to AB\overline{C}$
1	1	1	1	$\to ABC$

Then, to get the sum of f products equation, OR the fundamental products of the Table 4.13, i.e.

$$Y = \overline{A}BC + A\overline{B}C + AB\overline{C} + ABC$$

4.12.2 Logic circuit

The logic circuit for the above sum of products equation is shown in Fig. 4.14.

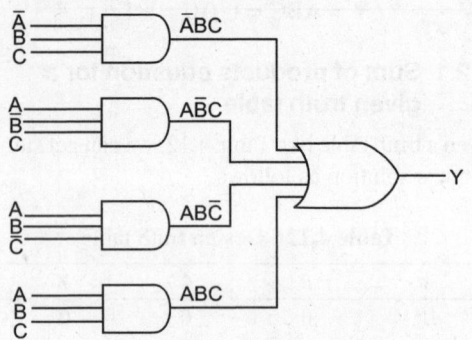

Fig. 4.16: Sum-of-Products circuit

Example 4.13: Given the following truth table. Write down the Boolean function and minimize using Boolean algebra so as to use the minimum number of gates

Table 4.14: (Ex. 4.13)

A	B	C	F
0	0	0	0
0	0	1	1
0	1	0	1
0	1	1	1
1	0	0	0
1	0	1	1
1	1	0	0
1	1	1	1

$$F = \overline{A}\,\overline{B}C + \overline{A}B\overline{C} + \overline{A}BC + A\overline{B}C + ABC$$

combine the terms so as to get $(A + \overline{A})$ common.

$$F = (\overline{A} + A)\overline{B}C + (\overline{A} + A)BC + \overline{A}B\overline{C}$$

$$= \overline{B}C + BC + \overline{A}B\overline{C}; \qquad (\because A + \overline{A} = 1)$$

$$= C + \overline{A}B\overline{C} \qquad (\because \overline{B} + B = 1)$$

$$= C + \overline{C}(\overline{A}B)$$

$$= C + \overline{A}B; \qquad (\because A + \overline{A}B = A + B)$$

Table 4.9 eqn. 10)

4.13 PRODUCT-OF-SUMS METHOD

Product-of-Sums method is similar to sum of products method. But with the product of sums method, the fundamental sum produces an output 0 for the corresponding input condition. Finally, by ANDing the sums, we get the POS equation corres-ponding to the truth table. This will become further clear by the following discussion.

4.13.1 Converting a Truth table to POS Equation

As an example, consider the truth table like Table 4.15. Then locate each output 0 in this truth table and write down its fundamental sum

Table 4.15

A	B	C	Y	
0	0	0	0	$\to A+B+C$
0	0	1	1	
0	1	0	1	
0	1	1	0	$\to A+\overline{B}+\overline{C}$
1	0	0	1	
1	0	1	1	
1	1	0	0	$\to \overline{A}+\overline{B}+C$
1	1	1	1	

In this table, the first output 0 appears for A = 0, B = 0 and C = 0. The fundamental sum for these inputs is A + B + C, because this produces an output zero for the corresponding input condition as

$$Y = A + B + C = 0 + 0 + 0 = 0.$$

Similarly, for the second output

$$Y = A + \overline{B} + \overline{C} = 0 + \overline{1} + \overline{1} = 0 + 0 + 0 = 0$$

(Note that B and C are complemented because this is the only way to set a logical sum of 0 for the given input conditions).

Similarly for the third output 0,

$$Y = \bar{A} + \bar{B} + C = \bar{1} + \bar{1} + 0 = 0 + 0 + 0 = 0$$

Note that each variable is complemented when the corresponding input variable is a 1, and the variable is uncomplemented when the corresponding input variable is -.

To get the product-of-sums equation, you have to AND all the fundamental sums obtained as described above, i.e.

$$Y = (A + B + C)(A + \bar{B} + \bar{C})(\bar{A} + \bar{B} + C) \quad \ldots(i)$$

This is the product-of-sums equation (POS equation) for Table 4.15.

4.13.2 Logic Circuit

A logic circuit for POS Eqn. (i) is shown in Fig. 4.17

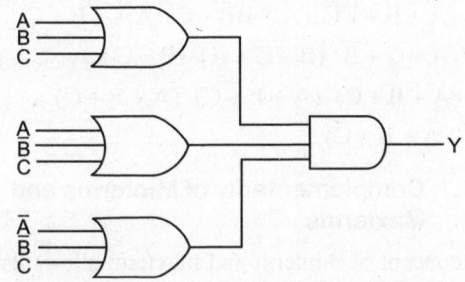

Fig. 4.17: Product-of-sums circuit (for equation (i)

Example 4.14: Develop a circuit to perform

$$Y = AB + D(B + \bar{C})$$

using NAND gates. You have regular and complementary inputs available.

Solution:

$$Y = \overline{\overline{(AB)}} + \overline{\{D(B + \bar{C})\}}'$$

$$= \left[\overline{(AB)} \cdot \overline{\{D(B + \bar{C})\}} \right]$$

(using De Morgan's theorem)

$$= \left[(\overline{AB}) \cdot \{D(\overline{BC})'\} \right]'$$

(Here a dash outside the square bracket denotes complement or inversion of the complete expression)

Or

$$Y = \left[(AB)' \cdot \{D(B'C)'\}' \right]'$$

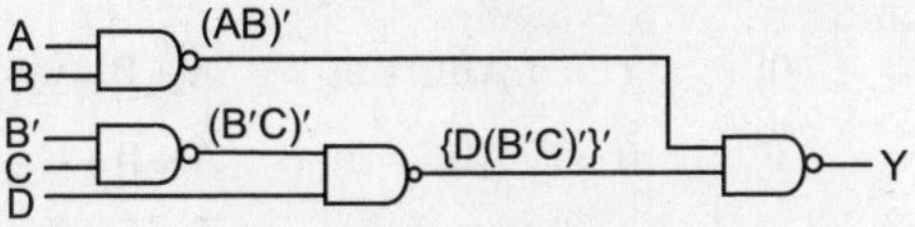

Fig. 4.18: (Example 4.14)

4.14 BOOLEAN FUNCTIONS

A Boolean function is an expression which is formed with binary variables, the two binary operators AND and OR, the unary operator NOT, parentheses and 'equal to' sign. For a given value of the variables, the value of the function can be either 0 or 1. For example consider the equation

$$W = X + \bar{Y} \cdot Z$$

Here the variable W is a function of X, Y and Z, i.e. $W = f(X, Y, Z)$. The right hand side of the equation is called an expression. The symbols X, Y and Z are called *literals* of this function.

The above is an example of a Boolean function represented as an algebraic equation. A Boolean function may also be represented in the form of a truth table. The number of rows in the table will be equal to 2^n where n is the number of literals (binary variables) used in the function. Table 4.16 shows the truth table for the Boolean function

$$W = X + \bar{Y} \cdot Z$$

Table 4.16 Truth table for the Boolean function $W = X + \bar{Y} \cdot Z$.

X	Y	Z	$\bar{Y}$	$\bar{Y} \cdot Z$	$X + \bar{Y} \cdot Z$
0	0	0	1	0	0
0	0	1	1	1	1
0	1	0	0	0	0
0	1	1	0	0	0
1	0	0	1	0	1
1	0	1	1	1	1
1	1	0	0	0	1
1	1	1	0	0	1

4.15 CANONICAL FORMS FOR BOOLEAN FUNCTIONS (Minterms and Maxterms)

A binary variable may appear either in its normal form (X) or in its complement form ($\bar{X}$). Consider now binary variables X and Y combined with an AND operator. Since each variable may appear in either form, There are four possible combinations

$$\bar{X} \cdot \bar{Y} \quad \bar{X} \cdot Y \quad X \cdot \bar{Y} \quad X \cdot Y$$

Each of these four AND terms is called a *minterm* or a *standard product*.

In a similar manner, n variables can be combined to form 2^n minterms. The 2^n different minterms may be determined by a method similar to the one shown in Table 4.17 for three variables.

Table 4.17: (a) Minterms and maxterms for three variables

Variables			Minterms		Maxterms	
x	y	z	Term	Designation	Term	Designation
0	0	0	$\bar{x}\cdot\bar{y}\cdot\bar{z}$	m_0	$x+y+z$	M_0
0	0	1	$\bar{x}\cdot\bar{y}\cdot z$	m_1	$x+y+\bar{z}$	M_1
0	1	0	$\bar{x}\cdot y\cdot\bar{z}$	m_2	$x+\bar{y}+z$	M_2
0	1	1	$\bar{x}\cdot y\cdot z$	m_3	$x+\bar{y}+\bar{z}$	M_3
1	0	0	$x\cdot\bar{y}\cdot\bar{z}$	m_4	$\bar{x}+y+z$	M_4
1	0	1	$x\cdot\bar{y}\cdot z$	m_5	$\bar{x}+y+\bar{z}$	M_5
1	1	0	$x\cdot y\cdot\bar{z}$	m_6	$\bar{x}+\bar{y}+z$	M_6
1	1	1	$x\cdot y\cdot z$	m_7	$\bar{x}+\bar{y}+\bar{z}$	M_7

The binary numbers from 0 to 2^n-1 are listed under the n variables. Each minterm is obtained from an AND term of the n variables, with each variable being primed if the corresponding bit of the binary number is 0, and unprimed, if it is a 1.

A symbol for each minterm is also shown in the table and is of the form m_j where j denotes the decimal equivalent of the binary number of the minterm designated.

Similarly, n variables forming an OR term, with each variable being primed or unprimed, provide 2^n possible combinations called maxterms or *standard sums*. The eight maxterms for three variables, together with their symbolic designation are shown in the table. Any 2^n maxterms for n variables may be determined similarly. Each maxterm is obtained from an OR term of the n variables, with each variable being unprimed, if the corresponding bit is a 0, and primed, if it is a 1. Note that each maxterm is the complement of its corresponding minterm and viceversa.

Now consider for example, the logical equation
$$Y = (\bar{A} + BC)(B + \bar{C}A) \quad \ldots(1)$$
(i) Sum of products form:
$$Y = A(B + \bar{C}A) + (BC)(B + \bar{C}A)$$
$$= AB + A\bar{C}A + BCB + BC\bar{C}A$$
$$= AB + A\bar{C} + BC \quad \ldots(2)$$
It can be converted into standard SOP form as
$$Y = AB(C + \bar{C}) + A\bar{C}(B + \bar{B}) + BC(A + \bar{A})$$
$$= ABC + AB\bar{C} + AB\bar{C} + A\bar{B}\bar{C} + ABC + \bar{A}BC$$
$$= \bar{A}BC + A\bar{B}\bar{C} + AB\bar{C} + ABC \quad \ldots(3)$$
(ii) Product of Sums form:

Using the theorem $A + BC = (A + B)(A + C)$, we can rewrite eqn (1) as

$$Y = (A + B)(A + C)(B + A)(B + \bar{C}) \quad \ldots(4)$$
$$= (A + B)(A + C)(B + \bar{C}) \quad \ldots(5)$$
Eqn. (5) be converted into standard POS form as
$$Y = (A + B + C\bar{C})(A + B\bar{B} + C)(A\bar{A} + B + \bar{C})$$
$$(A + C + \bar{B})\,(B + \bar{C} + A)\cdot(B + \bar{C} + \bar{A})$$
$$= (A + B + C)\cdot(A + B + \bar{C})\cdot(A + \bar{B} + C)$$
$$\cdot(\bar{A} + B + \bar{C}) \quad \ldots(6)$$

4.15.1 Complementarily of Minterms and Maxterms

The concept of minterm and maxterm allows us to introduce a very convenient shorthand notation to express logical functions. Table (4.17b) gives the minterms and maxterms for a three variable logical function where the number of minterms as well as maxterms is $2^3 = (8)$. While writing a particular minterm or maxterm, we always write it with the variables in an orderly way as evident from the table.

Table 4.17: (b) Minterms/maxterms for three variables

Variables			Minterms (m_i)	Maxterms (M_i)
A	B	C		
0	0	0	$\bar{A}\bar{B}\bar{C} = m_0$	$A + B + C = M_0$
0	0	1	$\bar{A}\bar{B}C = m_1$	$A + B + \bar{C} = M_1$
0	1	0	$\bar{A}B\bar{C} = m_2$	$A + \bar{B} + C = M_2$
0	1	1	$\bar{A}BC = m_3$	$A + \bar{B} + \bar{C} = M_3$
1	0	0	$A\bar{B}\bar{C} = m_4$	$\bar{A} + B + C = M_4$
1	0	1	$A\bar{B}C = m_5$	$\bar{A} + B + \bar{C} = M_5$
1	1	0	$AB\bar{C} = m_6$	$\bar{A} + \bar{B} + C = M_6$
1	1	1	$ABC = m_7$	$\bar{A} + \bar{B} + \bar{C} = M_7$

Each minterm is represented by m_i where i is the decimal equivalent of the natural binary number corresponding to the minterm with uncomplemented variables taken as 1s and the complemented variables taken as 0s. Similarly, each maxterm is represented by M_i where i is the decimal equivalent of the natural binary number corresponding to the maxterm with uncomplemented variables taken as 0s and complemented variables taken as 1s. Using these notations the standard SOP equation (3) (for example)

$$Y = \overline{A}BC + A\overline{B}\,\overline{C} + AB\overline{C} + ABC$$

can be written as

$$Y = m_3 + m_4 + m_6 + m_7 = \Sigma_m (3, 4, 6, 7) \quad …(7)$$

where $m_3 = \overline{A}BC$, $m_4 = A\overline{B}\,\overline{C}$, $m_6 = AB\overline{C}$ and $ABC = m_7$.

Similarly the standard POS equation.

$$Y = (A + B + C)\cdot(A + B + \overline{C})$$
$$\cdot(A + \overline{B} + C)\cdot(\overline{A} + B + \overline{C})$$

can be written as

$$Y = M_0, M_1, M_2, M_5 = \Pi M (0, 1, 2, 5) \quad …(8)$$

where

$$M_0 = (A + B + C), \; M_1 = (A + B + \overline{C}),$$
$$M_2 = (A + \overline{B} + C), \; M_5 = (\overline{A} + B + \overline{C})$$

Equations (7) and (8) are the shorthand forms of standard SOP and POS respectively. Since these two equations represent the same logical function

$$Y = (A + BC)(B + \overline{C}A)$$

therefore we notice that there is a complementary type of relationship between a function expressed in terms of minterms and in terms of maxterms. Hence, if a logical function is specified in terms of minterm (maxterm), its maxterm (minterm) representation can be determined by using this complementary property; for example, if

$$Y = \Sigma m (3, 4, 6, 7) \quad …(9)$$

Then $\quad Y = \Pi M (0, 1, 2, 5) \quad …(10)$

for the above three variable logical function.

4.16 SUM-OF-PRODUCTS IN TERMS OF MINTERMS OF A TRUTH TABLE

A sum-of-products (SOP) expression is a product term (minterm), or several product terms (minterms) logically added (ORed) together, e.g.

$$X \cdot \overline{X} + \overline{Y} \cdot Y$$

The following steps are followed to express a Boolean function in terms of SOP form:

1. Construct a truth table for the given Boolean function.
2. Form a minterm for each combination of the variables which produces a 1 in the function.
3. The desired expression is the sum (OR) of all the minterms obtained in step 2.

For example consider the Table 4.18

Table 4.18: Truth table for functions F_0

x	y	z	F_0
0	0	0	0
0	0	1	1
0	1	0	0
0	1	1	0
1	0	0	1
1	0	1	0
1	1	0	0
1	1	1	1

In case of function F_0, the following three combinations of the variables produce a 1: 001, 100 and 111.

The corresponding minterms are

$$\overline{x} \cdot \overline{y} \cdot z, \; x \cdot \overline{y} \cdot \overline{z} \quad \text{and} \quad x \cdot y \cdot z$$

Taking the sum (OR) of all these minterms, the function F_0 can be expressed in the sum-of-products form as

$$F_0 = \overline{x} \cdot \overline{y} \cdot z + x \cdot \overline{y} \cdot \overline{z} + x \cdot y \cdot z$$
$$(= m_1 + m_4 + m_7)$$

4.17 PRODUCT-OF-SUMS IN TERM OF MAXTERMS FOR A TRUTH TABLE

A product-of-sums (POS) expression is a sum term (maxterm), or several sum terms (maxterms), logically multiplied (ANDed) together., e.g.

$$(\overline{x} + y)\cdot(x + \overline{y}).$$

The following steps are followed to express a Boolean function in its POS form:

1. Construct a truth table for the given Boolean function.
2. Form a maxterm for each combination of the variables which produces a 0 in the function.
3. The desired expression is the product (AND) of all the maxterms obtained in step 2.

For example, in case of function F_0 of Table 4.18 the following five combinations of the variables produce a 0

000, 010, 011, 101, 110

Their corresponding maxterms are

$(x + y + z), (x + \overline{y} + z), (x + \overline{y} + \overline{z}),$

$(\overline{x} + y + \overline{z}), (\overline{x} + \overline{y} + z)$

Taking the product (AND) of all these maxterms, the function F_0 can be expressed in its POS form as

$F_0 = (x + y + z) \cdot (x + \overline{y} + z) \cdot (x + \overline{y} + \overline{z})$
$\cdot (\overline{x} + y + \overline{z}) \cdot (\overline{x} + \overline{y} + z).$

or $\quad F_0 = M_0 \cdot M_1 \cdot M_2 \cdot M_4$

Example 4.15: With the given truth table, draw a simplified set of logic gates to provide a closed circuit when F = 1.

Solution:

$F = \overline{A}B\overline{C} + \overline{A}BC + A\overline{B}\overline{C} + AB\overline{C} + ABC$

$\quad = B\overline{C} + BC + ABC$

or $\quad F = (\overline{A} + A)B\overline{C} + (\overline{A} + A)BC + A\overline{B}\overline{C}$

$\quad = B\overline{C} + BC + A\overline{B}\overline{C}$

$\quad = B + A\overline{B}\overline{C}$

$\quad = B + \overline{B}(A\overline{C}) = B + A\overline{C}$

$\quad\quad\quad$ (using A + A'B = A + B)

Table 4.19: (Example 4.15)

A	B	C	F
0	0	0	0
0	0	1	0
0	1	0	1
0	1	1	1
1	0	0	1
1	0	1	0
1	1	0	0
1	1	1	1

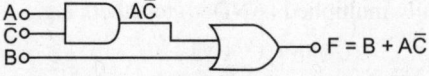

Fig. 4.19: Logic gates for F. (example 4.15)

4.18 TRUTH TABLES TO KARNAUGH MAPS

A part from algebraic simplification of Boolean expressions, there is yet another method known as Karnaugh map method.

A Karnaugh map is a visual display of the fundamental products needed for a sum-of-products solution for a given truth table. For example, consider Table 4.20.

Table 4.20

A	B	Y
0	0	0
0	1	0
1	0	1
1	1	1

Figure 4.20 shows steps for constructing Karnaugh map for this truth table. Begin by drawing

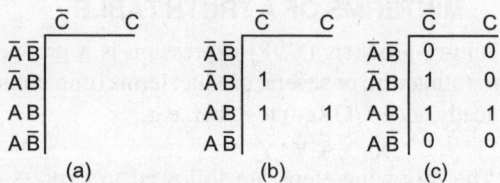

Fig. 4.20: Constructing a Karnaugh map

Figure 4.20a Put a1 for A = 1 and B = 0 (i.e. fundamental product $A\overline{B}$ (Fig. 4.20b). Then a 1 for A = 1 and B = 1 (fundamental product AB) (Fig. 4.20c) Finally enter OS in the remaining space (Fig. 4.20d).

4.18.1 Three-variable Maps

First draw the blank map (Fig. 4.21). The vertical column is labeled $\overline{A}\overline{B}, \overline{A}, B, AB, A\overline{B}$. On the horizontal side, label $\overline{C}$ and C. Then, look for output 1s in the table. Output 1s appear for ABC inputs of 010, 110 and 111. The fundamental products for these inputs are $\overline{A}B\overline{C}, AB\overline{C}$ and ABC. Enter 1s for these products on the Karnaugh map. Finally, enter 0s in the remaining spaces (Figs. 4.21a to c). Consider the Table 4.21

Table 4.21

A	B	C	Y
0	0	0	0
0	0	1	0
0	1	0	1
0	1	1	0
1	0	0	0
1	0	1	0
1	1	0	1
1	1	1	1

Fig. 4.21: Three variable Karnaugh map

4.18.2 Four variable maps

Many digital computer system use four bit numbers. So, logic circuits are often designed to handle 4 input variables (or their complements) consequently, one must know how to draw a four variable Karnaugh map. As an example, consider the truth Table 4.22. Start by drawing Fig. 4.22a.

Table 4.22

A	*B*	*C*	*D*	*Y*
0	0	0	0	0
0	0	0	1	1
0	0	1	0	0
0	0	1	1	0
0	1	0	0	0
0	1	0	1	0
0	1	1	0	1
0	1	1	1	1
1	0	0	0	0
1	0	0	1	0
1	0	1	0	0
1	0	1	1	0
1	1	0	0	0
1	1	0	1	0
1	1	1	0	1
1	1	1	1	0

Make the vertical column in the sequence $\bar{A}\bar{B}$, $\bar{A}B$, AB and $A\bar{B}$ and horizontal row in the sequence $\bar{C}\bar{D}$, $\bar{C}D$, CD and $C\bar{D}$. From the truth table, output 1s appear for ABCD inputs of 0001, 0110, 0111 and 1110. The fundamental products for these input conditions are $\bar{A}\bar{B}\bar{C}D$, $\bar{A}BC\bar{D}$, $\bar{A}BCD$ and $ABC\bar{D}$. Enter 1s for these in the Karnaugh map (Fig. 4.22b), and then 0s in the remaining spaces, and so we get Fig. 4.22c as the Karnaugh map for Table 4.22.

	$\bar{C}\bar{D}$	$\bar{C}D$	CD	$C\bar{D}$
$\bar{A}\bar{B}$				
$\bar{A}B$				
AB				
$A\bar{B}$				

(a)

	$\bar{C}\bar{D}$	$\bar{C}D$	CD	$C\bar{D}$
$\bar{A}\bar{B}$		1		
$\bar{A}B$			1	1
AB				1
$A\bar{B}$				

(b)

	$\bar{C}\bar{D}$	$\bar{C}D$	CD	$C\bar{D}$
$\bar{A}\bar{B}$	0	1	0	0
$\bar{A}B$	0	0	1	1
AB	0	0	0	1
$A\bar{B}$	0	0	0	0

(c)

Fig. 4.22: Four variable Karnaugh map for Table 4.22

We will consider five and six-variable maps in section 4.32.

4.19 PAIRS, QUADS AND OCTETS IN KARNAUGH MAP

1. Pairs

Consider Fig. 4.22. The map contains a pair of 1s adjacent to each other. The first 1 represents ABCD and the second $ABC\bar{D}$. As we move from the first 1 to the second 1, only one variable (here D) goes from uncomplemented to complemented form (other variables do not change). The sum-of-products equation is

$$Y = ABCD + ABC\bar{D}$$
$$= ABC(D + \bar{D})$$

Since D is ORed with its complement, the above equation simplifies to

$$Y = ABC$$

For easy identification we encircle a pair of adjacent 1s, and as a result, instead of two separate products ABCD and $ABC\bar{D}$, we visualize the

	$\bar{C}\bar{D}$	$\bar{C}D$	CD	$C\bar{D}$
$\bar{A}\bar{B}$	0	0	0	0
$\bar{A}B$	0	0	0	0
AB	0	0	1	1
$A\bar{B}$	0	0	0	0

(a)

	$\bar{C}\bar{D}$	$\bar{C}D$	CD	$C\bar{D}$
$\bar{A}\bar{B}$	0	0	0	0
$\bar{A}B$	0	0	0	0
AB	0	0	1	1
$A\bar{B}$	0	0	0	0

(b)

Fig. 4.23: A pair of adjacent 1s

Figure 4.23 shows more examples of pairs.

	$\bar{C}\bar{D}$	$\bar{C}D$	CD	$C\bar{D}$
$\bar{A}\bar{B}$	0	0	0	0
$\bar{A}B$	0	0	0	0
AB	0	0	0	1
$A\bar{B}$	0	0	0	1

(a) $Y = AC\bar{D}$

	$\bar{C}\bar{D}$	$\bar{C}D$	CD	$C\bar{D}$
$\bar{A}\bar{B}$	0	0	1	0
$\bar{A}B$	0	0	1	0
AB	0	0	0	0
$A\bar{B}$	0	0	0	0

(b) $Y = \bar{A}CD$

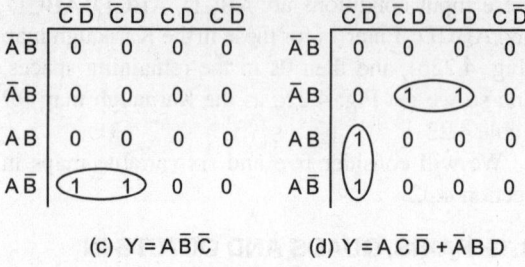

(c) $Y = A\bar{B}\bar{C}$ (d) $Y = A\bar{C}\bar{D} + \bar{A}BD$

Fig. 4.24: Examples of pairs

encircled pair as representing a single reduced product ABC.

2. Quads

A quad is a group of four 1s that are horizontally or vertically adjacent. When we encircle a quad, it leads to a simpler product by eliminating two variables and their complements for example, consider Fig. 4.25a. Visualize the four 1s as two pairs.

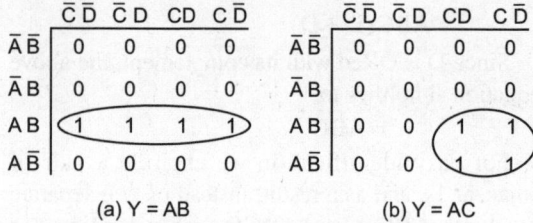

(a) $Y = AB$ (b) $Y = AC$

Fig. 4.25: Examples of quads

The first pair represents $AB\bar{C}$ and the second pair ABC. Consequently

$$Y = AB\bar{C} + ABC = AB(\bar{C} + C)$$

or $Y = AB$

i.e. the quad of Fig. 4.25a represents a product whose two variables and their complements have been eliminated (here C and D).

Similarly, for Fig. 4.25b

$$Y = AC$$

(here B and $\bar{B}$ and D and $\bar{D}$ have been eliminated)

3. Octets

An octet is a group of eight 1s (Fig. 4.26a). This may be visualized as a group of two quads, (Fig. 4.26b). Therefore, the equation is

$$Y = A\bar{C} + AC$$
$$= A(\bar{C} + C)$$

or $Y = A$

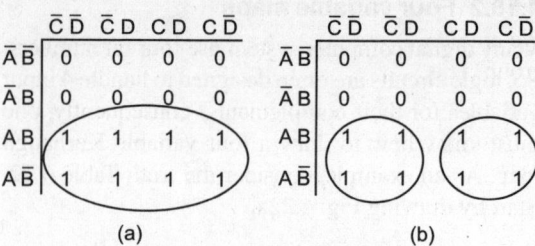

(a) (b)

Fig. 4.26: Examples of octets

Thus, when we encircle an octet, it eliminates three variables and their complements. A similar proof applies to any such other octet. So, without bothering about algebra, we may form pairs or quads or octets and obtain a simplified output.

4.20 KARNAUGH METHOD OF SIMPLIFICATION

We know that pair eliminates one variable and its complement, a quad eliminates two variables and their complements and an octet eliminates three variables and their complements. Thus, after drawing a Karnaugh map, encircle the octets first, the quads second and the pairs last. In this way we get greatest simplification.

Example 4.16: Suppose, we obtain from a truth table, a Karnaugh map of Fig. 4.27a

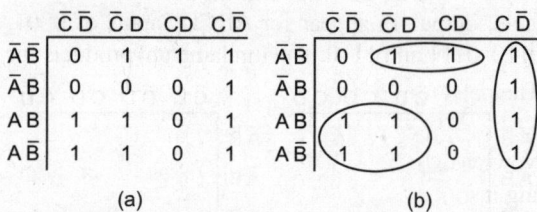

(a) (b)

Fig. 4.27: Encircling octets, quads and pairs

First we look for octets. There are none. Then quads. There are two, we encircle them. Finally we look for pair (there is one) and encircle it. We get Fig. 4.27b.

The pair represents $\bar{A}\bar{B}D$, the lower quad stands for $A\bar{C}$ and the quad on the right represents $C\bar{D}$. By ORing these, we get the Boolean equation corresponding to the Karnaugh map:

$$Y = \bar{A}\bar{B}D + A\bar{C} + C\bar{D}.$$

4.21 OVERLAPPING GROUPS

We are allowed to use the same 1 in more than one group. This is illustrated in Fig. 4.28.

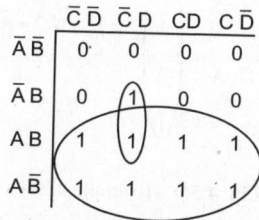

Fig. 4.28: Overlapping groups

For the octet, the fundamental product is A and for the pair, the fundamental product is $B\bar{C}D$. The simplified equation is

$$Y = A + B\bar{C}D.$$

Thus, we have to use the 1s more than once to get the largest groups as possible.

Rolling the Map: Consider the Fig. 4.29a.

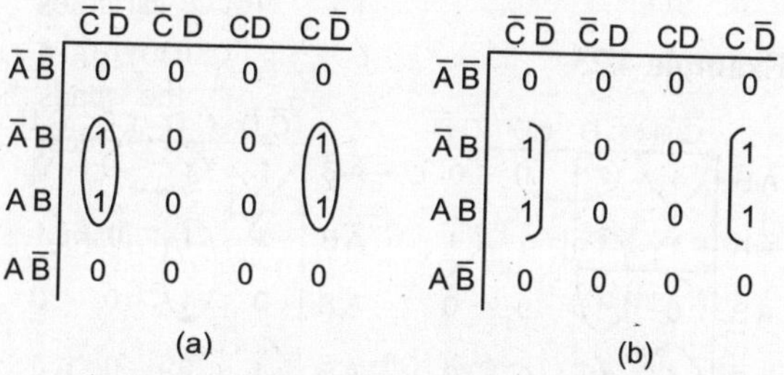

(a)   (b)

Fig. 4.29: Rolling the Karnaugh map

The pairs result in the equation

$$Y = B\bar{C}\bar{D} + BC\bar{D}$$

Now visualize picking up the Karnaugh map and rolling it so that the left side touches the right side.

Then we realize that the two pairs actually form a quad. To indicate this, Fig. 4.29b is drawn. From this viewpoint, for Fig. 4.29b, we have the equation

$$Y = B\bar{D}$$

Why rolling is valid? This may be seen as follows

$$Y = B\bar{C}\bar{D} + BC\bar{D}$$
$$= B\bar{D}(\bar{C} + C)$$
$$= B\bar{D}$$

Thus, rolled quad of Fig. 4.29b is equivalent to Fig. 4.29a.

4.22 ELIMINATING REDUNDANT GROUPS

After having finished encircling the groups, we have to eliminate any redundant group. This is a group whose all 1s are already used by other groups, i.e. a redundant group is an unnecessary group (Fig. 4.30).

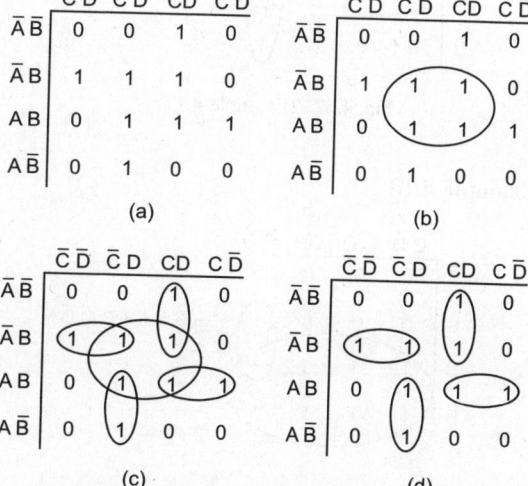

Fig. 4.30: Eliminating a redundant group

Summary of steps in Karnaugh Map Method

1. Enter a 1 on the Karnaugh map for each fundamental product that produces a 1 output in the truth table. Enter 0s elsewhere.
2. Encircle first octets, then quads, and finally pairs. Remember to roll and overlap to get the largest groups possible.
3. If any isolated ones remain, encircle each.
4. Eliminate any redundant group
5. Write the Boolean equation by ORing the products corresponding to the encircled groups.

Example 4.16

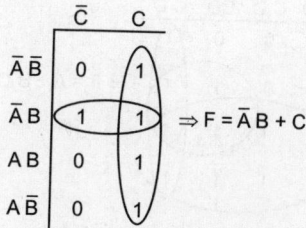

Fig. 4.31: (Example 4.16)

Example 4.17

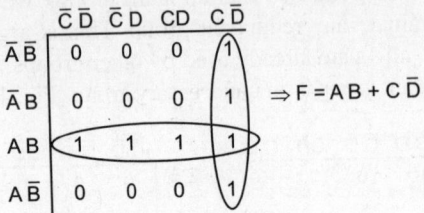

Fig. 4.32: (Example 4.17)

Example 4.18

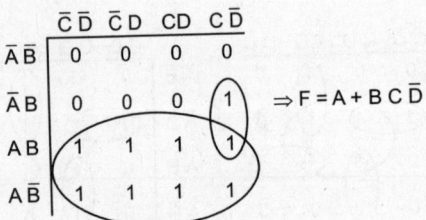

Fig. 4.33: (Example 4.18)

Example 4.19

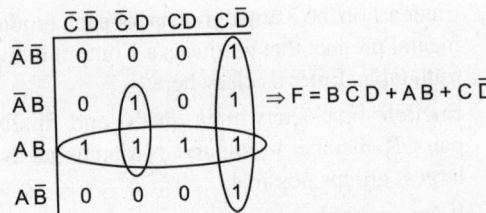

Fig. 4.34: (Example 4.19)

Example 4.20

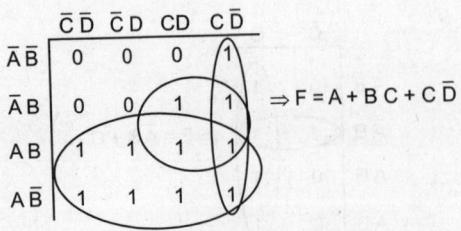

Fig. 4.35: (Example 4.20)

Example 4.21

$$\overline{C}\,\overline{D} \quad \overline{C}\,D \quad CD \quad C\,\overline{D}$$

	$\overline{C}\,\overline{D}$	$\overline{C}\,D$	CD	$C\,\overline{D}$
$\overline{A}\,\overline{B}$	0	1	1	1
$\overline{A}\,B$	0	0	0	1
$A\,B$	1	1	0	1
$A\,\overline{B}$	1	1	0	1

$\Rightarrow F = A\overline{C} + \overline{A}\,\overline{B}\,D + C\,\overline{D}$

Fig. 4.36: (Example 4.21)

Example 4.22

	$\overline{C}\,\overline{D}$	$\overline{C}\,D$	CD	$C\,\overline{D}$
$\overline{A}\,\overline{B}$	1	0	0	1
$\overline{A}\,B$	0	0	0	0
$A\,B$	0	0	0	0
$A\,\overline{B}$	1	0	0	1

$\Rightarrow F = \overline{B}\,\overline{D}$

(No isolated 1 should remain)

Fig. 4.37: (Example 4.22)

Example 4.23

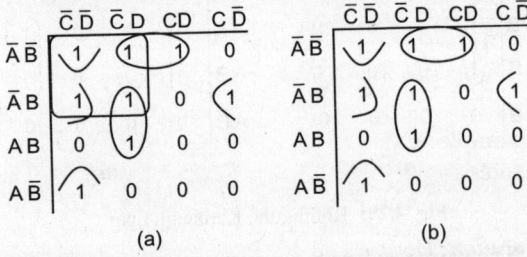

(a)   (b)

Fig. 4.38: As far as possible, no isolated 1 should remain. Therefore rolling is done. Figure 4.38a is wrong, because the quad is redundant. It is dropped and we get Fig. 4.38b, which is correct.

For Fig. 4.38b $F = \overline{A}\,\overline{B}D + B\overline{C}D + \overline{B}\,C\overline{D} + \overline{A}B\overline{D}$

Example 4.24: Write the simplified Boolean expression for the following map:

	$\overline{C}\,\overline{D}$	$\overline{C}\,D$	CD	$C\,\overline{D}$
$\overline{A}\,\overline{B}$	0	0	0	0
$\overline{A}\,B$	0	0	1	0
$A\,B$	1	1	1	1
$A\,\overline{B}$	0	1	1	1

Fig. 4.39: (Example 4.24)

Solution:

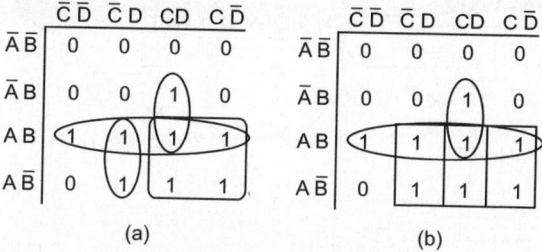

(a)   (b)

Fig. 4.40: (Example 4.24)

Grouping in Fig. 4.40a is not proper. Overlapping should be done to get the largest groups possible. Fig. 4.40b is correct. It gives

$$F = AB + AD + AC + BCD$$

Example 4.25: For the Karnaugh map given below, write the minimum sum-of-products expression.

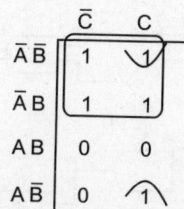

$\Rightarrow F = \bar{A}\,\bar{C}\,\bar{D} + AC\,\bar{D} + \bar{B}\,\bar{D}$

Fig. 4.41: (Example 4.25)

Example 4.26: Use Karnaugh map to simplify the expression

$$Y = \bar{A}\bar{B}\bar{C} + \bar{B}C + \bar{A}B$$

Solution: Here, we are not given the truth table from which to fill the K-map. Instead, we have to fill in the K-map by taking each of the product terms in the expression and placing 1s in the corresponding positions.

Fig. 4.42: (Example 4.26)

The first term $\bar{A}\bar{B}\bar{C}$ tells us to enter a1 for $\bar{A}\bar{B}\bar{C}$. The second term $\bar{A}B\bar{C}$ tells us to enter 1 in each position that contains a $\bar{B}C$ (This would be for $\bar{A}\bar{B}\bar{C}$ and $\bar{A}BC$). Similarly, the $\bar{A}B$ term tells us to place a 1 in the $\bar{A}BC$ and $\bar{A}B\bar{C}$ positions. Remaining positions are filled with zeros. Encircling quad and rolling 1s for $\bar{A}BC$ and $A\bar{B}C$ leads to the following simplified output

$$Y = \bar{A} + \bar{B}C.$$

Example 4.27: You have $F = (AB + C + D)(C' + D)$. Draw a Karnaugh map to yield F. By De Morgan's theorem determine the dual of the logic function and draw the logic circuit (i.e. for F = 0).

Solution:

$$F = ABC' + ABD + CD + D + C'D$$
$$= ABC' + ABD + D + C'D$$
$$= ABC' + ABD + D \qquad (\because A + AB = A)$$
$$= ABC' + D$$

The corresponding Karnaugh map is

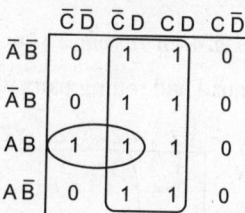

Fig. 4.43: (Example 4.27)

$$F = D + ABC'$$
$$F' = (ABC' + D)'$$
$$= (ABC')'(D') \qquad (\because (A + B)' = A'B')$$

Logic circuit

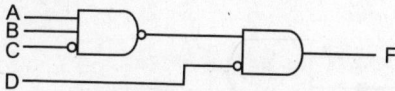

Fig. 4.44: (Example 4.27)

Example 4.28: Write the simplified logic expression for the circuit shown and also draw the simplified circuit.

$$F = Y'\{X' + Z[(W'Y)' + WX]\}$$
$$= Y'X' + Y'Z(W + Y') + WXY'Z$$
$$= X'Y' + WY'Z + Y'Z + WXY'Z$$

For K-map, the first term tells us to enter 1 in each position that contains a X'Y'. The second term permits to enter 1 for term containing WY'Z. The fourth term

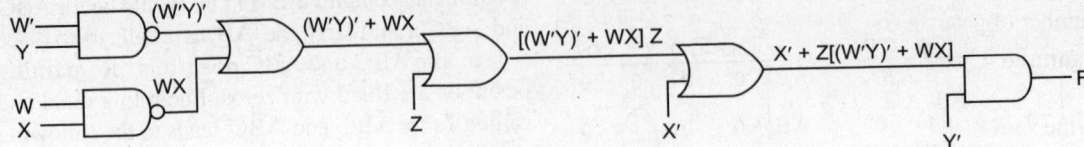

Fig. 4.45: (Example 4.28)

also tells to enter 1 for the same. The third term tells to enter 1s for the positions in column Y'Z. Thus we get the following K-map after entering 0s in remaining places

	Y'Z'	Y'Z	YZ	YZ'
W'X'	1	1	0	0
W'X	0	1	0	0
WX	0	1	0	0
WX'	1	1	0	0

Fig. 4.46: (Example 4.28)

Encircling quad and rolling pairs we get

	Y'Z'	Y'Z	YZ	YZ'
W'X'	1	1	0	0
W'X	0	1	0	0
WX	0	1	0	0
WX'	1	1	0	0

Fig. 4.47: (Example 4.28)

consequently $F = X'Y' + Y'Z$ and the simplified circuit is

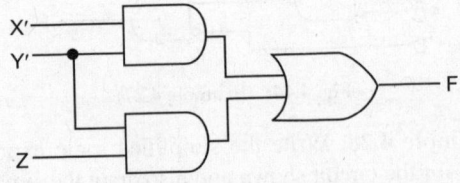

Fig. 4.48: (Example 4.28)

Example 4.29: A competitor performs the operation of the Fig. 4.49 shown with only five gates and is cutting you price, find and draw this circuit.

Karnaugh map is as shown in Fig. 4.50. Further reduction using Karnaugh map is not possible we will make use of De Morgan's theorem.

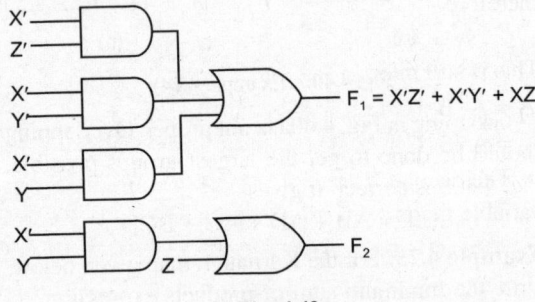

$F_1 = X'Z' + X'Y' + XZ$

F_2

Fig. 4.49

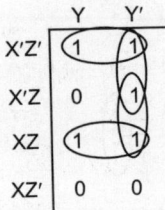

	Y	Y'
X'Z'	1	1
X'Z	0	1
XZ	1	1
XZ'	0	0

Fig. 4.50: Karnaugh map for Fig. 4.49

$$F_1 = [X'Z' + X'Y' + XZ)']'$$
$$= [(X'Z')' (X'Y')' (XZ)']'$$
$$= [(X + Z) (X + Y) (X' + Z')]'$$
$$= [(XZ' + X'Z) (X + Y)]' \qquad \ldots(i)$$

or $\quad F_1 = [XZ' + XZ'Y + X'ZY]'$
$$= (XZ' + X'ZY)' \qquad \ldots(ii)$$

Thus, the simplified circuit for F_1 is

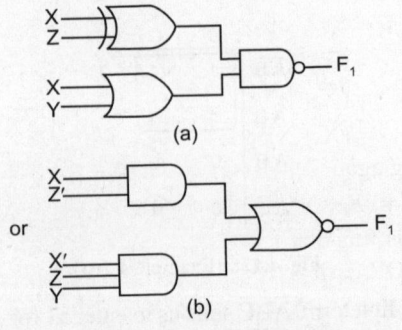

(a)

or

(b)

Fig. 4.51: (a) for eqn. (i) and (b) for eqn. (ii)

Thus using eqn. (i) or (ii), we have reduced the number of gates from 6 to 5 for F_1 and F_2.

Example 4.30: Express the Boolean function

$$F = A + \bar{B} \cdot C$$

in the sum-of-products (i.e. sum of minterms) form.

Solution: The function has three variables A, B and C. The first term in F is missing two variables, therefore

$$A = A \cdot (B + \bar{B}) = A \cdot B + A \cdot \bar{B}$$

This is still missing one variable, therefore

$$A = A \cdot B \cdot (C + \bar{C}) + A \cdot \bar{B}(C + \bar{C})$$
$$= A \cdot B \cdot C + A \cdot B \cdot \bar{C} + A \cdot \bar{B} \cdot C + A \cdot \bar{B} \cdot \bar{C} \quad \dots(i)$$

The second term in F, i.e. $\bar{B} \cdot C$ is missing one variable therefore,

$$\bar{B} \cdot C = \bar{B} \cdot C \cdot (A + \bar{A})$$
$$= A \cdot \bar{B} \cdot C + \bar{A} \cdot \bar{B} \cdot C \quad \dots(ii)$$

From eqn. (i) and (ii) we get

$$F = A + \bar{B} \cdot C$$
$$= A \cdot B \cdot C + A \cdot B \cdot \bar{C} + A \cdot \bar{B} \cdot C$$
$$+ A \cdot \bar{B} \cdot \bar{C} + A \cdot \bar{B} \cdot C + \bar{A} \cdot \bar{B} \cdot C$$

Note that, here, the term $A \cdot \bar{B} \cdot C$ appears twice and since $A + A = A$, we get, after rearranging the minterms in ascending order,

$$F = \bar{A} \cdot \bar{B} \cdot C + A \cdot \bar{B} \cdot \bar{C} + A \cdot \bar{B} \cdot C$$
$$+ A \cdot B \cdot \bar{C} + A \cdot B \cdot C$$
$$= m_1 + m_4 + m_5 + m_6 + m_7 \text{ (see table 4.17)}$$

Example 4.31: Express the Boolean function

$$F = X \cdot Y + \bar{X} \cdot Z$$

in the product-of-sums (maxterms) form.

Solution: First, we convert the function into OR terms by using

$$A + B \cdot C = (A + B) \cdot (A + C), \text{ i.e.}$$
$$F = X \cdot Y + \bar{X} \cdot Z$$
$$= (X \cdot Y + \bar{X}) \cdot (X \cdot Y + Z)$$
$$= (\bar{X} + X \cdot Y) \cdot (Z + X \cdot Y)$$
$$= (\bar{X} + X) \cdot (\bar{X} + Y) \cdot (Z + X) \cdot (Z + Y)$$

(using again $A + B \cdot C = (A + B) \cdot (A + C)$)

$$= (\bar{X} + Y) \cdot (X + Y) \cdot (Y + Z) \quad \dots(i)$$

This function has three variables (X, Y and Z). Each OR term is missing one variable, Therefore:

$$\bar{X} + Y = \bar{X} + Y + Z \cdot \bar{Z}$$
$$= (\bar{X} + Y + Z) \cdot (\bar{X} + Y + \bar{Z}) \quad \dots(ii)$$

$$X + Z = X + Z + Y \cdot \bar{Y}$$
$$= (X + Z + Y) \cdot (X + Z + \bar{Y})$$
$$= (X + Y + Z) \cdot (X + \bar{Y} + Z) \quad \dots(iii)$$

$$Y + Z = X, \bar{X} + Y + Z = (Y + Z + X \cdot \bar{X})$$
$$= (Y + Z + X) \cdot (Y + Z + \bar{X})$$
$$= (X + Y + Z) \cdot (\bar{X} + Y + Z) \quad \dots(iv)$$

From eqn. (i), (ii), (iii) and (iv) we get

$$F = (\bar{X} + Y + Z) \cdot (\bar{X} + Y + \bar{Z}) \cdot (X + Y + Z)$$
$$\cdot (X + \bar{Y} + Z) \cdot (X + Y + Z) \cdot (\bar{X} + Y + Z)$$
$$= (X + Y + Z) \cdot (X + \bar{Y} + Z)$$
$$\cdot (\bar{X} + Y + Z) \cdot (\bar{X} + Y + \bar{Z})$$
$$(\because A \cdot A = A \cdot B \cdot B = B)$$
$$= M_0 \cdot M_2 \cdot M_4 \cdot M_5 \text{ (Table 4.17)}$$

4.23 DESIGNING USING NAND GATES

NAND gates are widely used in modern computers and an understanding of their use is invaluable. Figure 4.52 shows an OR gate symbol with 'bubbles' (inverters) at each input. The two gate-symbols in Fig. 4.52 perform the same function on inputs, as shown; for the NAND gate yields $\bar{A} + \bar{B} + \bar{C}$ on the inputs A, B and C, as does the functionally equivalent gate.

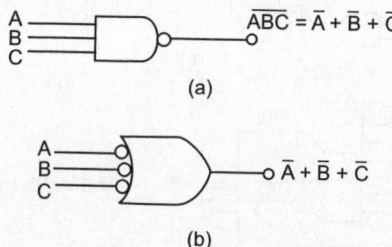

(a)

(b)

Fig. 4.52: (a) NAND gate and (b) Functionally equivalent gate

Some time, it is convenient to substitute a functionally equivalent gate in place of conventional NAND gate symbol. As an example of the use of an equivalent symbol to simplify the analysis of a NAND network, consider Fig. 4.53a. This shows a two-level NAND-to-NAND network with inputs A, B, C, D, E and F. Fig 4.53b shows the same network but with the rightmost NAND gate replaced by functionally equivalent block diagram symbol previous shown in Fig. 4.52. Note that the output function is the same for Fig. 4.53b as for Fig. 4.53a. Finally, note that the bubbles in Fig. 4.53b always occur in pairs, and so can be eliminated from the drawing from a functional viewpoint (since $\overline{X} = X$). This leads to Fig. 4.53c which is an AND-to-OR network. Thus, the NAND-to-NAND network in Fig. 4.53a yields the same function as the AND-to-OR network in Fig. 4.53c.

It is to be noticed here that the substitution of functionally equivalent symbols followed by removal of the "double bubbles" in Fig. 4.53b is a visual presentation of the following use of De Morgan's rule which can be compared with the transformation in the figure:

$$\overline{[(A \cdot B) \cdot \overline{(C \cdot D)} \cdot \overline{(E \cdot F)}]} = \overline{\overline{A \cdot B}} + \overline{\overline{C \cdot D}} + \overline{\overline{E \cdot F}}$$
$$= A \cdot B + C \cdot D + E \cdot F$$

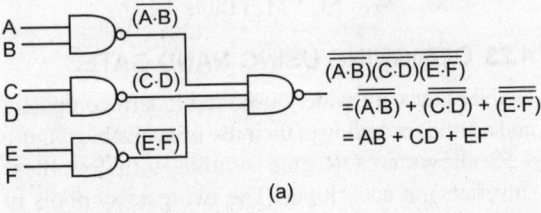

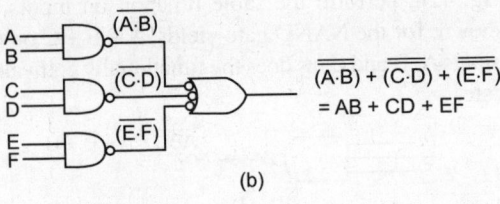

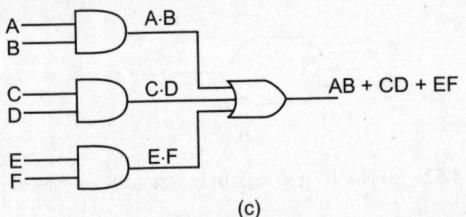

Fig. 4.53: NAND-to-NAND gate analysis (a) NAND network. (b) Network in (a) with equivalent gates (c) AND-to-OR network

The above analysis of a two-level NAND gate networks leads to the following design rule using NAND-to-NAND gates:

Design Rule: *To design a two-level NAND-to-NAND gate network, use the truth table procedure for a Sum-of-Products expression. Simplify this SOP expression by using K-map. Finally draw a NAND-to-NAND gate network in the two-level form and write the same*

inputs as would have been used in an AND-to-OR gate network, except use NAND gates in place of the AND and OR gates.

For example, let us design a NAND-to-NAND gate network for the truth Table 4.23 with three inputs A, B and C.

Table 4.23

A	B	C	F	Product terms
0	0	0	1	$\overline{A}\,\overline{B}\,\overline{C}$
0	0	1	1	$\overline{A}\,\overline{B}C$
0	1	0	0	$\overline{A}B\overline{C} = 0$
0	1	1	1	$\overline{A}BC$
1	0	0	0	$A\overline{B}\,\overline{C} = 0$
1	0	1	0	$A\overline{B}C = 0$
1	1	0	1	$AB\overline{C}$
1	1	1	1	ABC

$$\therefore \quad F = \overline{A}\,\overline{B}\,\overline{C} + \overline{A}\,\overline{B}C + \overline{A}BC + AB\overline{C} + ABC$$

Drawing K-map; we have

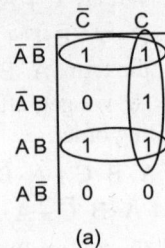

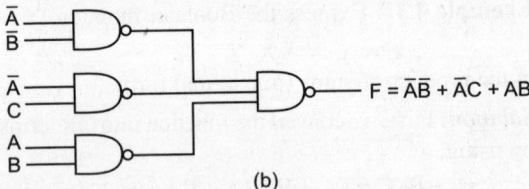

Fig. 4.54: (b) Two level NAND-to-NAND gate network for Table 4.23

simplified expression is

$$F = \overline{A}\,\overline{B} + \overline{A}C + AB.$$

The corresponding NAND-to-NAND gate network is shown in Fig. 4.53.

4.24 DESIGN USING NOR GATES

NOR gates are used often in computers because current IC technology yields NOR gates in efficient, fast circuit designs.

First we note that a symbol functionally equivalent to the NOR gate is as shown in Fig. 4.55 and satisfies the following De Morgan's rule:

$$\overline{A + B + C} = \overline{A} \cdot \overline{B} \cdot \overline{C}$$

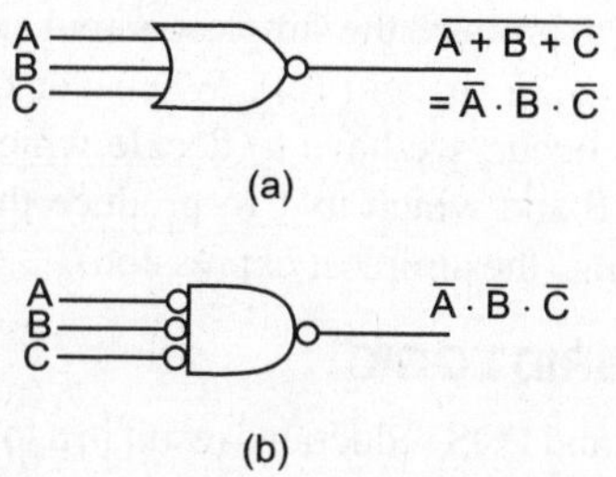

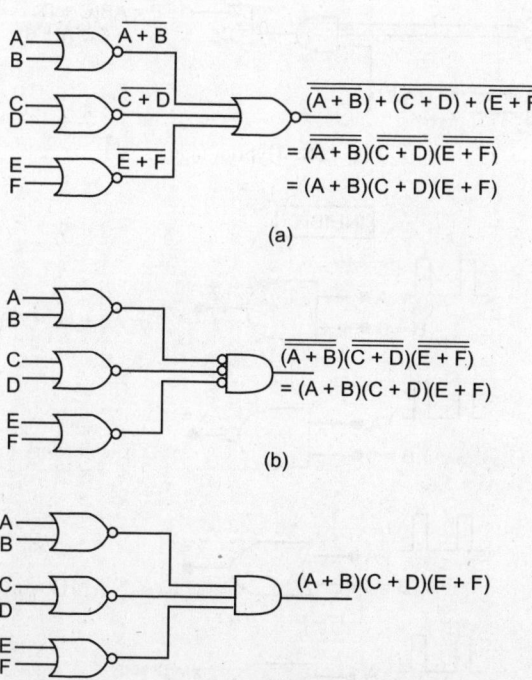

(a)

(b)

Fig. 4.55: (a) NOR gate and (b) functionally equivalent gate

Figure 4.56 shows a NOR-to-NOR gate network having the output function (A+B) (C+D) (E+F). To analyze this network, we substitute the functionally equivalent symbol for the right most NOR gate as shown in Fig. 4.56b. This yields the same function, but an examination of Fig. 4.56b shows that bubbles

occur in pairs. Since $\overline{\overline{X}} = X$, these can be eliminated as shown in Fig. 4.56c, which is for OR-to-AND network.

Figure 4.56 is a visual presentation of the use of following De Morgan's rule which can be compared with the transformation in the figure:

$$\overline{\overline{(A + B)} + \overline{(C + D)} + \overline{(E + F)}}$$
$$= \overline{\overline{(A + B)}} \; \overline{\overline{(C + D)}} \; \overline{\overline{(E + F)}}$$
$$= (A + B)(C + D)(E + F)$$

The above analysis of a two-level NOR gate networks leads to the following design rule using NOR-to-NOR gates.

Design Rule

To design a NOR-to-NOR gate network, use the procedures for designing OR-to-AND gate network. Simplify using K-map, as for OR-to-AND gate networks. Finally, draw the logic block diagram as for the OR-to-AND gate networks, but substitute NOR gates for the OR and AND gates.

For examples, let us design NOR-to-NOR gate network for the truth Table 4.24.

Fig. 4.56: NOR-to-NOR gate network analysis (a) NOR network, (b) network in (a) with equivalent gates (c) OR-to-AND network

Table 4.24

ABC	F	Product terms
0 0 0	1	$\overline{A}\overline{B}\overline{C}$
0 0 1	0	$\overline{A}\overline{B}\overline{C} = 0$
0 1 0	1	$\overline{A}B\overline{C}$
0 1 1	0	$\overline{A}BC = 0$
1 0 0	0	$A\overline{B}\overline{C} = 0$
1 0 1	1	$A\overline{B}C$
1 1 0	0	$AB\overline{C} = 0$
1 1 1	1	ABC

k-map

	$\overline{C}$	C
$\overline{A}\,\overline{B}$	1	0
$\overline{A}\,B$	1	0
$A\,B$	0	1
$A\,\overline{B}$	0	1

$$F = \overline{A}\,\overline{C} + AC$$
$$= (\overline{A} + C)(A + \overline{C})$$

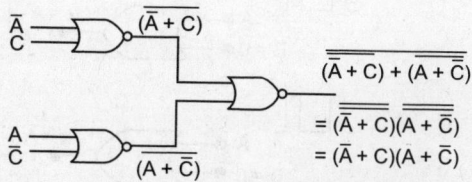

$$\overline{\overline{(\overline{A} + C)} + \overline{(A + \overline{C})}}$$
$$= \overline{\overline{(\overline{A} + C)}} \; \overline{\overline{(A + \overline{C})}}$$
$$= (\overline{A} + C)(A + \overline{C})$$

Fig. 4.57: Two level NOR-to-NOR gate network for Table 4.24

4.25 DON'T CARE CONDITIONS

Some logic circuits can be designed so that there are certain input conditions for which there are no specified output levels usually because these input conditions will never occur, i.e. there will be certain input combinations for which we need not care whether the output is HIGH or LOW. These are termed "don't care" conditions. This is illustrated in truth Table 4.25 of Fig. 4.57.

Table 4.25: Don't care condition has to be changed to 0 or 1 to produce K-map looping that yields the simplest expression.

A	B	C	Y
0	0	0	0
0	0	1	0
0	1	0	0
0	1	1	×
1	0	0	×
1	0	1	1
1	1	0	1
1	1	1	1

$\left.\begin{array}{c} \\ \\ \end{array}\right\}$ "Don't care"

$\begin{array}{c|cc} \overline{A}\overline{B} & 0 & 0 \\ \overline{A}B & 0 & \times \\ AB & 1 & 1 \\ A\overline{B} & \times & 1 \end{array} \Rightarrow \begin{array}{c|cc} \overline{A}\overline{B} & 0 & 0 \\ \overline{A}B & 0 & 0 \\ AB & 1 & 1 \\ A\overline{B} & 1 & 1 \end{array}$

$$Y = A$$

(a)   (b)   (c)

Here the output Y is not specified as either 0 or 1 for the conditions (A = 1, B = 0, C = 0) and (A = 0, B = 1, C = 1). Instead, a cross (×) is shown for these conditions (don't care condition). A don't care condition can arise for several reasons, the most common being that in some situations certain input combinations can never occur, and so there is no specified output. A circuit designer is free to make the output for any don't care condition either a 0 or a 1 in order to produce the simplest output expression. (e.g. Table 4.25 (b) and (c)). Whenever don't care conditions occur, we have to decide which ones to change to 0 and which to 1 to produce the best K-map looping (the simplest expression).

4.26 HYBRID LOGIC

Both SOP and POS reductions result in a logic circuit in which each input signal has to pass through two gates to reach the output. It is therefore called a two-level logic and has the advantage of providing uniform time delay between input signals and the output.

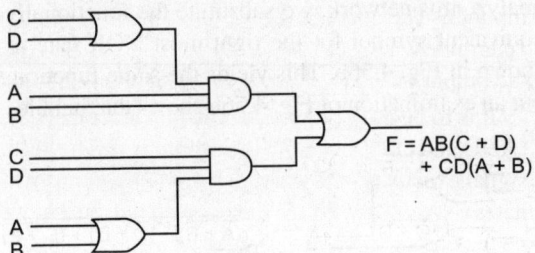

Fig. 4.58: Hybrid logic

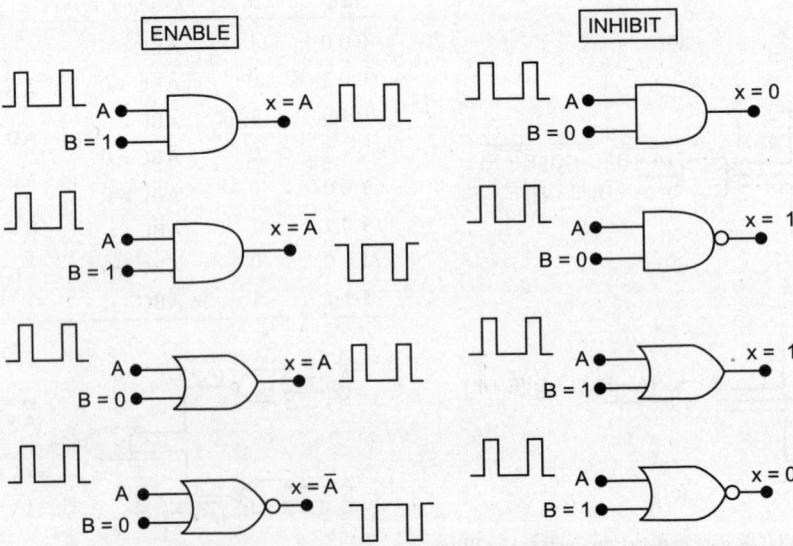

Fig. 4.59: Four basic gates can either enable or inhibit the passage of an input signal at A under control of logic level at control input B

However, there is a disadvantage that the minimal expression obtained by either SOP reduction or POS reduction may not be the actual minimal. In fact the actual minimal may be obtained by manipulating the minimal of SOP and POS forms into a hybrid form. For example, the expression ABC + ABD + ACD + BCD is in minimal SOP form and requires 16 inputs. It can however be reduced by factoring to AB (C + D) + CD (A + B) and implemented as shown in Fig. 4.58, with 12 inputs.

4.27 ENABLE AND INHIBIT CIRCUITS

Each of the basic logic gates can be used to control the passage of an input logic signal through to the output. This is depicted in Fig. 4.59 where a logical signal A, is applied to one input of each of the basic logic gates. The other input of each gate is the control input B. The logic level at this control input will determine whether the input signal is *enabled* to reach the output or *inhibited* from reaching the output.

It is to be noted that when the non inverting gates (AND and OR) are enabled, the output will follow. A signal exactly on the other hand, when the inverting gates (NAND and NOR) are enabled, the output will be the complement of the signal A.

Also note that AND and NOR gates produce a constant Low output when they are in the inhibited condition, whereas the NAND and OR gates produce a constant High output in the inhibited condition.

4.27.1 An Inhibit circuit with multiple inputs

A NOT circuit preceding one terminal (S) of an AND gate constitutes an inhibit circuit (Fig. 4.60a)

Table 4.26: Truth table for $Y = AB\overline{S}$ for eight possible input combinations

Inputs			Output
A	B	S	Y
0	0	0	0
0	1	0	0
1	0	0	0
1	1	0	1
0	0	1	0
0	1	1	0
1	0	1	0
1	1	1	0

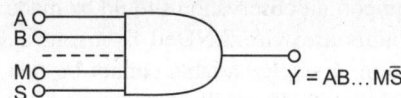

Fig. 4.60: The logic symbol for an AND with an enable terminal S

If A = 1, B = 1.....M = 1 then Y = 1 provided S = 0. However if S = 1, then the *coincidence* of A, B......M is inhibited and Y = 0. Such a configuration is also called an *anticoincidence* circuit. The terminal S is also called a strobe or an enable input. The enabling bit S = 0 allows the gate to perform its AND logic whereas the inhibiting bit S = 1 causes the output to remain at Y = 0 independent of the values of the input bits.

4.28 WIRED-OR AND WIRED-AND GATES

In certain integrated-circuit technologies (e.g. TTL circuits) it is possible to form OR and AND gates by means of a simple connection. Figure 4.60 shows a NAND-to-AND gate combination in which The AND gate is formed by simply connecting the NAND gate outputs. The wired AND gate in this figure requires no additional circuitry beyond that required for the NAND gates. This is shown by the dotted lines used in the AND symbol.

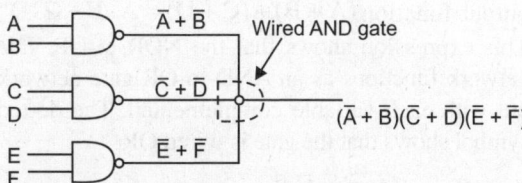

Fig. 4.61: NAND-to-Wired AND network. Here a NAND gate is equivalent to an OR with bubbled inputs

Only certain NAND gates can have their outputs connected in this way to serve as an AND gate. The designer of the NAND gates arranges for this feature and the manufacturer indicates on the specification sheet when this can be done. For example, when transistor-transistor logic (TTL) circuits are used, the specification sheets sometimes refer to the gates as having 'open collectors' which means they can be formed into NAND-to-AND nets by simply connecting their outputs. In effect, the circuits are so designed that the output level will be the lowest level any gate would if the gates were operated singly.

An important observation should be made here:

If inputs are wire-ANDed by using a simple connection, a single variable cannot be tied to the AND connection. This is illustrated in Fig. 4.61 which shows a design where a single variable B occurs in the minimal expression.

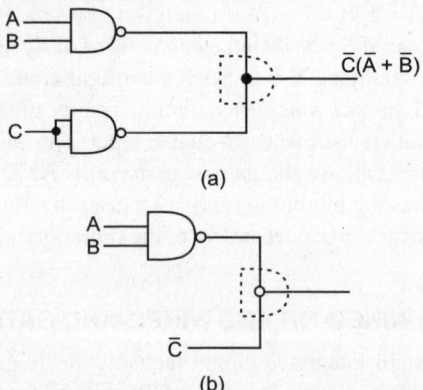

(a)

(b)

Fig. 4.62: (a) NAND-to AND gate design with a single variable C. (b) This design cannot be used in place of (a)

In this case, a single input NAND gate (i.e. inverter) must be used as shown in Fig. 4.62a.

Similar to NAND-to-Wired AND gates some NOR gates will form an OR gate at their output when they are connected. Fig. 4.62 shows a NOR-to-Wired OR net with output function $(\overline{A} + \overline{B}) + (\overline{C} + \overline{D}) = A \cdot B + C \cdot D$. This expression shows that the NOR-to-OR gate network functions as an AND-to-OR gate network but with each variable complemented. The dotted symbol shows that the gate is wired OR.

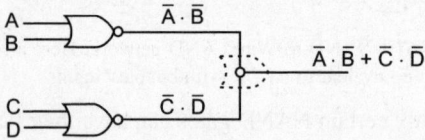

Fig. 4.63: NOR-to-wired OR gate network

Again note that only certain NOR gates can be connected at their outputs to form wired ORs. Some ECL circuits make this possible and the manufacturer mentions this on the specification sheets.

4.29 REPRESENTATION OF LOGICAL FUNCTIONS OR KARNAUGH MAP

We have described K-map method in section 4.18. For logical functions, the information contained in a truth table or available in SOP or POS form is represented on K-map. (It can be used upto six variables, beyond that, it becomes very cumbersome). Figure 4.64 shows the K-maps for two-, three-, and four variables.

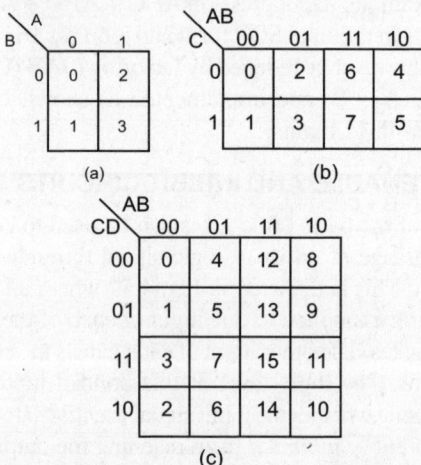

Fig. 4.64: Karnaugh maps for (a) two variables (b) three variables and (c) four variables

Representation of standard SOP form/POS form on K-map

A logical equation in standard SOP form can be represented on a K-map by simply entering 1s in the cells of the K-map corresponding to each minterm present in the equation.

Logical equation in standard POS form can be represented on K-map by entering 0s in the cells of K-map corresponding to each maxterm present in the equation.

Table 4.27: Minterms/maxterms for four variables

Variable	Minterm	Maxterm
A B C D	m_1	m_2
0 0 0 0	$\overline{A}\overline{B}\overline{C}\overline{D} = m_0$	$A + B + C + D = M_0$
0 0 0 1	$\overline{A}\overline{B}\overline{C}D = m_1$	$A + B + C + \overline{D} = M_1$
0 0 1 0	$\overline{A}\overline{B}C\overline{D} = m_2$	$A + B + \overline{C} + D = M_2$
0 0 1 1	$\overline{A}\overline{B}CD = m_3$	$A + B + \overline{C} + \overline{D} = M_3$
0 1 0 0	$\overline{A}B\overline{C}\overline{D} = m_4$	$A + \overline{B} + C + D = M_4$
0 1 0 1	$\overline{A}B\overline{C}D = m_5$	$A + \overline{B} + C + \overline{D} = M_5$
0 1 1 0	$\overline{A}BC\overline{D} = m_6$	$A + \overline{B} + \overline{C} + D = M_6$

(Contd.)

(a)

A⟍B	0	1
0	$\bar{A}\bar{B}$	$A\bar{B}$
1	$\bar{A}B$	AB

(b)

A⟍B	0	1
0	$A+B$	$\bar{A}+B$
1	$A+\bar{B}$	$\bar{A}+\bar{B}$

(c)

AB⟍C	00	01	11	10
0	$\bar{A}\bar{B}\bar{C}$	$\bar{A}B\bar{C}$	$AB\bar{C}$	$A\bar{B}\bar{C}$
1	$\bar{A}\bar{B}C$	$\bar{A}BC$	ABC	$A\bar{B}C$

(d)

AB⟍C	00	01	11	10
0	$A+B+C$	$A+\bar{B}+C$	$A+\bar{B}+C$	$\bar{A}+B+C$
1	$A+B+\bar{C}$	$A+\bar{B}+\bar{C}$	$\bar{A}+\bar{B}+\bar{C}$	$\bar{A}+B+\bar{C}$

(e)

AB⟍CD	00	01	11	10
00	$\bar{A}\bar{B}\bar{C}\bar{D}$	$\bar{A}B\bar{C}\bar{D}$	$AB\bar{C}\bar{D}$	$A\bar{B}\bar{C}\bar{D}$
01	$\bar{A}\bar{B}\bar{C}D$	$\bar{A}B\bar{C}D$	$AB\bar{C}D$	$A\bar{B}\bar{C}D$
11	$\bar{A}\bar{B}\bar{C}\bar{D}$	$\bar{A}BCD$	$ABCD$	$A\bar{B}CD$
10	$\bar{A}\bar{B}\bar{C}\bar{D}$	$\bar{A}BC\bar{D}$	$ABC\bar{D}$	$A\bar{B}C\bar{D}$

(f)

AB⟍CD	00	01	11	10
00	$A+B+C+D$	$A+\bar{B}+C+D$	$\bar{A}+\bar{B}+C+D$	$\bar{A}+B+C+D$
01	$A+B+C+\bar{D}$	$A+\bar{B}+C+\bar{D}$	$\bar{A}+\bar{B}+C+\bar{D}$	$\bar{A}+B+C+\bar{D}$
11	$A+B+\bar{C}+\bar{D}$	$A+\bar{B}+\bar{C}+\bar{D}$	$\bar{A}+\bar{B}+\bar{C}+\bar{D}$	$\bar{A}+B+\bar{C}+\bar{D}$
10	$A+B+\bar{C}+D$	$A+\bar{B}+\bar{C}+D$	$\bar{A}+\bar{B}+\bar{C}+D$	$\bar{A}+B+\bar{C}+D$

Fig. 4.65: Minterms and maxterms corresponding to each cell of K-maps (a) and (b) two variable maps (c) and (d) three variable maps and (e) and (f) four variable maps

Table 4.27 contd.

Variable	Minterm	Maxterm
$A\,B\,C\,D$	m_1	m_2
0 1 1 1	$\bar{A}BCD = m_7$	$A+\bar{B}+\bar{C}+\bar{D} = M_7$
1 0 0 0	$A\bar{B}\bar{C}\bar{D} = m_8$	$\bar{A}+B+C+D = M_8$
1 0 0 1	$A\bar{B}\bar{C}D = m_9$	$\bar{A}+B+C+\bar{D} = M_9$
1 0 1 0	$A\bar{B}C\bar{D} = m_{10}$	$\bar{A}+B+\bar{C}+D = M_{10}$
1 0 1 1	$A\bar{B}CD = m_{11}$	$\bar{A}+B+\bar{C}+\bar{D} = M_{11}$
1 1 0 0	$AB\bar{C}\bar{D} = m_{12}$	$\bar{A}+\bar{B}+C+D = M_{12}$
1 1 0 1	$AB\bar{C}D = m_{13}$	$\bar{A}+\bar{B}+C+\bar{D} = M_{13}$
1 1 1 0	$ABC\bar{D} = m_{14}$	$\bar{A}+\bar{B}+\bar{C}+D = M_{14}$
1 1 1 1	$ABCD = m_{15}$	$\bar{A}+\bar{B}+\bar{C}+\bar{D} = M_{15}$

4.30 MINIMIZATION OF LOGICAL FUNCTIONS SPECIFIED IN TERMS OF MINTERMS/MAXTERMS

Minimization of SOP form

For minimizing a given expression in SOP form or for a given truth table, we have to prepare the K-map first and then look for combinations of 1s on the K-map. We have to combine the 1s in such a way that the resulting expression is minimum. This will become clear from the following example.

Example 4.32: Minimize the four-variable logic function using K-map:

f (A, B, C, D) = Σm (0, 1, 2, 3, 5, 7, 8, 9, 11, 14).

Solution: The K-map of the given eqn. is shown in Fig. 4.66.

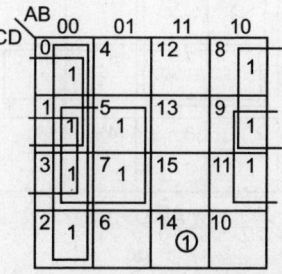

Fig. 4.66: K-map for eqn. of example 4.32

The equation is minimized using following steps.

1. Encircle 1 in cell (it cannot be combined with any other 1). The corresponding term is $AB C\bar{D}$.
2. There are at least two possible ways for every 1 forming groups of two adjacent ones. Therefore, for the time-being, ignore it.
3. For four adjacent ones, the groups are (8, 9, 0, 1), (11, 9, 1, 3), (5, 7, 3, 1), (2, 3, 10). Encircle these.

The corresponding terms are $\bar{B}\bar{C}$, $\bar{B}D$, $\bar{A}D$ and $\bar{A}\bar{B}$ respectively. (Obtained by encircling quads and rolling).

Since all the ones have been encircled, the minimized equation is

$$f(A, B, C, D) = AB C\bar{D} + \bar{B}\bar{C} + \bar{B}D + \bar{A}D + \bar{A}\bar{B}.$$

Minimization of POS form

For minimizing a given expression in POS form or for a given truth table, write zeros in the cells corresponding to maxterms for 0 outputs. The K-map is simplified by following the same procedure as used for SOP form (with ones replaced by zeros). Then, groups of zeros are formed. We now consider the following example.

Example 4.33: Minimize the logic function $f(A, B, C, D) = \Sigma m (0, 1, 2, 3, 5, 7, 8, 9, 11, 14)$ in POS form.

Solution: The given equation can b e expressed in standard POS form as

$$f(A, B, C, D) = \Pi M (4, 6, 10, 12, 13, 15) \quad \ldots(i)$$

The K-map corresponding to this eqn. is shown in Fig. 4.67.

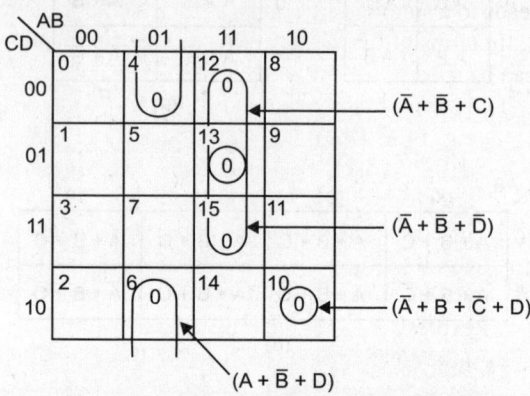

Fig. 4.67: K-map of eqn. (i) (example 4.33)

Enter zeros for the maxterms. Comparing with Fig. 4.65b we obtain the minimized POS expression

$$f (\bar{A} + B + \bar{C} + D) \cdot (\bar{A} + \bar{B} + C)$$
$$\cdot (\bar{A} + \bar{B} + \bar{D}) \cdot (A + \bar{B} + D)$$

4.31 DON'T CARE CONDITIONS IN KARNAUGH MAPS

We have described don't care condition in section 4.25. These correspond to cases in which certain combinations of input variables do not occur or the outputs corresponding to these input variables do not matter. In such situations, there is a flexibility for the designer to assume a 0 or a1 as output conditions are represented on the K-map by putting a X mark in the

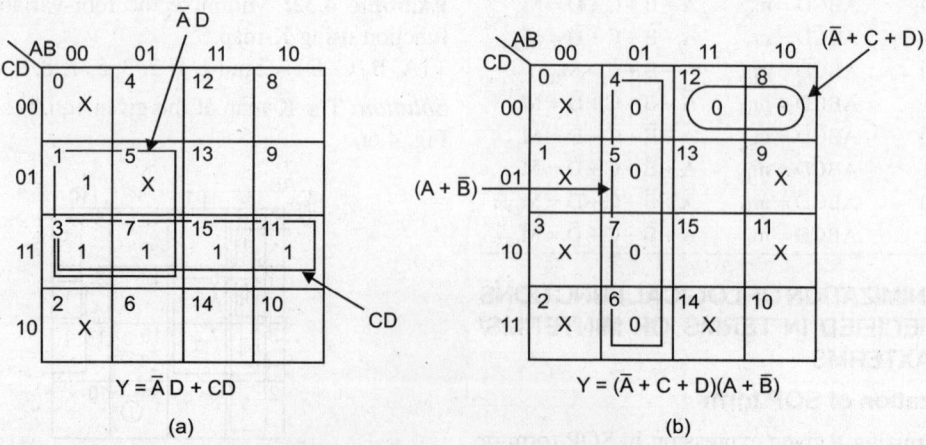

Fig. 4.68: K-maps with don't care conditions

corresponding cell. This X mark in a cell may be assumed to be a 0 or a1 depending upon which one leads to a simpler expression.

In terms of the minterms/maxterms, the function can be specified as

1. In terms of minterms and don't care conditions, e.g. f (A, B, C, D) = Σm (1, 3, 7, 11, 15) + d (0, 2,5). Its K-map and minimized expression are given in Fig. 4.68a.
2. In terms of maxterms and don't care conditions, e.g. if (A, B, C, D) = ΠM (4, 5, 6, 7, 8, 12) · d (1, 2, 3, 9, 11, 14). Its K-map and the minimized expressions are given in Fig. 4.68b.
3. In terms of truth table, we have already discussed earlier in article 4.25.

4.32 FIVE-AND SIX-VARIABLE K-MAPS

A five and a six variable K-maps are shown in Figs. 4.69 and 4.70 respectively. In a *five-variable map*, we have to consider the two four variable maps superimposed on one another; not 'hinged' or mirror imaged.

A *six-variable map* has four four-variable maps. The adjacencies between the entries in each four-variable map are visualized in the normal way.

The use of five and six variable K-maps are illustrated with the help of subsequent examples.

A = 0

DE \ BC	00	01	11	10
00	0	4	12	8
01	1	5	13	9
11	3	7	15	11
10	2	6	14	10

A = 1

DE \ BC	00	01	11	10
00	16	20	28	24
01	17	21	29	25
11	19	23	31	27
10	18	22	30	26

Fig. 4.69: A five-variable K-map

Example 4.34: Simplify the logic expression
F(A, B, C, D, E) = Sm (0, 5, 6, 8, 9, 10, 11,
16, 20, 24, 25, 26, 27, 29, 31) ...(i)

Solution: The K-map is shown in Fig. 4.71

The simplified expression is

$$F = \bar{A}\bar{B}C\bar{D}E + \bar{A}\bar{B}CD\bar{E} + A\bar{B}\bar{D}\bar{E} + \bar{C}\bar{D}\bar{E} + ABE + B\bar{C}$$

Example 4.35: Simplify the six-variable logic expression of
F = (A, B, C, D, E, F) = Σm (0, 5, 7, 8, 9, 12, 13, 23, 24, 25, 28, 29, 37, 40, 42, 44, 46, 55, 56, 57, 60, 61) ...(ii)

A = 0, B = 0

EF \ CD	00	01	11	10
00	0	4	12	8
01	1	5	13	9
11	3	7	15	11
10	2	6	14	10

A = 1, B = 0

EF \ CD	00	01	11	10
00	32	36	44	40
01	33	37	45	41
11	35	39	47	43
10	34	38	46	42

A = 0, B = 1

EF \ CD	00	01	11	10
00	16	20	28	24
01	17	21	29	25
11	19	23	31	27
10	18	22	30	26

A = 1, B = 1

EF \ CD	00	01	11	10
00	48	52	60	56
01	49	53	61	57
11	51	55	63	59
10	50	54	62	58

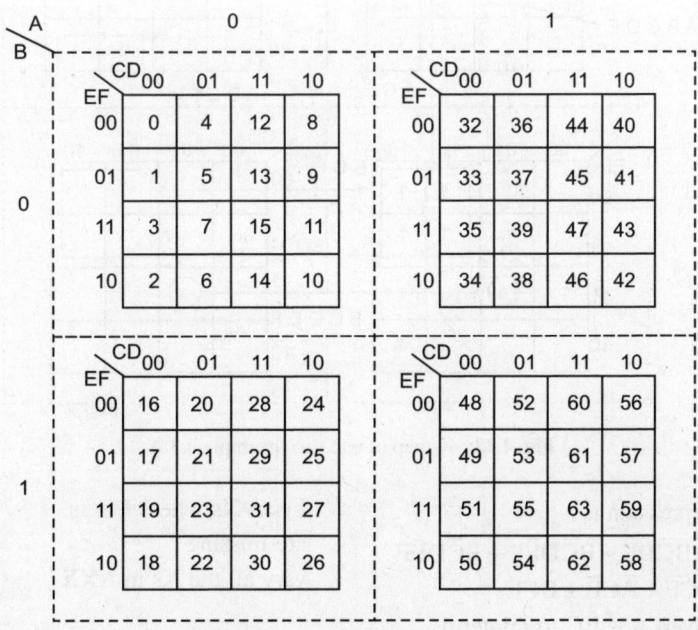

Fig. 4.70: A six variable K-map

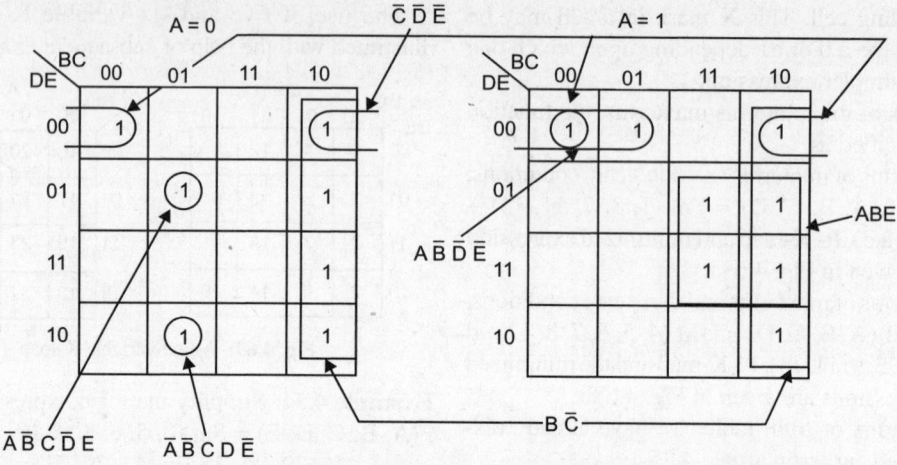

Fig. 4.71: K-map of eqn. (i) (example 4.34)

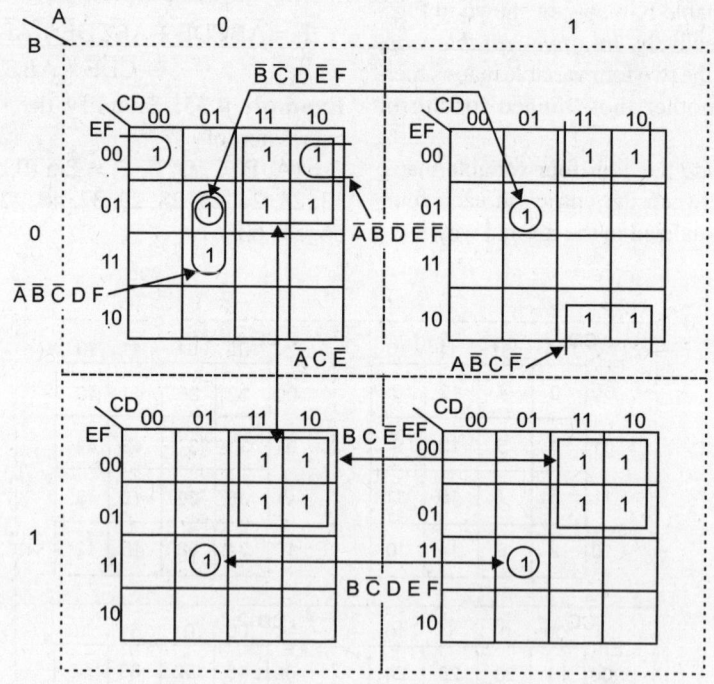

Fig. 4.72: K-map of eqn. (ii) (example 4.35)

The simplified expression is

$$F = \overline{A}\overline{B}\overline{D}\overline{E}\overline{F} + \overline{A}\overline{B}CDF + \overline{B}\overline{C}D\overline{E}F + B\overline{C}D\overline{E}F$$
$$+ B\overline{C}DEF + A\overline{B}C\overline{F} + \overline{A}C\overline{E} + BC\overline{E}$$

Example 4.36: Convert A + BC to minterms.

Solution: Write down terms

Insert Xs where letters are missing	A + BC
	AXX XBC
vary all the Xs in AXX	$A\overline{B}\overline{C}$, $A\overline{B}C$,
	$AB\overline{C}$, ABC
vary all the Xs in XBC	$\overline{A}BC$, ABC

Therefore

$$A + BC = A\overline{B}\overline{C} + A\overline{B}C + AB\overline{C}$$
$$+ ABC + \overline{A}BC + ABC.$$
$$= A\overline{B}\overline{C} + A\overline{B}C + AB\overline{C} + ABC + \overline{A}BC.$$

Example 4.37: Find the minterms for AB + ACD

Solution:

AB XX generates $AB\overline{C}\overline{D}$, $AB\overline{C}D$, $ABC\overline{D}$, ABCD

AXCD generates $A\overline{B}CD$, ABCD

Hence,

$$AB + ACD = AB\overline{C}\overline{D} + AB\overline{C}D + ABC\overline{D} + A\overline{B}CD$$
$$+ ABCD$$

Example 4.38: Find the minterms designation of $A\overline{B}\overline{C}\overline{D}$.

Solution:

Original term :	:	$A\overline{B}\overline{C}\overline{D}$
Substitute 1s for non barred letters and 0s for barred letters }	:	1 0 0 0
Express as decimal:	:	m_8
subscript of m		
Therefore	:	$A\overline{B}\overline{C}\overline{D} = m_8$.

Example 4.39: Reduce the expression
$$F = \Sigma m\ (0, 1, 2, 3, 6, 7, 13, 15)$$
by mapping and implement in NAND logic.

Solution: Plot on a four variable map, as shown in Fig. 4.73a. The next process is to read the map.

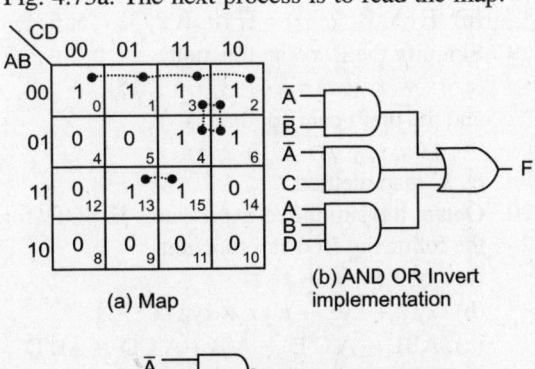

(a) Map

(b) AND OR Invert implementation

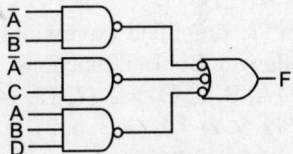

(c) NAND Invert implementation

Fig. 4.73: Pertaining to Ex. 4.39

Reading the quad m_0, m_1, m_2 and m_3 yields $\overline{A}\overline{B}$ (for C and D can both be 1s and 0s). Reading the quad m_2, m_3, m_6 and m_7 produces $\overline{A}C$ and reading m_{13} and m_{15} gives ABD. Consequently

$$F = \overline{A}\overline{B} + \overline{A}C + ABD.$$

It then can be implemented in AOI logic and then NAND logic as shown in Fig. 4.73b and 4.73c.

Example 4.40: Reduce the expression
$$F = \Sigma m\ (2, 3, 5, 7, 9, 11, 12, 13, 14, 15)$$
by mapping and implement in NOR logic

Solution: Plot on a four variable map, as shown in Fig. 4.74a.

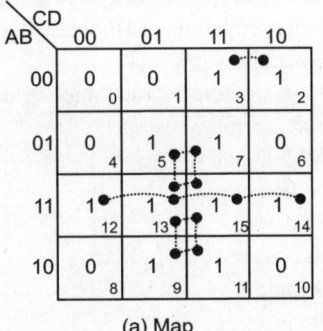

(a) Map

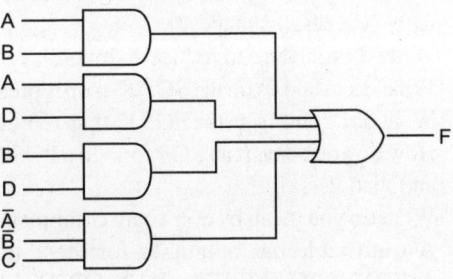

(b) AND-OR-Invert logic

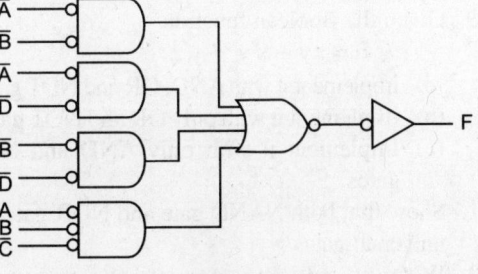

(c) NOR Logic

Fig. 4.74 Pertaining to Ex. 4.40

Reading the map produces the following terms:

m_2 and m_3 yield $\overline{A}\overline{B}C$

Quad m_5, m_7, m_{13}, m_{15} yields BD

Quad m_{12}, m_{13}, m_{14}, m_{15} yields AB

Quad m_9, m_{11}, m_{13}, m_{15} yields AD

Hence

$$F = \overline{A}\overline{B}C + BD + AB + AD$$

Figure 4.74b shows the AOI implementation and Fig. 4.74c shows the NOR implementation of the function.

EXERCISE

1. Demonstrate by mean of truth tables, the validity of the following theorems of Boolean algebra
   (a) The associative laws
   (b) Demorgan's theorems for three variables
2. Realize X-OR operation using
   (a) Only NAND gates
   (b) Only NOR gates
   (c) AOI logic
3. How do you convert AOI logic to
   (a) NAND logic
   (b) NOR logic
4. What is hybrid logic? What are its main advantages and disadvantages?
5. Write a procedure to reduce K-maps.
6. What do 1s and 0s on the SOP K-map represent? What do 0s and 1s on the POS K-map represent?
7. How do you convert an SOP form to a POS form and vice versa?
8. What do you mean by don't care combinations?
9. A truth table has output 1s for these inputs ABCD = 0011, ABCD = 0110, ABCD = 1001 and ABCD = 1110. Draw the Karnaugh map showing the fundamental products.
10. Given the Boolean function
    $$F = xy + x'y' + y'z$$
    (a) Implement it with AND, OR and NOT gates.
    (b) Implement it with only OR and NOT gates.
    (c) Implement it with only AND and NOT gates.
11. Show that both NAND gate and NOR gate are universal gates.
12. Without reducing, convert the following expressions to NAND logic:
    (a) $(A + B)(C + D)$

(b) $(A + \overline{B}C)D$
(c) $AB + CD(A\overline{B} + CD)$
(d) $A + BC + ABC$

13. Without reducing, convert the following expressions to NOR logic.
    (a) $X + Y + XY$
    (b) $X\overline{Y} + X + (\overline{X + Y})$
14. Draw the logic diagram and construct the truth table for the following expression
    $$X = A + B + \overline{CD}$$
15. Given the logic equation
    $$F = ABC + B\overline{C}D + \overline{A}BC$$
    (a) Make a truth table
    (b) Simplify using K-map
    (c) Realize F using NAND gates only
16. (a) Make a K-map for the function
    $$F = AB + A\overline{C} + C + AD + A\overline{B}C + ABC$$
    (b) Express F in standard SOP form
    (c) Minimize it and then realize the minimized expression using NAND gates only.
17. Express the following functions in a sum of minterms and a product of maxterms
    (a) $F(A, B, C, D) = D(A' + B) + B'D$
    (b) $F(x, y, z) = (xy + z)(y + xz)$
18. Convert the following to the other canonical from
    (a) $F(A, B, C, D) = \Sigma(0, 2, 6, 11, 13, 14)$
    (b) $F(A, B, C, D) = \Pi(0, 1, 2, 3, 4, 5, 6, 12)$
19. Simplify the Boolean function
    $$F(w, x, y, z) = \Sigma(1, 3, 7, 11, 15)$$
    and the don't care conditions
    $$d(w, x, y, z) = \Sigma(0, 2, 5)$$
    by K-map method.
20. Obtain the simplified expressions in SOPs for the following Boolean functions
    (a) $xy + x'y'z' + x'yz'$
    (b) $xy'z + xyz' + x'yz + xyz$
    (c) $ABD + A'C'D' + A'B + A'CD' + AB'D'$
21. Obtain the simplified expressions in SOPs for the following Boolean functions
    (a) $F(A, B, C, D) = \Sigma(7, 13, 14, 15)$
    (b) $F(x, y, z) = \Sigma(2, 3, 6, 7)$
22. Obtain the simplified expressions in POSs
    (a) $F(A, B, C, D) = \Pi(0, 1, 2, 3, 4, 10, 11)$
    (b) $F(x, y, z) = \Pi(0, 1, 4, 5)$

23. Simplify the Boolean function F in sum of products using the don't care conditions (d):
    (a) $F = y' + x' z'$
        $d = xy + yz$
    (b) $F = B' C' D' + BCD' + ABCD'$
        $d = B'CD' + A'BC'D$

24. Given the logic function
    $$Y = \overline{(AB + \overline{A}\overline{B})} + \overline{A}B$$
    Assuming the complements are not available, simplify the function using De Morgan's theorem and realize using minimum number of NAND gates.

25. Prove the following identities using Boolean theorems
    (a) $(A + B)(\overline{A} + C) = AC + \overline{A}B$
    (b) $(A + C)(A + D)(B + C)(B + D) = AB + CD$

26. Prove the following using De Morgan's theorem:
    (a) $AB + CD = \overline{\overline{AB}.\overline{CD}}$
    (b) $(A + B).(C + D) = \overline{\overline{(A + B)} + \overline{(C + D)}}$
    and hence prove
    (i) An AND-OR configuration is equivalent to a NAND-NAND configuration
    (ii) An OR-AND configuration is equivalent to a NOR-NOR configuration

27. Is it possible to INHIBIT (or DISABLE) AND, OR, NAND, NOR gates? If yes, how?

28. For the logic operation
    $$Y = A\overline{B} + \overline{A}B$$

    (a) Obtain the truth table
    (b) Name the operation performed
    (c) Realize this operation using AND, OR, NOT gates
    (d) Realize this operation using only NAND gates

29. For the logic equation
    $$f = AB'D + A'BC + BC'D'$$
    (a) Obtain the standard SOP equation
    (b) Make a truth table
    (c) Simplify using K-map
    (d) Realize the simplify expression obtained in (c) using only NAND gates

30. Prove the following
    (a) $A \oplus B = \overline{A} \oplus \overline{B}$
    (b) $\overline{A \oplus B} = A \oplus \overline{B} = \overline{A} \oplus B$

31. Using Boolean algebra verify
    (a) $\overline{\overline{A} + B} + \overline{\overline{A} + \overline{B}} = A$
    (b) $AB + AC + BC' = AC + BC'$
        (**Hint:** multiply the first term on the l.h.s. by $C + C' = 1$)
    (c) $(AB + BC + CA)' = A'B' + B'C' + C'A'$
    (d) $(A + B)(B + C)(C + A) = AB + BC + CA$
    (e) $AB + B'C' + AC' = AB + B'C'$
        (**Hint:** A term may be multiplied by $B + B' = 1$)
    (f) $A'BC + AB'C + ABC' + ABC = AB + BC + CA$
    (g) $AB'C + A'BC + ABC = AC + AB$

5

Logic Families

5.1 INTRODUCTION

As a result of advances in micro electronics, the digital IC technology in less than four decades has rapidly advanced from small scale integration (SSI), through medium scale integration (MSI), large scale integration (LSI), very large scale integration (VLSI) to ultra large scale integration (ULSI) and the technology is now entering giant scale integration (GSI) with millions of gate equivalent circuits integrated on a single chip. The use of ICs has rendered reduced size, reduced cost, improved reliability and reduced power consumption. Besides this there are some limitations. ICs cannot handle very large voltages or currents and also electrical devices like precision resistors, inductors, transformers and large capacitors cannot be implemented on small chips. Thus operations that require high power levels or devices that cannot be integrated are still handled by discrete components.

ICs have been fabricated using various technologies which we will consider in the subsequent sections. Prior to that, it will be first worthwhile to discuss IC specification terminology.

5.2 DIGITAL IC SPECIFICATION TERMINOLOGY

The digital IC nomenclature and terminology is fairly standardized. The most specific terms are defined below:

5.2.1 Threshold Voltage

The threshold voltage is defined as that voltage at the input of a gate which causes a change in the state of the output from one logic level to the other.

5.2.2 Propagation Delay

A pulse through a gate takes a certain amount of time to propagate from input to output. This interval of time is known as the propagation delay of the gate. It is the average transition delay time t_{pd}, expressed by

$$t_{pd} = \frac{(t_{PLH} + t_{PHL})}{2}$$

where t_{PLH} is the signal delay time when the output goes from a logic 0 to a logic 1 state and t_{PHL} is the signal delay time when the output goes from a logic 1 to a logic 0 state.

5.2.3 Power Dissipation

Every logic gate draws some current from the supply for its operation. The current drawn in HIGH state is different from that drawn in LOW state. The power dissipation P_D of a logic gate is the power required by the gate to operate with 50% duty cycle at a specified frequency and is expressed in milli watts. This means that 1 and 0 periods of the output are equal. The power dissipation of a gate is given by

$$P_D = V_{CC} \times 1_{CC} \text{ (avg)}/n$$

where V_{CC} is the gate supply voltage, I_{CC} (avg) is the average current drawn from the supply by the enter IC and n is the number of gates in the IC. Now,

$$I_{CC}(avg) = \frac{I_{CCH} + I_{CCL}}{2}$$

where I_{CCH} is the current drawn by the IC when all the gates in IC are in HIGH state and I_{CCL} is the current drawn by the IC when all the gates in the IC are in LOW state. The total power consumed by an IC is equal to the product of the power dissipated by each gate and the number of gates in that IC.

5.2.4 Fan-in

The fan-in of a logic gate is defined as the number of inputs that the gate is designed to handle.

5.2.5 Fan-out

The fan-out of a logic gate is defined as the maximum number of standard loads that the output of the gate can drive without impairing its normal operation. A standard load is usually specified as the amount of current needed by an input of another gate of the same IC family. If a gate is made to drive more than this number of gate inputs, the performance of the gate is not guaranteed and it may malfunction.

Fan-outs may be HIGH state fan-out, i.e. the fan-out of the gate when its output is a logic 1 or it may be LOW state fan-out, i.e. the fan-out of the gate when its output is a logic 0. The smaller of these two numbers is taken as the actual fan-out.

The fan-out of a gate affects the propagation delay time as well as saturation. The driving gate sinks current when it is in LOW state and sources current when it is in HIGH state.

$$\text{HIGH state fan-out} = \frac{I_{OH}(\max)}{I_{IH}}$$

where I_{OH} (max) is the maximum current that the driver gate can source when it is in a 1 state and I_{IH} is the current drawn by each driven gate from the driver gate.

$$\text{LOW state fan-out} = \frac{I_{OL}(\max)}{I_{IL}}$$

where I_{OL} (max) is the maximum current that the driver gate can sink when its output is a logic 0 and I_{IL} is the current drawn from each driven gate by the driver gate. Figure 5.1 depicts the current sourcing and current sinking for the TTL 7400 NAND gate.

5.2.6 Voltage and Current Parameters

The definitions of voltage and current levels corresponding to the logic 0 and logic 1 states are as follows:

V_{IH} **(min) (HIGH level input voltage):** It is the minimum voltage level required at the input of a gate for that input to be treated as logic 1. Any voltage below this level will not be accepted as a logic 1 by the logic circuit.

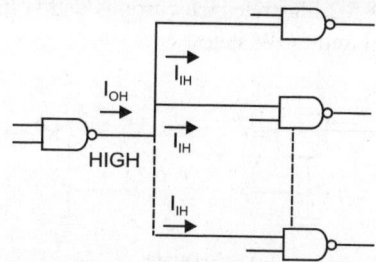

(a) Current sourcing in HIGH state

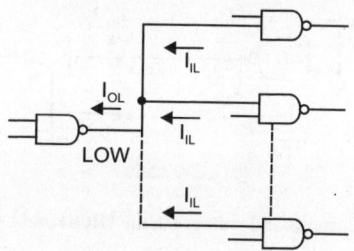

(b) Current sinking in LOW state

Fig. 5.1: HIGH state and LOW state fan-outs for TTL 7400 NAND gate

V_{OH} **(min) (HIGH level output voltage):** It is the minimum voltage level required at the output of a gate for that output to be treated as a logic 1. Any voltage output below this level will not be accepted as a logic 1 output.

V_{IL} **(max) (LOW level input voltage):** It is the maximum voltage level that can be treated as a logic 0 at the input of the gate. Any voltage above this level will not be treated as a logic 0 input by the logic gate.

V_{OL} **(max) (LOW level output voltage):** It is the maximum voltage level that can be treated as a logic 0 at the output of the gate. Any voltage above this level will not be treated as a logic 0 output.

I_{IH} **(HIGH level input current):** The current that flows into an input when a specified high voltage is applied to that input.

I_{IL} **(LOW level input current):** The current that flows into an input when a specified low voltage is applied to that input.

I_{OH} **(HIGH level output current):** The current that flows from an output in a logic 1 state under specified load conditions.

I_{OL} **(LOW level output current):** The current that flows from an output in a logic 0 state under specified load conditions.

Figure 5.2 illustrates the currents and voltages in the HIGH and LOW states.

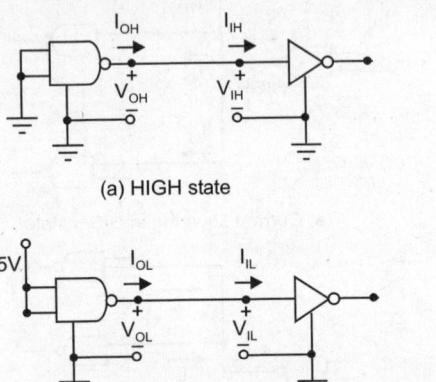

(a) HIGH state

(b) LOW state

Fig. 5.2: Currents and voltages in the HIGH and LOW states

5.2.7 Noise Margin

The input and output voltage levels are shown again in Fig. 5.3. Stray electric and magnetic fields may induce unwanted voltages, known as noise, on the connecting wires between logic circuits. This may cause the voltage at the input to a logic circuit to drop below V_{IH} or rise above V_{IL}, and may produce undesired operation.

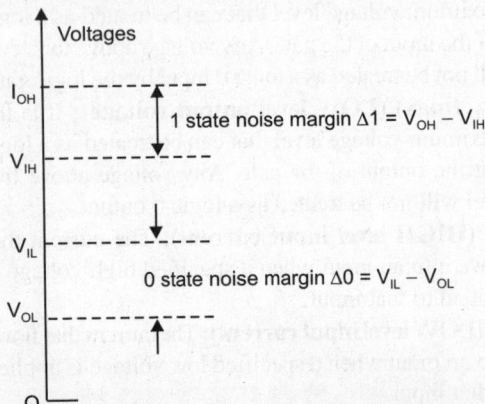

Fig. 5.3: Voltage levels and noise margins of ICs

The circuits ability to tolerate noise signals is referred to as the *noise immunity* a quantitative measure of which is called *noise margin*. Noise margins are illustrated in Fig. 5.3.

ac noise margins

The noise margins defined above are referred to as *dc noise margins*. Strictly speaking, the noise is generally thought of as an ac signal with amplitude and pulse width. For high speed ICs, a pulse width of a few microseconds is extremely long in comparison to the propagation delay time of the circuit and therefore may be treated as dc as far as the response of the logic circuit is concerned. As the noise pulse width decreases and approaches the propagation delay time of the circuit, the pulse duration is too short for the circuit to respond. Then, a large pulse amplitude will be required to produce a change in the circuit output. This means that a logic circuit can effectively tolerate a large noise amplitude if the noise is of a very short duration. This is referred to as *ac noise margin* and is substantially greater than the dc noise margin. It is generally supplied by the manufacturers in the form of a curve between noise margin and noise pulse width.

5.2.8 Operating Temperatures

The IC gates being semiconductor devices, they are temperature sensitive. However, they are designed to operate satisfactorily over a specified temperature range. The range specified for commercial applications is 0 to 70°C and for industrial applications is 0 to 85°C.

5.2.9 Speed-Power Product

A common means for measuring and comparing the overall performance of an IC family is the *speed power product*. It is obtained by multiplying the gate propagation delay by the gate power dissipation. A low value of speed power product is desirable. The smaller its value, the better the overall performance. It is the figure of merit of an IC family. It has units of energy and is expressed in Pico joules.

5.2.10 Power Supply Requirements

The supply voltages and the amount of power required by an IC are important characteristics required to choose the proper power supply.

5.2.11 Flexibilities Available

Various flexibilities are available in different IC logic families and these must be considered while selecting a logic family for a particular job. Some of the flexibilities available are

1. **The breadth of the series:** Types of different logic functions available in the series.
2. **Popularity of the series:** The cost of manufacturing depends upon the number of ICs manufactured. When a large number of ICs of one type are manufactured, the cost per function will be very small and it will be easily available because of multiple sources.
3. **Wired-logic capability:** The output can be connected together to perform additional logic without any extra hardware.
4. **Availability of complement outputs:** This eliminates the need for additional inverters.
5. **Type of output:** Passive pull up, active pull up, open collector/ drain and tristate, etc.

Example 5.1: Given below are the specification of a low power Schottky TTL NOR gate. Calculate its propagation delay, power dissipation, fan-out, fan-in, noise margin and also find its figure of merit.

Parameter	Values
V_{CC}	5V
I_{CCH}	1.6 m A (feeding 4 gates)
I_{CCL}	2.8 m A (feeding 4 gates)
V_{OH} (min)	2.7 V
V_{OL} (min)	0.4 V
V_{IH} (max)	2.0 V
V_{IL} (max)	0.8 V
I_{OH} (max)	4.0 mA (out)
I_{OL} (max)	8.0 mA (in)
I_{IH} (max)	0.02 mA (in)
I_{IL} (max)	0.4 mA (in)
$t_{PLH} t_{PHL}$	10 ns10 ns

Solution: (i) Propagation delay,

$$t_p = \frac{t_{PLH} + t_{PHL}}{2} = 10\,ns$$

(ii) Power dissipation $P_D = V_{CC} I_C$ (ON)

$$I_C \,(ON) = \frac{I_{CCH} + I_{CCL}}{2} = \frac{1.6 + 2.8}{2} = 2.2\,mA$$

P_D (4 gates) = $5 \times 2.2 = 11$ m W

P_D/gate = $11/4 = 2.75$ mW.

(iii) Case 1: Output is HIGH:

$$\text{Fan-out} = \frac{I_{OH}}{I_{IL}} = \frac{0.4\,mA}{0.02\,mA} = 20$$

Case 2: Output is LOW:

$$\text{Fan-in} = \frac{I_{OL}}{I_{IL}} = \frac{8\,mA}{0.4\,mA} = 20$$

(iv) Noise margin : $(NM)_H = V_{OH} - V_{IH} = 0.7\,V$

$$(NM)_L = V_{IL} - V_{OL} = 0.4\,V$$

(v) Figure of merit = $P_D\,t_p = 2.75 \times 10 = 27.5$ mW/ns

5.3 DIGITAL LOGIC FAMILIES

In the first chapter, we have acquainted ourselves with various levels of integration for digital integrated circuits which we have also mentioned in the introduction of this chapter. We will now consider various digital logic families. The digital IC families can be divided into two broad categories.

(A) Bipolar logic families

(B) Unipolar logic families

(A) Bipolar logic families: There are two types of bipolar families

(a) Saturated

(b) Non-saturated

In saturated bipolar logic, the transistors are driven into saturation and cut off. These are further classified as

(i) Resistor-Transistor Logic (RTL)

(ii) Direct Coupled Transistor Logic (DCTL)

(iii) Diode Transistor Logic (DTL)

(iv) High Threshold Logic (HTL)

(v) Transistor Transistor Logic (TTL)

(vi) Integrated Injection Logic (IIL)

In non saturated bipolar logic families, transistor is not driven into saturation but operates between cut off and active states. These are classified as

(i) Schottky TTL

(ii) Emitter Coupled Logic (ECL)

(B) Unipolar logic families: These logic families use MOSFET design for their internal circuitry. The common type of MOS families are

(i) P-Channel MOS (PMOS)

(ii) N- Channel MOS (NMOS)

(iii) Complementary MOS (CMOS)

5.4 DIODE LOGIC OR CIRCUIT

In a *diode-logic* (DL) system, the logical gates are implemented by using diodes. A diode OR for negative logic is shown in Fig. 5.4.

We first consider the case where the supply voltage V_R has a value equal to the voltage $V(0)$ of the 0 state for dc logic.

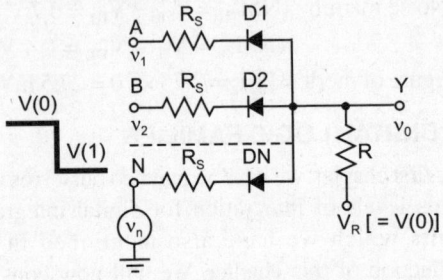

Fig. 5.4: A diode OR circuit for negative logic

If all inputs are in the 0 state, the voltage across each diode is $V(0) - V(0) = 0$. Since, in order for a diode to conduct, it must be forward-based by at least the cut in voltage V_γ, therefore none of the diodes conducts and the output voltage is $v_0 = V_0$ and γ is in the 0 state.

Suppose now input A is changed to the 1 state, which for negative logic is at the potential $V(1)$, less positive than the 0 state, then D_1 will conduct. The output becomes

$$v_0 = V(0) - [V(0) - V(1) - V\gamma]\frac{R}{(R + R_s + R_f)}$$

where R_s is the generator-source resistance and R_f is the diode forward resistance. R is chosen much larger than $R_s + R_f$. Then

$$v_0 \approx V(1) + V_g$$

Hence, the output voltage exceeds the more negative level $V(1)$ by V_γ (~0.2V for Ge or 0.6 V for S_i). Further, the step in output voltage is smaller by V_γ than the change in input voltage. If we assume $R \gg R_s$ and ideal diodes, i.e. $R_f = 0$ and $V_\gamma = 0$, then the output for input A excited is $v_0 = V(1)$ and the circuit has performed the following operation; if A = 1, B = 0, ... N = 0, then Y = 1 which is the logic for the OR operation. If two or more inputs are in the 1 state, the diodes connected to these inputs conduct and other diodes remain reverse biased. The output is $V(1)$ and again, the OR function is satisfied.

A positive logic OR gate uses the same configuration as that in Fig. 5.4, except that all diodes must be reversed. The output now is equal to the most positive level $V(1)$.

For Fig. 5.4 a second mode of operation of the OR circuit is possible if V_R is set equal to a voltage more positive than $V(0)$ by at least V_g. Then all diodes conduct in the 0 state and $v_0 \approx V(0)$ if $R \gg R_s + R_f$. If one or more inputs are excited, then the diode connected to the most negative $V(1)$ conducts, the output equals this value of $V(1)$ and all other diodes are back-biased. Clearly, the OR function has been satisfied.

5.5 DIODE LOGIC AND CIRCUIT

A diode logic (DL) AND circuit for negative logic is shown in Fig. 5.5a. To understand the operation of the circuit, assume initially that all source resistances are zero the diodes are ideal. If any input is at the 0 level $V(0)$, the diode connected to this input conducts and the output is clamped at the voltage $V(0)$ or Y = 0. However if all inputs are at the 1 level $V(1)$, then all diodes are reverse based and $v_0 = V(1)$ or Y = 1. Clearly the AND operation has been implemented. The AND gate is also called a *coincidence circuit*.

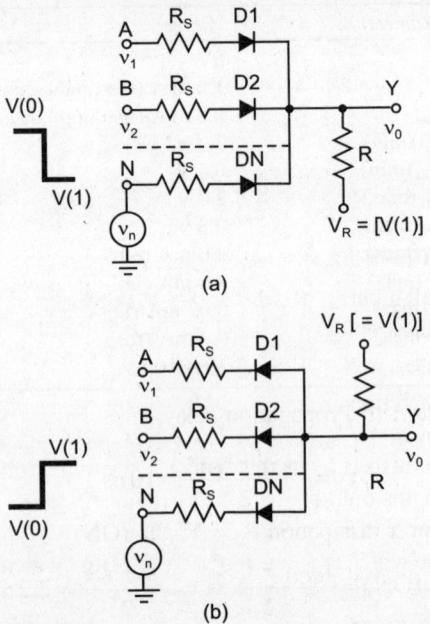

Fig. 5.5: A diode logic (DL) AND circuit for (a) negative logic and (b) positive logic

For positive logic AND gate, the configuration is similar to Fig. 5.5a except that all diodes are reversed as indicated in Fig. 5.5b. Figure 5.5b may be compared with Fig. 5.4. Note that V(1) in Fig. 5.4 equals V(0) in Fig. 5.5b since both represent the lower binary level. Thus, the two circuits are identical, i.e. a negative logic OR gate is the same circuit as a positive logic AND gate. This result is independent of the hardware used to implement the circuit.

If all the inputs are low, all the diodes will be conducting and output Y will be low. Similarly when one or more inputs are low output Y is low. On the other hand, if all the inputs are high, none of the diodes conduct and output Y is high.

5.6 EMITTER FOLLOWER LOGIC FOR AN OR CIRCUIT

An implementation of OR circuit using an emitter coupled OR gate is indicated in Fig. 5.6, where an emitter follower is used for each input and the output is taken from the common emitter resistor R. The bottom of this resistor goes to a supply voltage equal to V(0). For negative logic the transistor must be p-n-p type as shown.

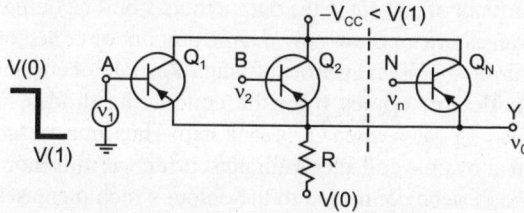

Fig. 5.6: An OR circuit for negative logic. (This circuit is also a positive logic AND gate)

If all inputs are at the 0 level V(0) then each base-to-emitter voltage is at

$$V(0) - V(0) = 0V$$

and each transistor is virtually at cut off. The output is then $v_0 = V(0)$ or $Y = 0$. On the other hand, if any input is at the 1 level, then, because of emitter follower action, the output is $v_0 = V(1)$, (neglecting the small V_{BE}) or $Y = 1$. The V_{BE} of all transistors other than the one which is excited is $V(0) - V(1)$. Since this is a positive voltage (for negative logic), these p-n-p transistors are off. Here, OR logic is obeyed and the circuit has a higher input impedance than the diode logic gate. If more than one input is excited simultaneously then the output follows the most negative value V(1). Note that $-V_{CC}$ must to more negative than V(1). For positive logic, n-p-n transistors

are to be used and V_{CC} must be more positive than V(I).

5.7 AN INVERTER CIRCUIT USING TRANSISTORS

The transistor circuit of Fig. 5.7 implements a NOT circuit (i.e. inverter) for positive logic. It has a 0 state of $V(0) = V_{EE}$ and a 1 state of $V(1) = V_{CC}$. If the input is low $v_i = V(0)$, then the parameters are chosen so that the transistor is off, and hence $v_0 = V_{CC} = V(1)$. On the other hand, if the input is high $v_i = V(1)$, then the circuit parameters are such that transistor Q is in saturation and then $v_0 = V_{EE} = V(0)$, if we neglect the collector-to-emitter saturation voltage V_{CE} sat.

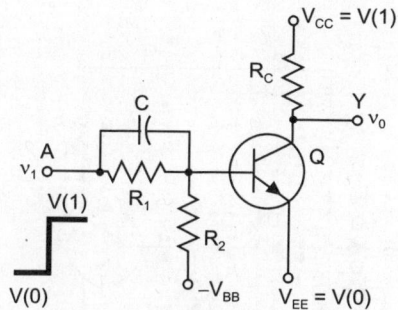

Fig. 5.7: A NOT circuit using transistor for positive logic

5.8 THE NAND AND NOR DIODE TRANSISTOR LOGIC (DTL) GATES

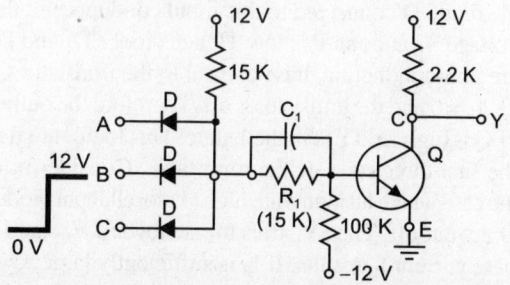

Fig. 5.8: A three input positive logic NAND (or negative logic NOR) gate

Here, the NAND has been implemented by placing a transistor NOT circuit after the diode AND, as shown in Fig. 5.8 this is known *as diode-transistor logic* (DTL) gate.

5.9 MODIFIED (INTEGRATED-CIRCUIT) DTL NAND GATE

Most logic gates are fabricated as an integrated circuit (IC). In IC fabrication, large values of resistance (above 30K) and of capacitance (above 100 pF) cannot be fabricated economically, whereas transistors and diodes may be constructed comparatively inexpensively. With these facts in view, the NAND gate of Fig. 5.8 is modified for IC implementation by eliminating the capacitor C, reducing the resistance values drastically and using diodes or transistors to replace resistors wherever possible. With these changes, only a single 5V supply is sufficient. The resulting circuit is Fig. 5.9.

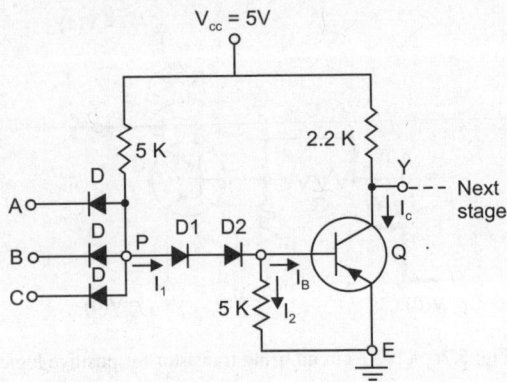

Fig. 5.9: An integrated positive logic DTL NAND gate

The operation of this circuit can be understood as follows. If at least one of the inputs is low (0 state), the diode D connected to this input conducts and the voltage V_p at point P is low. Hence diodes D_1 and D_2 are non-conducting, base current to the transistor Q, i.e. $I_B = 0$ and the transistor is off. Therefore the output of Q is high, and Y is in the 1 state. This logic satisfies the first three rows of the truth table. Consider now the case where all inputs are high (1) so, all input diodes D are cut off. Then V_p tries to rise toward V_{CC} and a base current I_B results. If I_B is sufficiently large, Q is driven into saturation and output Y falls to its low (0) state. This satisfies fourth row of the truth table.

The main disadvantages of DTL gates are

(i) Low fan-out capability
(ii) Low noise margin which can be increased by addition of optional diodes (D_1 or D_2) but it further decreases fan-out capability.

A modified DTL NAND gate with increased Fan-out: A modified DTL NAND gate is shown in Fig. 5.10

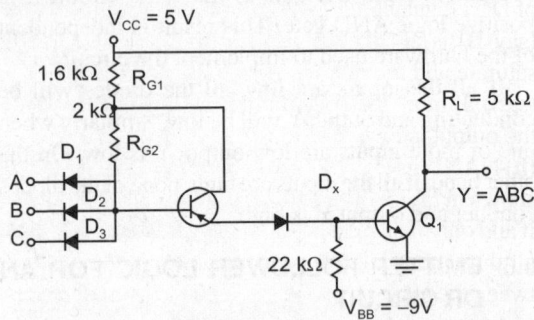

Fig. 5.10: Modified DTL NAND gate

The fan-out of the DTL gate can be increased by replacing one of the diodes in the base circuit of Q_1 by a transistor Q_2. The transistor Q_2 is maintained in the active region when output of Q_1 is saturated. As a result, this modified DTL circuit can supply a large amount of current to the output transistor. The output transistor can draw now a large amount of collector current before it goes out of saturation. Part of current (collector) comes from the conducting diodes in loading gates when Q_1 is saturated. Thus an increase in allowable collector saturated current permits more loads to be connected to the output which increases fan-out of the gate.

5.10 DIRECT COUPLED TRANSISTOR LOGIC (DCTL)

(a) NOR gate: A three input NOR DCTL gate for positive logic is shown in Fig. 5.11. The transistors are connected in parallel. Assume all inputs are at the logical 0 level. The input to any one transistor comes from a saturated transistor in the output of the

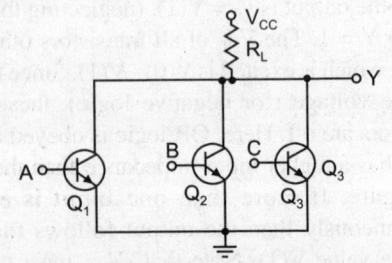

Fig. 5.11: DCTL NOR gate

preceding gate and is typically about 0.3V. As this voltage is less than the cut in voltage of Q_1, the Q_1 does not conduct and the output voltage equals V_{CC}.

Thus with all inputs at logical 0 level, the output of DCTL gate is at logical 1 level. If any of the inputs (say B) is at logical 1 level, the transistor Q_2 will saturate and the output drops to logical 0 level (V_{CE} (sat)). If more than one input is at the logical 1 level, the output is low.

(b) NAND gate: The series connected NAND gats is shown in Fig. 5.12. A logical 0 output can be produced if and only if all transistors are forward biased. Unless all three transistors are forward biased, the path between the supply voltage V_{CC} and the ground is open and no current can pass through load resistance R_L.

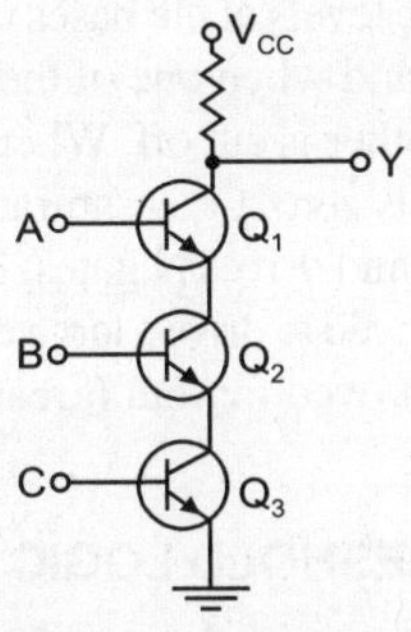

Fig. 5.12: DCTL NAND gate

The advantages of this logic circuit are ease of manufacture in IC form and requirement of low power supply, low breakdown voltage for the transistors and low power dissipation.

5.11 RESISTOR TRANSISTOR LOGIC (RTL)

The basic RTL NOR gate is shown in Fig. 5.13. It is modified version of DCTL circuit in which variations of V_{BE} (sat) caused current hogging problem. This problem is minimized here by using resistor R in series with the transistor base. With the addition of this resistor R, the base current depends on supply voltage V_{CC}, load resistor R_C and the base resistor R as long as V_{CC} is much greater than V_{BE} (sat).

In order to reduce storage delay time, the resistor R is bypassed by a small capacitor C. This converts the input network to a compensated attenuator. The product CR should match the base time constant. When the capacitor is included, it is referred to as resistor capacitor transistor logic (RCTL).

The major disadvantages of RTL are its low noise immunity and its low speed.

A typical commercially available integrated circuit RTL NOR gate has $R_C = 640\Omega$ and $R_b = 450\Omega$ as shown. A low power RTL NOR gate is also available having $R_C = 3.6k\Omega$ and $R_b = 1.5kW$. A disadvantage of this low power gate is that the larger resistors result in slower operation, i.e. the propagation delay time is larger. The reason for this longer delay is that all stray capacitors and capacitors inherent in the active devices must charge and discharge through these large resistors.

Pull-up Resistors: The collector resistor R_c in the RTL gate is often called a *passive pull-up resistor*. The reason for this terminology can be seen with

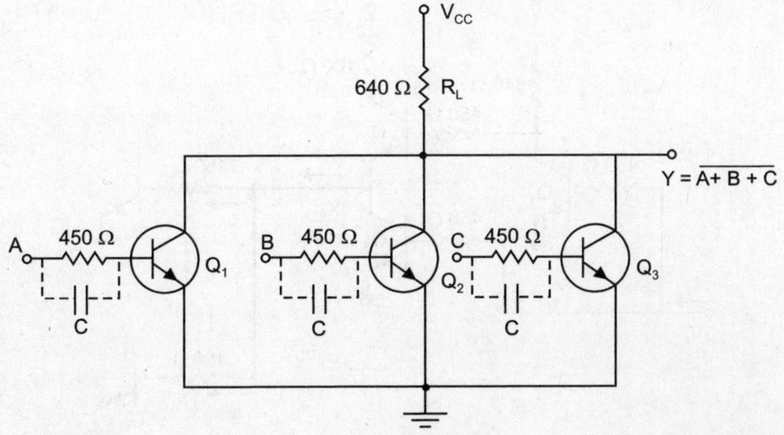

RTL NOR gate

Fig. 5.13: RTL NOR gate

reference to Fig. 5.14 where an RTL gate drives a capacitive load C. This capacitance is due in part to stray capacitance and in part to the capacitance associated with the base-emitter junctions of gates driven by the gate shown.

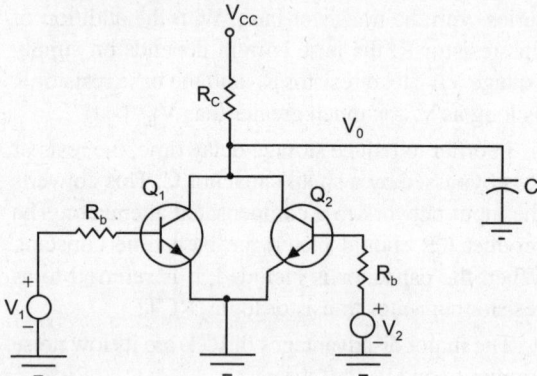

Fig. 5.14: A two-input RTL gate with a capacitive load

Assume initially that V_1 is high and V_2 low, so that Q_1 is saturated, Q_2 cut off and V_0 in the 0 state. Now let V_1 fall to the 0 state. Q_1 now cuts off and V_0 rises towards V_{CC} with a time constant $i = R_CC$. We say that V_0 is pulled up to V_{CC} by the *pull up* resistor R_C. Since R_C is passive, we say that the gate here employs passive pull-up.

5.12 AN RTL BUFFER AND ACTIVE PULL UP

The RTL gate discussed above employed a passive pull up R_c. This limited the output current available to drive other gates, to a value less than V_{CC}/R_C. The buffer is an RTL gate using an *active pull up to* achieve a very low output impedance. Thus, the output-current capability of the buffer is significantly greater than that of the ordinary gate. As a result, while the ordinary RTL gate has a fan-out of 5, the buffer has a fan-out of 25. Further, due to low output impedance, the rise time when driving a capacitive load is significantly less than that obtained with passive pull up.

A typical RTL buffer is shown in Fig. 5.15. The transistor Q_2 is the active pull up which replaces the passive pull up resistor R_c in the ordinary NOR gate. The transistor Q_1 serves as a logic inverter. When V_1 is at logic 0, the output of Q_1 is at logic 1 and vice versa.

Thus, the logic levels at the bases of Q_2 and Q_3 are always different and when one of these transistors is conducting, the other is cut off. When Q_2 is "off", its collector current is zero, Q_3 is saturated. When Q_3 is "off", Q_2 is "on" and drives N gates. Note that buffer is generally employed to drive a large number of gates, and provides improved fan-out (greater than that for an RTL gate).

5.13 HIGH THRESHOLD LOGIC (HTL)

In industrial environment, the noise level may be high due to presence of inductive loads. By using higher supply voltage V_{CC} (15V in place of 5V) and a zener diode D_z of 6.9V in place of D_x (in Fig. 5.16.), the circuit is converted into high noise immunity gate as

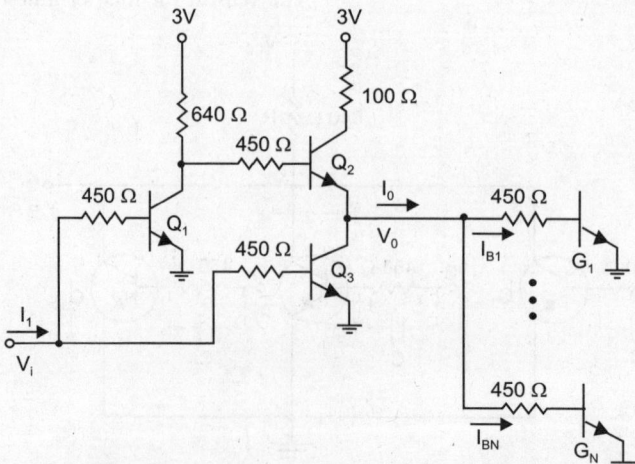

Fig. 5.15: An RTL buffer inverter driving N gates

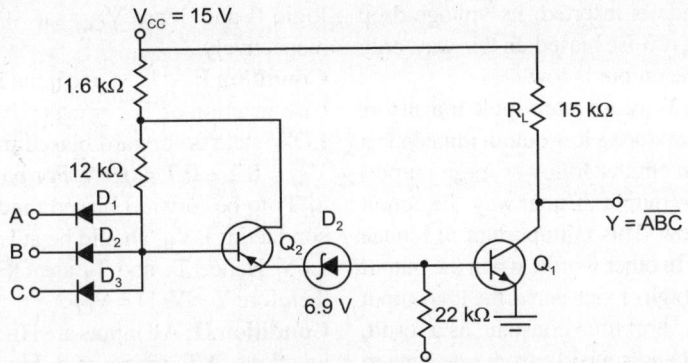

Fig. 5.16: A high threshold logic NAND or NOR gate

shown in Fig. 5.16. The noise margin obtained with this circuit is ~7V.

In order for output transistor Q_1 to conduct, the emitter of Q_2 must rise to a potential of $V_{BE} = +6.9V$. The low level for gate remains at 0.2V, but high level is at 15V. With the input of 0.2V, the base of Q_2 is at 0.9V and Q_2 is cut off. The noise voltage should be greater than 7.5 V to change the state of Q_1. With all inputs at 15V, the output transistor Q_1 is saturated.

5.14 STANDARD TTL NAND GATE

Figure 5.17 shows a standard TTL NAND gate. Q_1 is a two-input transistor where each emitter acts like a diode. Therefore, Q_1 and the 4kΩ resistor act like a 2-input AND gate. The rest of the circuit inverts the signal and the overall circuit acts like a 2-input NAND gate.

The output transistors Q_3 and Q_4 form a totem pole connection, typical of most TTL devices. Either one or the other is on. When Q_3 is on, the output is high, when Q_4 is on, the output is low. The advantage of a totem pole connection is its low output impedance.

Ideally, the input voltages A and B are either low (grounded) or high (+5V). If A or B low, Q_1 saturates. This reduces the base voltage of Q_2 to almost zero. Q_2 cuts off, forcing Q_4 to cut off. Under these conditions Q_3 acts like an emitter-follower and couples a high voltage to the output.

On the other hand, when both A and B are high, the collector diode of Q_1 goes into forward conduction, this forces Q_2 and Q_4 into saturation, producing a low output at Y.

Incidentally, without diode D_1 in the circuit, Q_3 would conduct slightly when the output is low. To

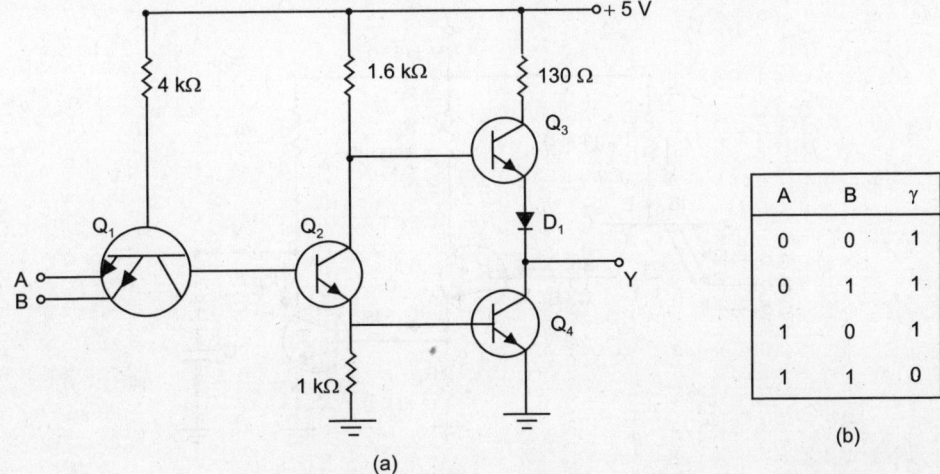

A	B	Y
0	0	1
0	1	1
1	0	1
1	1	0

(b)

(a)

Fig. 5.17: (a) Standard TTL NAND gate (b) Truth table

prevent this, the diode is inserted; its voltage drop keeps the diode of Q_3 reverse biased. In this way, only Q_4 conducts when the output is low.

Totem-pole output: Why are totem-pole transistors used? Because they produce a low output impedance. Either Q_3 acts like an emitter follower (high output) or Q_4 is saturated (low output). Either way, the output impedance is very low. This is important to reduce the switching speed. In other words, when the output changes from low to high or vice versa, the low output impedance implies a short time constant, as a result, output voltage can change quickly from one state to the other.

Propagation Delay Time and Power Dissipation: A standard TTL gate has a power dissipation of about 10mW. It may vary from this value because of signal levels, tolerances, etc but on the average, it is 10 mW per gate.

The propagation delay time is the amount of time it takes for the output of a gate to change after the inputs have changed, the propagation delay time of a TTL gate is ~10ns.

Operation of TTL NAND Gate: The operation of TTL gate of Fig. 5.18 is similar to the operation of the DTL NAND gate as far as the steady-state operation is concerned, as will be evident from conditions I and II discussed below. It is the condition III that differentiates the operation of TTL from the DTL and makes it the fastest of all saturating logic families.

For the operation discussed below, we assume that the load gates are not present and the voltages for

logic 0 and 1 are V_{CE}, sat ~0.2V and V_{CC} = 5V respectively.

Condition I: At least one input is LOW. The emitter-base junction of T_1 corresponding to the input in the LOW state is forward biased making voltage at B_1, $V_{B1} = 0.2 + 0.7 = 0.9$ V. For base-collector junction of T_1 to be forward biased (and for T_2 and T_3 to be conducting), V_{B1} should be at least $0.6 + 0.5 + 0.5 = 1.6$V. Hence T_2 and T_3 are OFF. Since T_3 is OFF, therefore $Y = V(1) = V_{CC}$.

Condition II: All inputs are HIGH. The emitter-base junctions of T_1 are reverse -biased. If we assume T_2 and T_3 are ON, then $V_{B2} = V_{C1} = 0.8 + 0.8 = 1.6$V. Since B_1 is connected to V_{CC} (+5V) through R_{B1}, the collector-base junction of T_1 is forward biased. The transistor T_1 is operating in the active inverse mode, making I_{C1} flow in the reverse direction. This current flows into the base of T_2 driving T_2 and T_3 into saturation. Therefore $Y = V(0) \approx 0.2$ V.

From above conditions I and II, it appears that T_1 is acting like back-to-back diodes. The importance of T_1 will become clear from condition III.

Condition III: Let the circuit be operating under condition II when one of the inputs suddenly goes to V(0). The corresponding emitter-base junction of T_1 starts conducting and V_{B1} drops to 0.9 V.

T_2 and T_3 will be turned off when the stored base charge is removed. Since $V_{C1} = V_{B2} = 1.6$V, therefore the collector-base junction of T_1 is back-biased making T_1 operate in the normal active region. This large collector current of T_1 is in a direction which

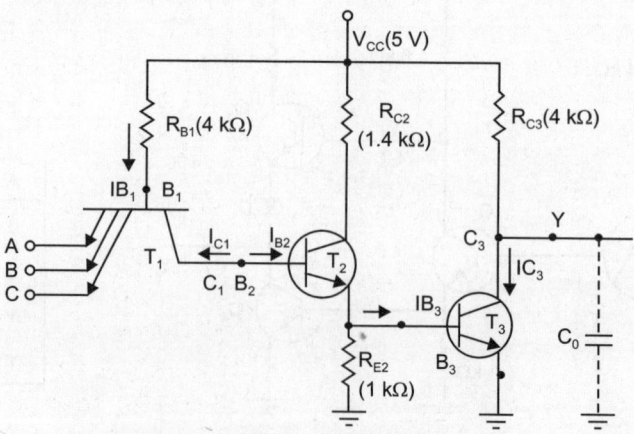

Fig. 5.18: A 3-input TTL NAND gate

helps in the removal of stored base charge in T_2 and T_3 and improves the speed of the circuit. The resistor R_{C3} is called passive pull-up resistor. The speed of the circuit can be improved by decreasing R_{C3} which decreases the time constant $(R_{C3} \times C_0)$ with which the output capacitance charges from 0 to 1 logic level. This would, however increase dissipation and would make it more difficult for T_3 to saturate.

Totem-Pole Output

It is possible in TTL gates to hasten the charging of output capacitance without increase in corresponding power dissipation by using an output circuit arrangement shown in Fig. 5.19. This is referred to as an *active pull-up* or *totem-pole output*.

Advantage of Totem-Pole Output: Totem-Pole transistors are used because they produce a low output impedance. Either T_4 acts as an emitter follower (high output), or T_3 is saturated (low output). When T_4 is conducting, the output impedance is ~70Ω, when T_3 is saturated, the output impedance is ~10Ω (This can be calculated from the data sheet's information). Either way, the output impedance is low, this enables the output voltage to change quickly from one state to the other because any stray output capacitance is rapidly charged or discharged through the low output impedance.

If one of the inputs drops to LOW, T_2 and T_3 go to cut off. The output voltage cannot change instantaneously (being the voltage across C_0) and because

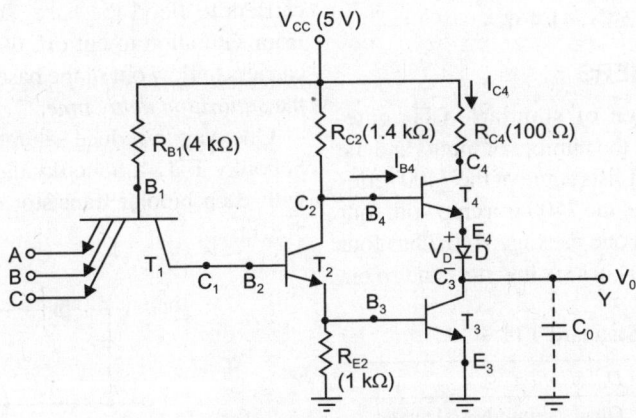

Fig. 5.19: A TTL NAND gate with totem-pole output driver

The operation of the circuit can be described as follows:

For output Y to be in LOW state, transistor T_4 and diode D are cut off. When the output makes a transition from LOW to HIGH corresponding to any input going to LOW, transistor T_4 enters saturation and supplies current for the charging of the output capacitor with a small time constant. This current decreases and eventually becomes 0 under steady state condition when $Y = V(1)$.

Diode D is used in the circuit to keep T_4 in cut off when the output is at logic 0. Corresponding to this, T_2 and T_3 are in saturation, therefore
$$V_{C2} = V_{B4} = V_{BE3, sat} + V_{CE2, sat} = 0.8 + 0.2 = 1.0V.$$

Since $V_0 = V_{CE3, sat} \approx 0.2$ V, the voltage across the base emitter junction of T_4 and diode D equals 1.0 - 0.2 = 0.8 V, which means T_4 and D are cut off.

of T_2 going to cut off, the voltage at the base of T_4 rises, driving it to saturation.

As soon as T_2 is cut off,
$$V_{B4} = V_{BE 4, sat} + V_D + V_0 = 0.8 + 0.7 + 0.2 = 1.7V.$$
Therefore,
$$I_{B4} = \frac{V_{CC} - V_{B4}}{R_{C2}} = \frac{5 - 1.7}{1.4} = 2.36 \text{mA}$$

and
$$I_{C4} = \frac{V_{CC} - V_{CE4, sat} - V_D - V_0}{R_{C4}}$$
$$= \frac{5 - 0.2 - 0.7 - 0.2}{0.1} = 39 \text{mA}$$

Hence, T_4 is in saturation if $h_{FE} > \dfrac{39}{2.36}$ or 16.5

The output voltage V_0 rises exponentially towards V_{CC} with the time constant $= (R_{C4} + R_{CS4} + R_f) \, \text{Co}$

where R_{CS4} is the saturation resistance of T_4 and R_f is the forward resistance of the diode.

As V_0 increases, the base and collector currents of T_4 are decreased and eventually T_4 just comes out of conduction at steady state. Therefore

$$V(1) = V_{CC} - V_\gamma (T_4) - V_\gamma \text{ (diode)}$$
$$= 5 - 0.5 - 0.6 = 3.9 \text{ V}$$

Now, if the output is at $V(1)$ and all the inputs go to HIGH, T_2 goes ON. Consequently T_4 and D go OFF and T_3 conducts. The capacitor Co discharges through T_3 and as V_0 approaches $V(0)$, T_3 enters into saturation.

From above, it is clear that the maximum current is drawn from the supply when the output makes a transition from $V(0)$ to $V(1)$ and equals

$$I_{C4} + I_{B4} = 39 + 2.4 = 41.4 \text{ mA}.$$

5.15 DEVICE NUMBERS

By varying the design of standard TTL gate, manufacturers can alter the number of inputs and the logic function. Table 5.1 lists some of the 7400 series TTL gates. For example, the 7400 is a chip with four 2-input NAND gates in one package, 740 2 has four 2-input NOR gates, 7404 has six inverters and so on.

Table 5.1: Standard TTL ICs

Device Number	Description
7400	Quad 2-input NAND gates
7402	Quad 2-input NOR gates
7404	Hex inverter
7408	Quad 2-input AND gates
7410	Triple 3-input NAND gates
7411	Triple 3-input AND gates
7420	Dual 4-input NAND gates
7421	Dual 4-input AND gates
7427	Triple 3-input NOR gates
7430	8-input NAND gate
7486	Quad 2-input XOR gate

5.15.1 High Speed TTL

The circuit of Fig. 5.17 (a) is a standard TTL. By decreasing the resistance, a manufacturer can lower the internal time constants, this decreases the propagation delay time. This, however, increases the power dissipation. This variation is known as high speed TTL. Devices of this type are numbered 74H00, 74H01, 74H02, etc. A high speed TTL gate has a

power dissipation around 22 mW and a propagation delay time of ~6ns.

5.15.2 Low Power TTL

By increasing the internal resistances, a manufacturer can reduce the power dissipation of TTL gates. Devices of this type are called low-power TTL and are numbered 74L00, 74L01, etc. These devices are slower than standard TTL. A low power TTL gate has a power dissipation of ~1mW and a propagation delay time ~35 ns.

5.15.3 Schottky TTL

With standard TTL, high speed TTL and low power TTL, the transistors go into saturation causing extra carriers to flood the base. To switch the transistor from saturation to cut off, one has to wait for extra carriers to flow out of the base, the delay is known as the *saturation delay time*.

One way to reduce saturation delay time is with Schottky TTL: a Schottky diode is fabricated along with each bipolar transistor of a TTL circuit (Fig. 5.20).

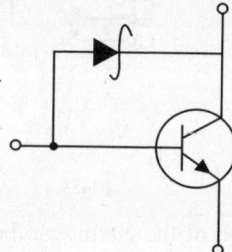

Fig. 5.20: Using Schottky diode to prevent transistor saturation

Because the Schottky diode has a forward voltage of only 0.4V, it prevents the transistor from saturating fully. This almost eliminates saturation delay time and thus provides better switching speed. These devices are numbered 74S00, 74S01, etc.

Schottky TTL devices are very fast, capable of operating reliably at 100 MHz. The 74S00 has a power dissipation of ~20mW per gate and a propagation delay time of ~3ns.

5.15.4 Low Power Schottky TTL

By increasing internal resistances together with using Schottky diode, manufacturers have come up with the best compromise between low power and high speed: low power Schottky TTL. These devices are numbered 74LS00, 74LS01, etc. A low power Schottky TTL gate has a power dissipation of ~ 2mW and a propagation delay time of ~ 10 ns.

The power dissipation and delay time values of the above mentioned TTL gates are summarized in Table 5.2. Of the five TTL types, standard TTL and low-power Schottky TTL have emerged as the favourites of the digital designers.

Table 5.2: The power dissipation and delay time values of TTL gates

Type	Power (mW)	Delay time (ns)
Low power	1	35
Low power Schottky	2	10
Standard	10	10
High speed	22	6
Schottky	20	3

5.16 5400/7400 TTL SERIES

TTL 5400/7400 series is the most popular and commonly used series of digital ICs. 7400 devices are used for commercial applications whereas the 5400 devices are used for military applications. The only difference in these two series are in the temperature and power supply range. The temperature

range is 0°C to 70°C for the 7400 series and –55°C to 125°C for the 5400 series. The supply voltage range is 5 ± 0.25V for the 7400 series and 5 ± 0.5V for the 5400 series. There are seven different series of TTL 54-/74- logic family given in Table 5.3.

Table 5.3: 54/74-TTL ICs with numbering scheme

Series	Prefix	Examples
Standard TTL	74-	7402, 74193
High power TTL	74H-	74H02, 74H193
Low power TTL	74L-	74L02, 74L193
Schottky TTL	74S-	74S02, 74S193
Low power Schottky TTL	74LS-	74LS02, 74LS193
Advanced Schottky TTL	74AS-	74AS02, 74AS193
Advanced low power Schottky TTL	74ALS-	74ALS02, 74ALS193

(*l*) Table 5.4 summarizes various specifications of 54/74 TTL logic families.

Unused Inputs: Electrically open input degrades noise immunity as well as the switching speed of a circuit. For example, it two inputs of a three input NAND gate are being used and the third is allowed to float, unpredictable result will occur if the third input picks up electrical noise from adjoining circuitry. Unused inputs on AND and NAND gates

Table 5.4: Specification of TTL IC families (*)

Parameter		5400 7400	54H00 74H00	54L00 74L00	54S00 74S00	54LS00 74LSOO	54AS00 74ASOO	54ALSOO 74ALSOO	Units
V_{IH}		2	2	2	2	2	2	2	Volts
v_H	54 Series	0.8	0.8	0.7	0.8	0.7	0.8	0.8	Volts
	74 Series	0.8	0.8	0.7	0.8	0.8	0.8	0.8	
V_{OH}	54 Series	2.4	2.4	2.4	2.5	2.5	3	3	Volts
	74 Series	2.4	2.4	2.4	2.7	2.7	3	3	
	54 Series	0.4	0.4	0.3	0.5	0.4	0.5	0.4	Volts
	74 Series	0.4	0.4	0.4	0.5	0.5	0.5	0.5	
I_{IH}		40	50	10	50	20	20	20	μA
I_{IL}		–1.6	–2.0	–0.18	–2.0	–0.36	–0.5	–0.1	mA
I_{OH}		–400	–500	–200	–1000	–400	–2000	–400	μA
I_{OL}	54 Series	16	20	2	20	4	20	4	mA
	74 Series	16	20	3.6	20	8	20	8	mA
$I_{cc}(I)$		8	16.8	0.8	16	1.6	3.2	0.85	mA
$I_{cc}(0)$		22	40	2.04	36	4.4	17.4	3	mA
t_pHL		15	10	60	5	15	4	8	ns
t_pLH		22	10	60	4.5	15	4.5	11	ns

(*) After R.P. Jain, Modern Digital Electronics, Tata McGraw-Hill, New Delhi (1997).

should be tied high and on OR and NOR gates, should be tied to ground. An example of this is a three input AND gate that is using only two of its inputs. Also, the outputs of unused gates on an IC should be forced high to reduce the I_{CC} supply current and thus reduce power requirements. To do this, tie AND and OR inputs HIGH and the NAND and NOR inputs LOW.

Advantages and Disadvantages of TTL: The main advantages of TTL integrated circuits are:

  (i) High speed operation
  (ii) Immunity to static electricity
  (iii) Good driving/interfacing
  (iv) Single power supply

    The main disadvantages are

  (i) High power consumption
  (ii) Large size for given function.

5.17 EMITTER COUPLED LOGIC

The switching time of logic circuits can be improved significantly by establishing one of the two states in the active region instead of the saturation region of the transistor. The other state remains in the cut off region. This eliminates the problem of storage time. The logic using this is known by different names, viz. emitter coupled logic (ECL), current mode logic (CML) or current controlled logic (CCL). ECL is extremely fast, with propagation delay times as low as 0.8 ns.

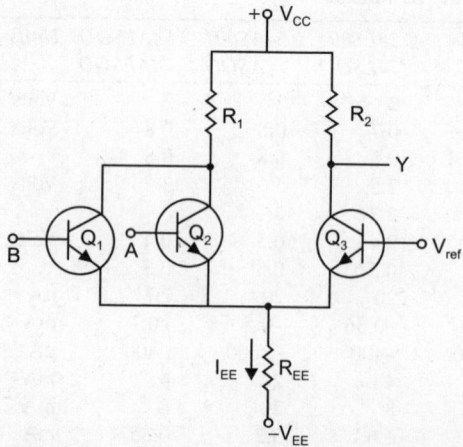

Fig. 5.21: Basic ECL gate

Figure 5.21 shows a logic ECL circuit. Two inputs are marked A and B with output Y. A bias voltage v_{ref}

is applied to the bias of Q_3 to establish bias level for gate operation.

The current through resistor R_E is switched from Q_2 to Q_3 as circuit changes state. The circuit is basically a differential amplifier. With Q_1 and Q_2 off, Q_3 is turned on by the reference voltage, resulting in low output. If either input (A or B) drives Q_1 or Q_2 on, the current I_{EE} switches from Q_3 resulting in high output at Y. It can be verified that the gate functions as positive OR gate.

The symbol of an ECL OR/NOR gate is shown in Fig. 5.22

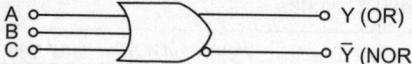

Fig. 5.22: Symbol for a 3-input OR/NOR gate

Fan-out: If all the input are LOW, the input transistors are out off. Therefore the input resistance is very high. On the other hand, if an input is HIGH, the input resistance is that of an emitter follower which is also high. Therefore the input impedance is always high.

The output resistance is either that of an emitter follower which is always low. Because of the low output impedance and high input impedance, the fan-out is large.

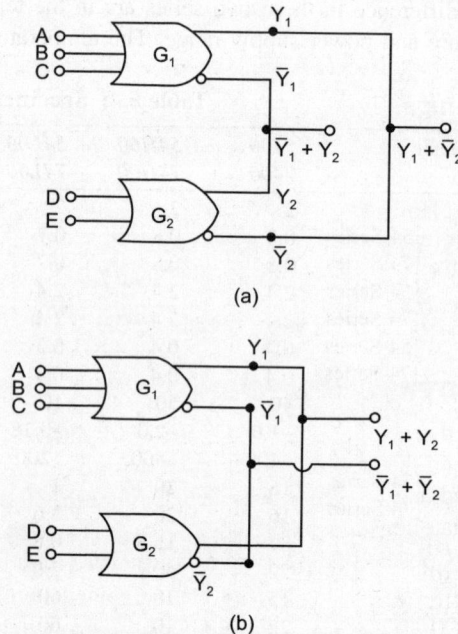

Fig. 5.23 Wired-OR connections of ECL gates

Wired-OR-Logic: The outputs of two or more ECL gates can be connected to obtain additional logic without using additional hardware, e.g. wired-OR configurations shown in Fig. 5.23.

Open Emitter Output: Similar to open collector outputs in TTL, open emitter output are available in ECL which is useful for wired-OR applications.

Unconnected Inputs: If any input of an ECL gate is left unconnected, the corresponding E-B junction of the input transistor will not be conducting. Hence it acts as if a logical 0 level voltage is applied to that input. Therefore in ECL ICs, all unconnected inputs are treated as logical 0s.

ECL Families: There are two popular ECL families: 10K series and 100K series. The 100K series is the fastest of all logic families and has a propagation delay time less than 1ns. Their voltage specifications are given in Table 5.5.

Table 5.5: Voltage specification of ECL series

Series	Supply voltage V_{EE} (volts)	V_{OL} (volts)	V_{OH} (volts)	V_{IL} (volts)	V_{IH} (volts)
10K	5.2	−1.7	−0.9	−1.4	−1.2
100K	4.5	−1.7	−0.9	−1.4	−1.2

Advantages and Disadvantage of ECL

The noise immunity of ECL is poor. Further, because of relatively low resistor values, the power dissipation is high. However, the unsaturated transistor for operation results in switching speeds better than Schottky TTL. Fan out can be greater than 20 because of high current driving capability of emitter follower outputs.

5.18 INTERFACING ECL AND TTL

It is often necessary to mix logic circuits of different families in the design of a digital system to realize the speed and power requirements by choosing the appropriate logic families for different parts of the system. Consider the interfacing between TTL and ECL gates. The logic levels in the two systems are entirely different and therefore level shifting circuits are required to be interposed between TTL and ECL gates. For TTL to ECL and ECL to TTL interfacing,

two level translator ICs available. Interfacing using these ICs are described below:

TTL to ECL Translator: The MC 10 H 124 is a quad TTL to ECL translator IC. It is a 16-pin IC and its logic diagram is shown in Fig. 5.24. It uses two power supplies, one positive and another negative for the generation of proper logic levels for ECL and TTL.

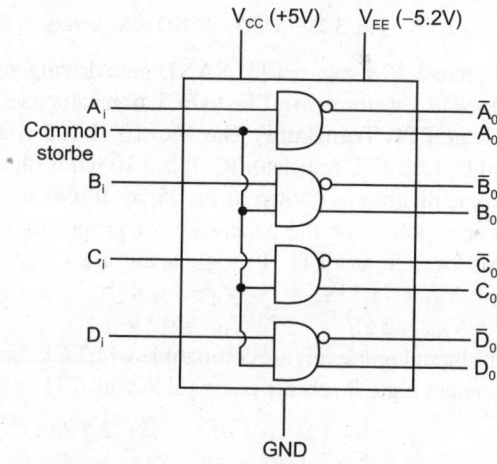

Fig. 5.24: Logic diagram of MC 10 H 124 TTL to ECL translator (After R.P. Jain, *Modern Digital Electronics* Tata McGraw-Hill, (1997)

The logic levels of the translator circuit are

$$V_{IH} = 2V, \qquad V_{IL} = 0.8V$$
$$V_{OH} = -0.98V \qquad V_{OL} = -1.63V$$

For TTL ICs, we have

$$V_{OH} = 2.4V \qquad V_{OL} = 0.4V$$

Comparing the output logic levels of TTL and the input logic levels of the translator IC, we observe

$$V_{IH} \text{ (translator)} < V_{OH} \text{ (TTL)}$$
$$V_{IL} \text{ (translator)} > V_{OL} \text{ (TTL)}$$

which shows that the input logic levels of the translator are compatible with the output logic levels of TTL.

Similarly, from Table 5.5

$$V_{IH} \text{(ECL)} = -1.2 \text{ V}$$
$$V_{IL} \text{(ECL)} = -1.4 \text{ V}$$

Comparing the output logic levels of the translator with the input logic levels of ECL, we find

$$V_{IH} \text{ (ECL)} < V_{OH} \text{ (translator)}$$
$$V_{IL} \text{ (ECL)} > V_{OL} \text{ (translator)}$$

which shows that the output logic levels of the translator are compatible with the input logic levels of ECL.

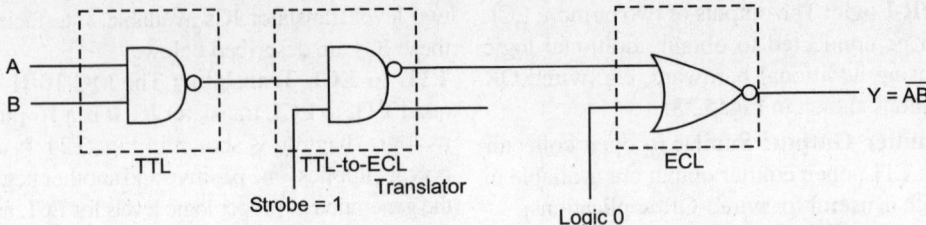

Fig. 5.25: A TTL NAND gate driving an ECL NOR gate through a TTL to ECL translator

Figure 5.25 shows a TTL NAND gate driving an ECL NOR gate through a TTL to ECL translator gate. **ECL to TTL Translator:** The MC 10 H 125 is a quad ECL to TTL translator IC. It is a 16 pin IC and its logic diagram is shown in Fig. 5.26. It uses two power supplies for the generation of proper logic levels for ECL and TTL. Its logic levels are

$$V_{IH} = -1.13V, \qquad V_{IL} = -1.48V$$
$$V_{OH} = 2.5V \qquad V_{OL} = 0.5V$$

Its input logic levels are compatible with ECL and the output logic levels are compatible with TTL.

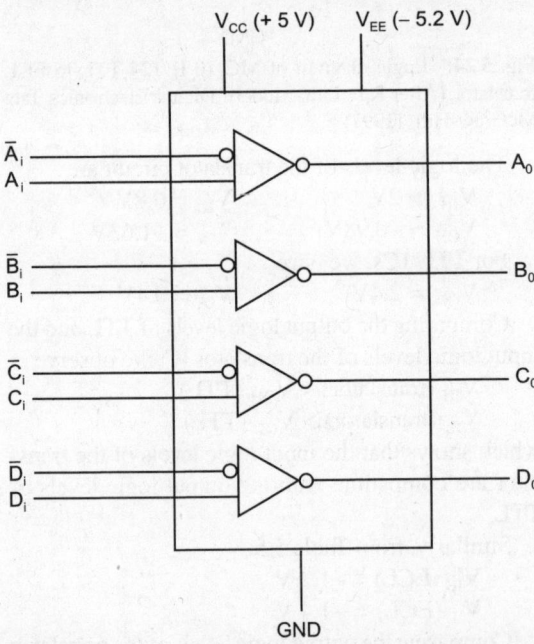

Fig. 5.26: Logic diagram of MC 10 H 125 ECL to TTL translator

5.19 TTL AND-OR-INVERT GATE

The Boolean equation

$$Y = AB + CD$$

corresponds to an AND-OR circuit shown in Fig. 5.27a. Using De Morgan's theorem, an equivalent NAND-NAND network is shown in Fig. 5.25b. 1s there any TTL device with the output given by the above equation? Yes, there are some AND-OR gates but they are not easily derived from the basic NAND gates.

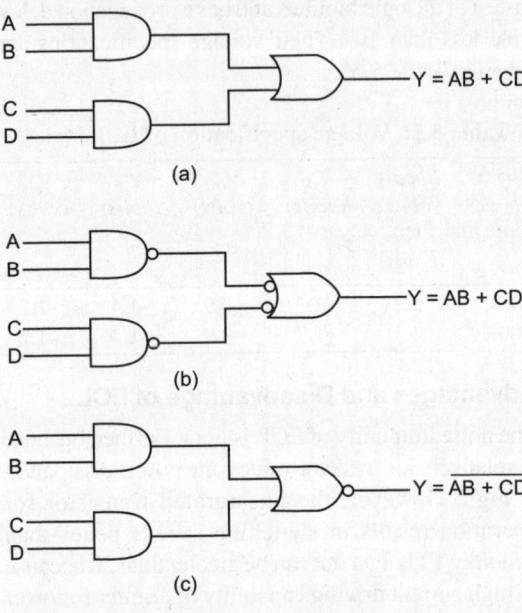

Fig. 5.27: (a) AND-OR circuit (b) NAND-NAND circuit (c) AND-OR-INVERT circuit

The gate that is easy to derive and comes close to expression like the above one is the AND-OR-INVERT gate shown in Fig. 5.25c and the output is

$$Y = \overline{(AB + CD)}$$

Figure 5.28 shows the schematic diagram of a TTL AND-OR-INVERT gate

Q_1, Q_2, Q_3 and Q_4 form the basic 2-input NAND gate of the 7400 series. By adding Q_5 and Q_6, we convert the basic NAND gate to an AND-OR-

INVERT gate. Q_1 and Q_5 act like 2-input AND gates; Q_2 and Q_6 produce OR-ing and inversion. Because of this, the circuit is logically equivalent to Fig. 5.27c.

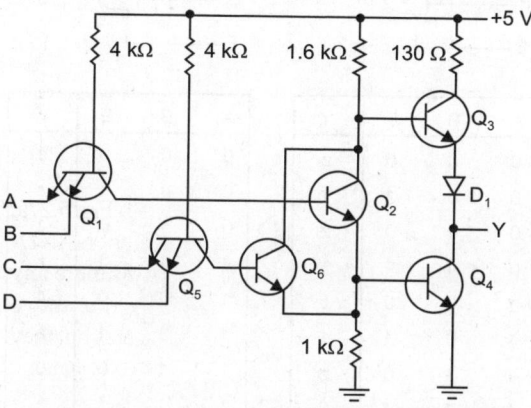

Fig. 5.28: Schematic TTL AND-OR-INVERT gate

In Table 5.6 listing the AND-OR-INVERT gates available in the 7400 series, 2-wide means two AND gates across, 4-wide means four AND gates across and so on. For instance, the 7454 is a 2-input 4-wide AOI gate like Fig. 5.29a; each AND gate has two inputs and there are four AND gate (4-wide)

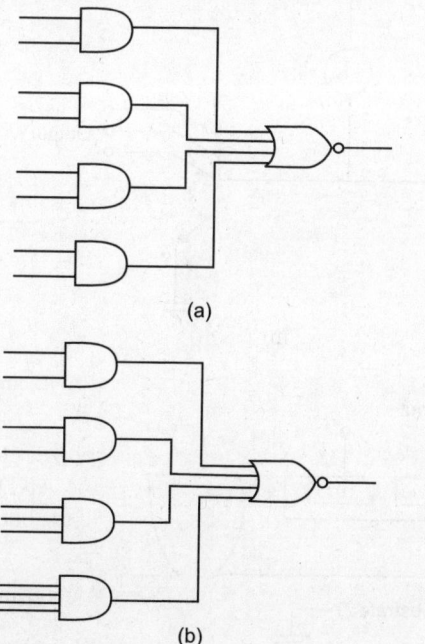

(a)

(b)

Fig. 5.29: Examples of AND-OR-INVERT circuits. (a) 2-input 4-wide AOI. (b) 2-2-3-4-input 4-wide AOI

Fig. 5.29b shows the 7464; it is a 2-2-3-4-input 4-wide AOI gate.

An output given by equation
$$Y = AB + CD,$$

We can connect the output of a 2-input 2-wide AOI gate to another inverter. This gives us the equivalent of an AND-OR circuit (Fig. 5.27a or a NAND-NAND network Fig. 5.27b.

Table 5.6: AND-OR-INVERT gates

Device	Description
7451	Dual 2-input 2-wide
7454	2-input 4-wide
7459	Dual 2-3 input 2-wide
7464	2-2-3-4 input 4-wide

5.20 TRISTATE TTL

In a memory chip, while paralleling chips to increase the number of words, it is necessary to connect in parallel the output bit lines of each of the other chips. Then, it is necessary to devise for a memory chip an output stage which allows the output logic level of a selected chip to be unaffected by connection of the output terminals of other chips which are not selected. For this purpose, tristate TTL is used, shown in Fig. 5.30 which is the standard TTL gate modified through the addition of a disabling input. It is known as tristate because it allows three possible output states. High, low and high impedance.

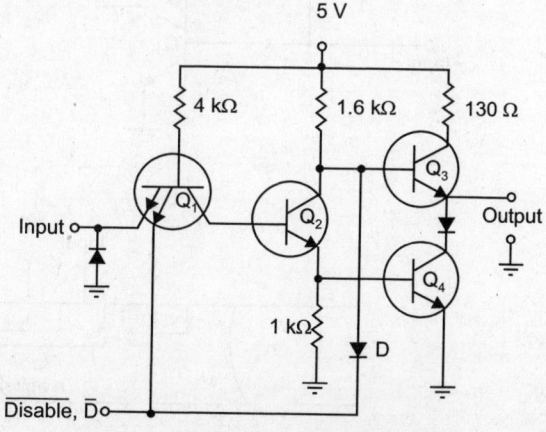

Fig. 5.30: A TTL tristate output stage

When the $\overline{\text{disable}}$ or $\overline{D}$ is at logic 1, diode D does not conduct nor does the emitter of Q_1 which is connected to this line. Hence with $\overline{D} = 1$ a straight forward output stage is provided with all of the usual advantages of active pull-up. However when $\overline{D} = 0$, both Q_3 and Q_4 will be off and the output terminal will be entirely isolated. The impedance seen looking back into the output terminal will be very high, being less than infinite only because of leakage and stray capacitance. Thus this circuit is described as having a tristage stage, the available states being logic 0, logic 1 and the third state, a nominally isolated high impedance state.

When the memory chips are paralleled for the sake of increasing the number of words, the tristate output stage is enabled and disabled by the chip select input and only the output stages on the selected chip are enabled.

In tristate logic, the high impedance state works according to the external enable input ($\overline{D}$ or E). The enable input ($\overline{D}/E$) decides whether the gate is enable or is in the high impedance state. When enable, it can be 0 or 1 depending upon inputs conditions. Fig. 5.31a shows the circuit symbol of a tristate NAND with active HIGH enable input with its truth table. The one shown in Fig. 5.31b has an active LOW enable input.

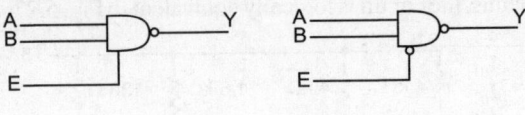

A	B	E	Y
0	0	0	z
0	0	1	1
0	1	0	z
0	1	1	1
1	0	0	z
1	0	1	1
1	1	0	z
1	1	1	0

(a)

A	B	E	Y
0	0	0	1
0	0	1	z
0	1	0	1
0	1	1	z
1	0	0	1
1	0	1	z
1	1	0	0
1	1	1	z

(b)

Fig. 5.31: Tristate NAND with (a) active-HIGH and (b) active LOW enable input

The output and input current specifications of tristate logic family (TSL) are given in Table 5.7.

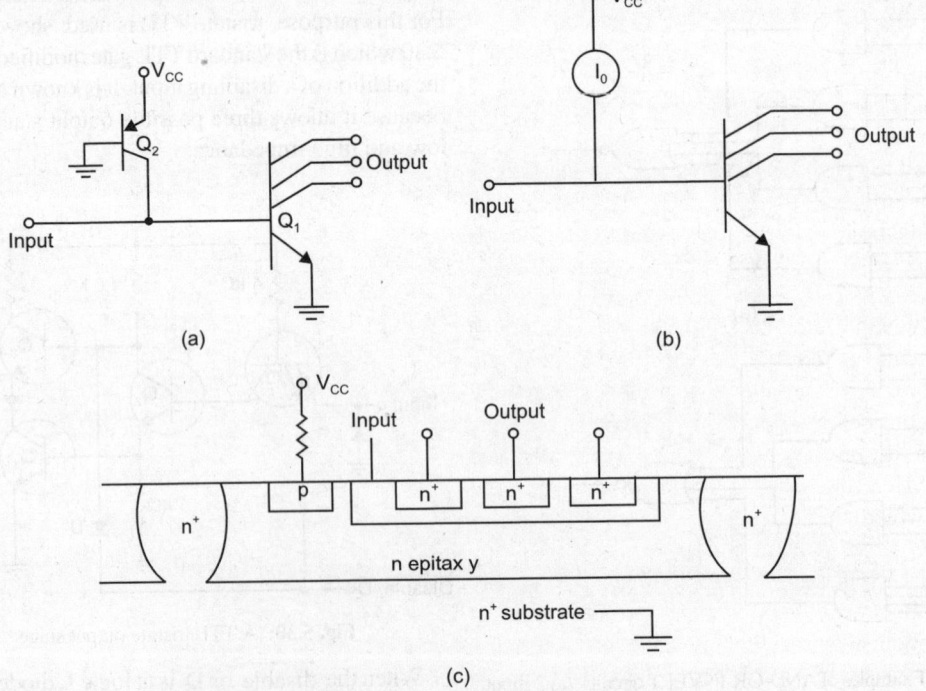

Fig. 5.32: (a) Basic I^2L gate (b) Circuit diagram (c) Cross section

Table 5.7: Current specifications of TSL family

Parameter	Control input	
	LOW (D)	HIGH (E or $\bar{D}$)
I_{IH}	40μA	40μA
I_{IL}	–1.6mA	–1.6mA
I_{OH}	40μA	–5.2mA
I_{OL}	–40μA	16mA

5.21 INTEGRATED INJECTION LOGIC (IIL)

In IIL technology, the components are merged together, i.e. one semiconductor region is part of two or more devices. Because of this, it is also known as merged transistor logic (MTL).

IIL or I^2L basic gate is similar in operation to the RTL gate with following differences:

(i) Base resistor used in RTL is removed.

(ii) Collector resistor is replaced by pnp-transistor that acts as load for I^2L gate.

(iii) I^2L gate makes use of multiple collectors instead of individual transistors as in RTL.

The basic I^2L gate is shown in Fig. 5.32a It has an npn transistor Q_1 with multiple collector for output. The base circuit has pnp transistor Q_2 connected to V_{CC} supply. Since collector of pnp and base of npn are both p regions, they can be termed together as a common p-region. Similarly, the base of pnp, an n-region is connected to the ground, can be formed with the emitter of npn, which is also an n-region connected to ground. This integrating or merging of two transistors is shown in Fig. 5.32c. I^2L gate operation cannot be analyzed when standing alone, It must be connected to other gates to make any meaning.

Figure 5.33 shows the circuit diagram of a two input I^2L, NOR gate. It is a type of bipolar saturated logic circuit (i.e. where transistors are entered into saturation). It does not contain resistances; as a result, there is considerable reduction in the size of the logic gate. It is very much used in very large scale integrated circuits. Here transistors Q_1 and Q_3 act as current sources which supply current to the base of transistors Q_2 and Q_4.

The I^2L NOR gate shown in Fig. 5.33. is simply two inverters with their output connected together. If either or both the inputs are HIGH, the corresponding output transistor is ON and the output is a current sink. So, the output is LOW, if both the inputs are

LOW, both the output transistors are OFF and so, the output is HIGH. This is a NOR operation.

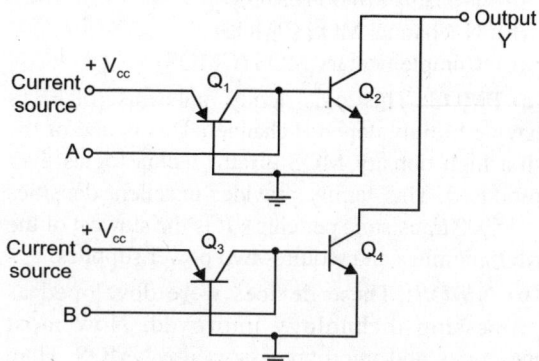

Fig. 5.33: IIL two input NOR gate

The I^2L NAND gate is shown in Fig. 5.34. It is simply an inverter with inputs connected directly together at the inverter input. If either input A or input B or both the input A and B are LOW (current sinks), the injected current flows into those inputs and Q_2 remains OFF (HIGH). If both the inputs are HIGH, the injected current turns on Q_2 making the output LOW. Thus NAND operation is performed. The transistor Q_1 acts as a current injector transistor because when its emitter is connected to an external power source, it can supply current to the base of Q_2.

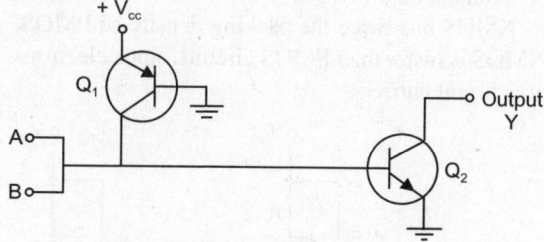

Fig. 5.34: Two input I^2L NAND gate

5.22 UNIPOLAR LOGIC FAMILIES

Unlike bipolar IC families, MOS logic IC families use field effect transistor design (FET) for internal circuitry. MOS families have become very popular due to high density of fabrication and low power dissipation. However, their lower speed and suscepti-bility to static electricity have been their primary drawback.

The following are more common types of MOS logic families:

(a) P-channel MOS (PMOS)

(b) N- channel MOS (PMOS)

(c) Complementary MOS (CMOS)

(a) PMOS: This logic family makes use of FETs having heavily doped P-channel. This is one of the first high density MOS circuit technologies ever produced. This family provides excellent densities (~15000 transistors per chip). It is the slowest of the MOS families and requires two power supplies.

(b) NMOS: These devices were developed as processing technology improved. Now most memories and microprocessors use NMOS. Here n-channel MOS field effect transistors are used. These devices are faster as compared to PMOS. These also provide larger currents for a given geometry than that can be obtained with PMOS technology.

(c) CMOS: It is a popular MOS logic. CMOS family combines both P and N-channel devices. Therefore it has speed and packing density somewhere between PMOS and NMOS. Power dissipation is very small. These devices are gradually becoming popular and replacing TTL.

5.23 PMOS AND NMOS CIRCUITS

These circuits have high packing density and are more economical than CMOS.

NMOS has twice the packing density of PMOS. NMOS is faster than PMOS circuits, since electrons are current carriers.

Here we will describe source of the NMOS basic circuits keeping in mind that PMOS circuits are similar except for voltage polarities. These circuits are widely used in LSI and VLSI (microprocessors, memories, ROMs etc).

NMOS logic circuits are shown in Fig. 5.35. For NMOS, supply voltage V_{DD} is positive to allow positive current flow from drain to source. The two voltage levels are function of threshold voltage V_T. The low level lies between 0 and V_T and high level ranges between V_T and V_{DD}. The inverter circuit shown in Fig. 5.35a uses two NMOSFETS where Q_1 acts as load and Q_2 as active device. The load resistor MOS has its gate connected to V_{DD}, thus maintaining it in conduction. As input voltage is low (below V_T), Q_2 turns off. As Q_1 is always on, the output voltage is V_{DD}. When input voltage is high, Q_2 turns on and current flows from V_{DD} to Q_2 through load resistor Q_1.

The NAND gate shown in Fig. 5.35b uses MOSFETs in series. Inputs A and B must both be high for output to be low. If either of the input is low, the corresponding transistor is turned off and output high.

The NOR gate shown in Fig. 5.35c uses MOSFETs in parallel. If either input is high, the corresponding transistor conducts and output is low. If all inputs are low, all active transistors are off and output, i.e. high.

5.24 CMOS CIRCUITS

The different CMOS circuits are

(a) CMOS Inverter: The most basic CMOS inverter is shown in Fig. 5.36a. It consists of a channel and an

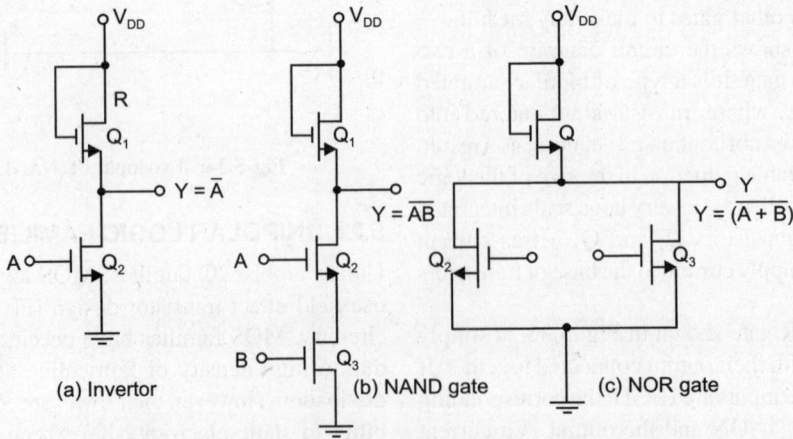

Fig. 5.35: NMOS logic circuits (a) Inverter (b) NAND gate (c) NOR gate

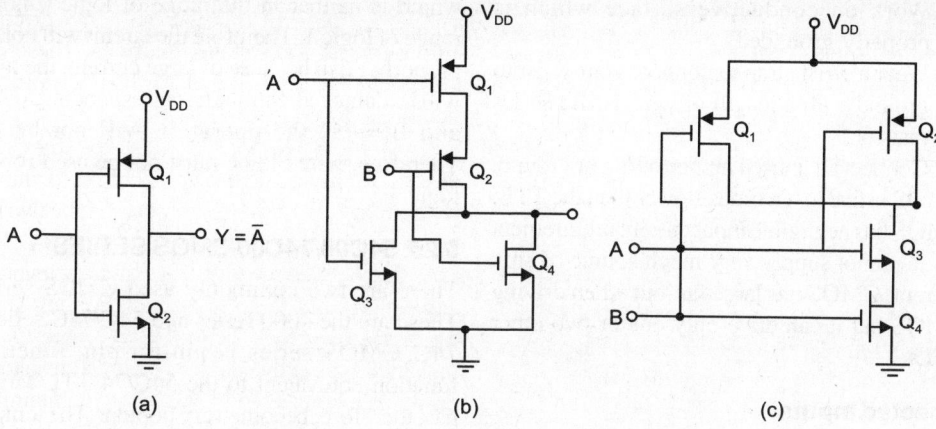

Fig. 5.36: (a) CMOS inverter (b) two input NOR gate (c) two input NAND gate

n channel enhancement type MOSFET. When a logical 0 (zero volts) is applied to the inverter input, the gate to source voltage V_{GS} of the p channel device is equal to the supply voltage V_{DD} and the p channel is on. Thus low impedance path exists between output and V_{DD} and a very high impedance path from the output to ground. As a result, V_0 approaches the logical 1 level, V_{DD} under normal loading conditions when $v_i = V_{DD}$, the p-channel device is off and the n-channel device is on. Then V_0 approaches zero, the logic zero level.

As one transistor is always off, the quiescent power consumption is the product of supply voltage and leakage current and is typically in the low micro watt range.

(b) CMOS NOR gate: A two input NOR gate is shown in Fig. 5.36 (b). It consists of an inverter, two n channel units in parallel and two p-channel units in series. Each of the two inputs is connected to the gate of one n-type transistor and one p-type transistor. When either A or B is positive, the p-type transistor turns off and disconnects the output from the V_{DD} supply while the n-type transistor turns on and is connected to ground. Thus the output is a low negative voltage. When both A and B are at ground potential, p-type transistors are on and both n-type transistors are off. In this case, the output is a high positive voltage since it is coupled to the V_{DD} terminal.

Three and four input NOR gates can be easily formed by placing three or four n-type transistors in parallel and a corresponding number of p-type transistors in series arrangement.

(c) CMOS NAND gate: A two input NAND gate consists of an inverter with two p-type transistors in parallel and two n-type transistors in series. The output becomes negative only if both inputs are positive, in which case, the p-type units are turned off and the n-type units are turned on. This condition couples the output to ground. If either input is negative, the associated transistor (n-type) is turned off and the associated p-type transistor is turned on. Then the output is coupled to the V_{DD} terminal and is high. The number of inputs can be extended as in case of NOR gates.

Advantages of CMOS

1. CMOS consume less power as compared to standard TTL.
2. The wide supply voltage range offers greater design flexibility over TTL where power supplies are required to be tightly regulated.
3. CMOS has higher margin for noisy signals.

Disadvantages of CMOS

1. The most important feature of CMOS is the thin layer of silicon dioxide insulating the input from the substrate. The infinite input impedance allows build up of electrostatic charge where a build up of 100V potential can damage the layer. To avoid this burn through, following guide lines should be observed while handling these devices:
   (a) Store ICs in conductive foam.

(b) Work on conductive surface which is properly grounded

(c) Wear a wrist strap to connect your wrist to ground with a length of wire 1 MΩ series resistor.

(d) Connect all unused inputs to V_{DD} or ground.

2. CMOS is five to ten times slower than the TTL.

3. CMOS has negligible input current requirement but it cannot supply very much source or sink current. CMOS has large fan-out when driving CMOS but it can drive only one or two other TTLs.

Unconnected Inputs

The unused inputs must be connected to either the supply voltage terminal or one of the used inputs provided that the fan-out of the single source is not exceeded. This is highly unlikely for CMOS circuit because of their high fan-out.

Wired-logic: Figure 5.37 shows two CMOS inverters with their outputs connected together. In this circuit,

(i) When $A = B = V(0)$
T_1 and T_1' are cut off and $Y = V(1) = V_{CC}$.

(ii) When $A = B = V(1)$
T_1 and T_1' are ON and $Y = V(0) = 0$.

(iii) When $A = V(1)$ and $B = V(0)$
T_1 and T_1' are ON whereas T_1' and T_2 are OFF.

Therefore a large current I will flow as shown in the figure. This will make voltage at Y equal to $V_{CC}/2$

which is neither in the range of logic 0 nor in the range of logic 1. Therefore the circuit will not operate properly. Also because of large current, the transistor will be damaged. Similarly, corresponding to $A = V(0)$ and $B = (1)$ the operation will not be proper. Therefore, wired logic must not be used for CMOS logic.

5.25 54C00/74C00 CMOS SERIES

There are two commonly used CMOS series ICs. These are the 4000 series and 54/74C series. 54C/74C CMOS series is pin-for-pin, function-for-function equivalent to the 54C/74 TTL family and has, therefore, become very popular. The temperature range for 54C series is –55°C to +125°C and for 74C series is –40°C to 85°C. It has a wide supply voltage range 3V to 15V. A person can take full advantage of his knowledge of 54/74 TTL series for the effective use of 54C/74C CMOS series.

There have been significant improvements in 54C/74C series. The 74HC/74HCT have higher speed and better current carrying capabilities. 74 HC is known as *high speed* CMOS and 74 HCT is known as *high speed TTL compatible* CMOS series. 74AC/74 ACT are very fast and have very high current sinking capabilities. These are known as *advanced* CMOS and *advanced TTL compatible* CMOS respectively. The 74 HC/74HCT/74AC/74ACT series can be operated at supply voltages in the range of 2-6 volts.

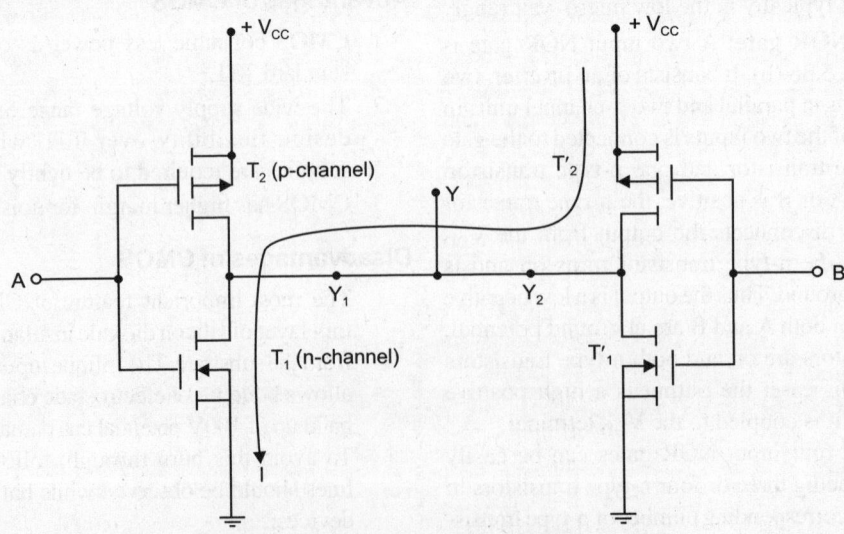

Fig. 5.37: CMOS inverters with outputs connected

The voltage and current parameters of various 74 CMOS series with 5V supply voltage are given in Table 5.8.

Table 5.8: Specifications of CMOS IC families (*)

Parameter	Load	74C	74HC	74HCT	74AC	74ACT	Units
V_{IH}		3.5	3.85	2.0	3.85	2.0	volts
V_{IL}		1.5	1.35	0.8	1.35	0.8	volts
V_{OH}	CMOS	4.5	4.4	4.4	4.4	4.4	volts
	TTL		3.84	3.84	3.76	3.76	volts
V_{OL}	CMOS	0.5	0.1	0.1	0.1	0.1	volt
	TTL		0.33	0.33	0.37	0.37	volt
I_{IH}		1	1	1	1	1	μA
I_{II}		−1	−1	−1	−1	−1	μA
I_{OH}	CMOS	−0.1	−0.02	−0.02	−0.05	−0.05	mA
	TTL		−4.0	−4.0	−24.0	−24.0	mA
I_{OL}	CMOS	0.36	0.02	0.02	0.05	0.05	mA
	TTL		4.0	4.0	24.0	24.0	mA

(*) After R.P. Jain, Modern Digital Electronics, Tata McGraw-Hill, (1997).

5.26 INTERFACING CMOS AND TTL

To achieve optimum performance in a digital system, devices form more than one logic family can be used, taking advantages of the superior characteristics of each family for different parts of the system. For example, CMOS logic ICs can be used in those parts of the system where low power dissipation is required, while TTL can be used where we require high speed of operation. Thus, it becomes necessary to examine the interface between TTL and CMOS devices.

The 74C series of CMOS ICs can be operated for any supply voltage in the range of 3V to 15V, whereas the 74 HC/74 HCT/74 AC/74 ACT series have the supply voltage range of 2V to 6V. Since the supply voltage used for all the 74 series TTL ICs is 5V, therefore it is necessary to operate CMOS devices at +5V, to make it compatible with TTL devices.

CMOS Driving TTL

Figure 5.38 shows a CMOS gate driving NTTL gates. For such an arrangement to operate properly the following conditions are required to be satisfied:

$$V_{OH} (CMOS) \geq V_{IH} (TTL) \qquad \ldots(i)$$
$$V_{OL} (CMOS) \leq V_{IL} (TTL) \qquad \ldots(ii)$$
$$I_{OH} (CMOS) \geq NI_{IH} (TTL) \qquad \ldots(iii)$$
$$I_{OL} (CMOS) \leq -NI_{IL} (TTL) \qquad \ldots(iv)$$

From the specifications (Table 5.8), we observe that

(i) The conditions of equations (iii) and (iv) are always satisfied for 74 HC/74 HCT/ 74 AC/ 74 ACT series. The value of N is different for different series. (The value of N when 74 ACT is driving 74 ALS is 240).

(ii) The conditions of equation (i) and (ii) are always satisfied. The noise margins when 74 ACT is driving 74 ALS are

$$\Delta 1 = 3.76 - 2.0 = 1.76 \text{ V}$$
$$\Delta 0 = 0.8 - 0.37 = 0.43 \text{ V}$$

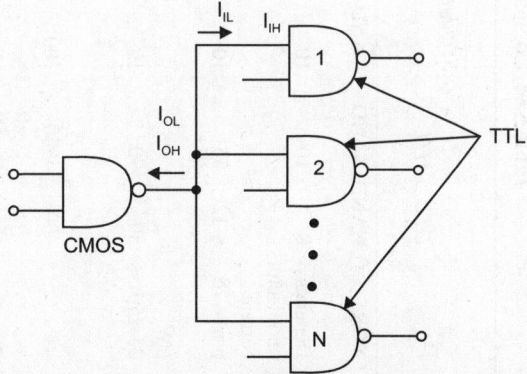

Fig. 5.38: A CMOS driving N TTL gates

In case of 74 C series, the condition of equation (iii) is satisfied for small values of N but the condition of equation (iv) is not satisfied even for N = 1, except in case of 74 L and 74 ALS TTL series. This difficulty can be overcome by using CMOS buffers having an adequate available output current.

Table 5.9: Comparison of various digital IC logic families

Logic Family Parameter	RTL	r²L	DTL	HTL	TTL				ECL	MOS	CMOS
					Standard High power high speed H	Low power low speed	Schrottky low power LS	Schrottky S,^			
Basic gate	NOR	NOR	NAND	NAND	NAND	NAND	NAND	NAND	NOR	NAND	NOR or NAND
Fan-out	5	Depends on injector current	8	10	IU	20	20	10	25	20	> 50
Power dissipation mW per gate	12	6-nW-70 μW	8-12	55	22	1	2	19	40-55	0.2-10	0.01 stalic; 1 at 1 MHz
Propagation delay ns per gate	12	25-250	30	90	6	33	9.5	3	2	300	70
Speed-power product (pJ)	144	<1	300	4950	132	33	19	57	100	60	0.7
Noise immunity	Nominal	Poor	Good	Excellent	Very good	Very good	Very good	Very good	Poor	Nominal	Very good
Clock rate for FFs	8	–	72	4	50	3	45	125	>60	2	10
Available Functions	High	LSI only	Fairly High	Nominal	Very high	Very high	Very high	Very high	High	Low	High

If 74 C series gate is driving 74 L series gate, the condition of equation (iv) is satisfied for $N = 2$ and in case of 74 ALS gates for $N = 3$.

TTL Driving CMOS

Figure 5.39 shows a TTL gate driving N CMOS gates. For such an arrangement to operate properly, the following conditions are to be satisfied:

$$V_{OH} \text{ (TTL)} \geq V_{IH} \text{ (CMOS)} \qquad \ldots\text{(v)}$$
$$V_{OL} \text{ (TTL)} \leq V_{IL} \text{ (CMOS)} \qquad \ldots\text{(vi)}$$
$$-I_{OH} \text{ (TTL)} \geq NI_{IH} \text{ (CMOS)} \qquad \ldots\text{(vii)}$$
$$I_{OL} \text{ (TTL)} \leq -NI_{IL} \text{ (CMOS)} \qquad \ldots\text{(viii)}$$

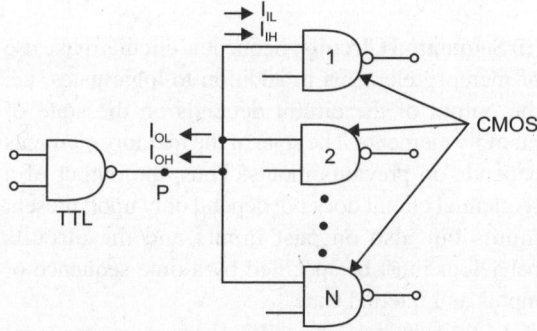

Fig. 5.39: A TTL gate driving N CMOS gates

All the above conditions are always satisfied in case of 74 HCT and 74 ACT series for high values of N. This shows that these two CMOS series are TTL compatible.

5.27 INTERFACING CMOS AND TTL ECL

Using MC10H124 TTL to ECL translator, and MC10H125 ECL to TTL translator ICs, it is possible to interface CMOS and ECL logic families.

The input of MC10H124 translator is compatible to the output logic voltages of CMOS and therefore, this can be used for CMOS to ECL interfacing. Similarly, the output of MC10H125 translator is compatible to the input logic voltages of CMOS (74 HCT and 74 ACT) families which makes it possible to be used it for ECL to CMOS interfacing. Table 5.9 gives a comparison of various digital IC logic families.

EXERCISE

1. What are the different characteristics of digital ICs?
2. Describe two features of TTL which make it faster than DTL. How is Schottky TTL made faster than standard TTL?
3. What is positive and negative logic system? Explain.
4. Compare the advantages and disadvantages of MOS logic over bipolar logic.
5. What are the advantages of CMOS over (a) NMOS (b) TTL?
6. Explain the statement: "CMOS has lower power dissipation than NMOS".
7. Write short notes on (a) HTL (b) RTL (c) I^2L
8. What do you mean by interfacing? Explain its need. How will you interface TTL to CMOS?
9. Name five series of TTL circuits.
10. What are advantages of ECL? Explain how ECL gate works.
11. Write short notes on (a) CMOS logic (b) PMOS logic.
12. Draw the circuit diagrams for the following: (a) Schottky TTL NAND gate (b) ECL Three input gate (c) I^2L circuit.
13. With the help of a neat diagram, explain the working of
    (a) A CMOS inverter
    (b) A two-input CMOS NAND gate
    (c) A two-input CMOS NOR gate
14. What are the characteristics of ECL family?
15. What is wired ANDing?
16. Name the TTL outputs which can be wired ANDed and which cannot be.
17. Why should'nt logic devices with totem-pole outputs be wired ANDed?
18. How do open-collector outputs differ from totem-pole outputs?
19. Why do open-collector outputs need a pull-up resistor?
20. What factors are involved in determining the value for the pull-up resistor?
21. What are the three possible output states of a tristate IC?

6

Combinational and Arithmetic Logic Circuits

6.1 COMBINATIONAL AND SEQUENTIAL CIRCUITS

In the previous chapters, we have studied the operation of the basic logic gates. All logic circuits can be realized using these basic gates. Logic circuits can be classified as:

(i) Combinational circuits
(ii) Sequential circuits

(i) Combinational Circuits: Combinational circuits consist of logic gates whose output at any time depends on the combination of logic levels present at the input. A combinational circuit has no memory characteristic and so, the output depends only on the current values of the inputs. A combinational circuit performs specific data processing function specified logically by a set of Boolean functions.

The block diagram of a combinational circuit is shown in Fig. 6.1. The circuit has N inputs and M outputs. For N inputs, there are 2^N possible combination of binary input values. For each input combination, there is only one output. This combinational circuits can be described by M Boolean expressions, one for each outputs, i.e. each output is expressed in terms of N inputs.

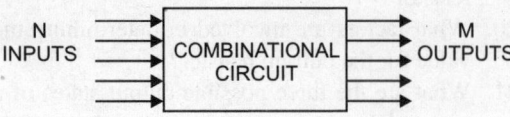

Fig. 6.1: Block diagram of a combinational circuit

(ii) Sequential Circuits: Sequential circuits make use of memory elements in addition to logic gates, i.e. the output of the circuit depends on the state of memory elements. The state of the memory elements depends on previous inputs. Thus, the output of a sequential circuit does not depend only upon present inputs but also on past inputs and the circuits behaviour must be specified by a time sequence of inputs and internal state.

In this chapter, we will consider combinational circuits such as adders/sub tractors 1-bit comparators and parity generators/checkers. Code converters, multiplexers and demultiplexers will be discussed in Chapter 10.

6.2 ADDERS

Addition is the most fundamental operation. Simple addition comprises four possible elementary operations:

$$0 + 0 = 0,\ 0 + 1 = 1,\ 1 + 0 = 1,\ 1 + 1 = 10$$

The first three operations produce one bit, but when both addend and augend are equal to 1, the binary sum consists of two digits. The higher significant bit of the sum is called a carry. When the addend and augend contain more digits, the carry is added to the next higher order pair of bits. A combinational circuit that performs addition of two bits is called half adder, whereas one that performs addition of three bits is called full adder.

6.2.1 The Half Adder

A half adder circuit adds two binary digits, giving a sum bit and a carry bit. The sum (S) bit and the carry (C) bit, according to the rules of binary addition are given by Table 6.1

Table 6.1

Inputs		Outputs	
A	B	S	C
0	0	0	0
0	1	1	0
1	0	1	0
1	1	0	1

The sum (S) is the XOR of A and B, therefore
$$S = A\overline{B} + \overline{A}B = A \oplus B$$
The carry (C) is the AND of A and B, therefore
$$C = AB$$

A half adder can, therefore be realized by using one XOR gate and one AND gate as shown in Fig. 6.2.

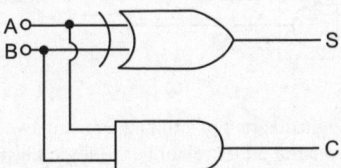

Fig. 6.2: Logic diagram for a half adder

A half adder can also be realized in universal logic by using either only NAND gates or only NOR gates as explained below.

NAND Logic

$$S = A\overline{B} + \overline{A}B$$
$$= A\overline{B} + A\overline{A} + \overline{A}B + B\overline{B}$$
$$= A(\overline{A} + \overline{B}) + B(\overline{A} + \overline{B})$$
$$= A \cdot \overline{AB} + B \cdot \overline{AB}$$

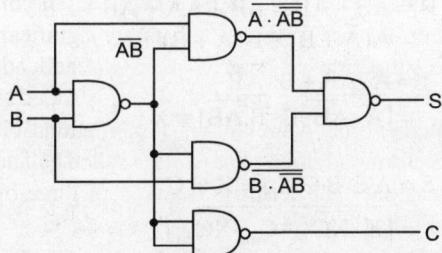

Fig. 6.3: Half adder using NAND logic

$$= \overline{\overline{(A \cdot \overline{AB})(B \cdot \overline{AB})}}$$
$$C = AB = \overline{\overline{AB}}$$

NOR Logic

$$S = A\overline{B} + \overline{A}B$$
$$= A\overline{B} + A\overline{A} + \overline{A}B + B\overline{B}$$
$$= A(\overline{A} + \overline{B}) + B(\overline{A} + \overline{B})$$
$$= (A + B)(\overline{A} + \overline{B})$$
$$= \overline{\overline{(A + B)(\overline{A} + \overline{B})}}$$
$$= [\overline{(A + B) + (\overline{A} + \overline{B})}]$$

Note: Here a dash outside the bracket denotes inverse of the complete expression inside.

$$C = AB = \overline{\overline{AB}} = (\overline{A} + \overline{B})'$$

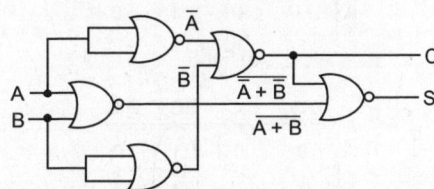

Fig. 6.4: Half adder using NOR logic

6.2.2 Full Adder

A half adder can only add two inputs and has no provision to add carry coming from the lower order bits when multibit addition is performed. For this purpose, a third input terminal is added and this circuit is used to add augend, addend and carry generated from previous addition. Thus a *full adder* is a combinational circuit that performs sum of three input bits. Two of the inputs, A and B represent the two bit to be added. The third input c_{in} represents the carry

Table 6.2

Inputs			Sum	Carry
A	B	C_{in}	(S)	(C_{out})
0	0	0	0	0
0	0	1	1	0
0	1	0	1	0
0	1	1	0	1
1	0	0	1	0
1	0	1	0	1
1	1	0	0	1
1	1	1	1	1

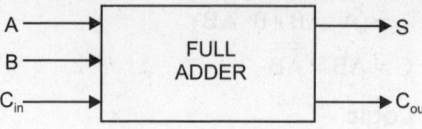

Fig. 6.5: Block diagram

from the previous lower significant position. The two outputs are denoted by S (the sum) and c_{out} (carry). The S gives the value of LSB of the sum and the C gives the value of the output carry. The truth table and the block diagram are shown above.

The K-maps for the output S and C_{out} are given in Fig. 6.6.

	$\bar{B}\bar{C}_{in}$	$\bar{B}C_{in}$	BC_{in}	$B\bar{C}_{in}$
$\bar{A}$	0	1	0	1
A	1	0	1	0

(a) For S

	$\bar{B}\bar{C}_{in}$	$\bar{B}C_{in}$	BC_{in}	$B\bar{C}_{in}$
$\bar{A}$	0	0	1	0
A	0	1	1	1

(b) For C_{out}

Fig. 6.6: Karnaugh maps for S and C_{out}

Consequently, the minimized expressions are given by

$$S = \bar{A}\bar{B}C_{in} + \bar{A}B\bar{C}_{in} + A\bar{B}\bar{C}_{in} + ABC_{in}$$
$$C_{out} = AB + AC_{in} + BC_{in}$$

Therefore

$$S = \bar{C}_{in}(\bar{A}B + A\bar{B}) + C_{in}(AB + \bar{A}\bar{B})$$
$$= \bar{C}_{in}(\bar{A}B + A\bar{B}) + C_{in}\overline{(AB + \bar{A}\bar{B})}$$
$$= \bar{C}_{in}(AB + \bar{A}\bar{B}) + C_{in}(AB + \bar{A}\bar{B})$$

(*see* example 4.5)

$$= C_{in} \oplus [\overline{AB + \bar{A}\bar{B}}] \quad (\because A \oplus B = \bar{A}B + A\bar{B})$$
$$= C_{in} \oplus (\bar{A}B + A\bar{B})$$

$$(\because \overline{(AB + \bar{A}\bar{B})} = \bar{A}B + A\bar{B})$$

or $S = C_{in} \oplus [A \oplus B]$
$$= A \oplus B \oplus C$$

From the truth table

$$C_{out} = \bar{A}BC_{in} + A\bar{B}C_{in} + AB\overline{C_{in}} + ABC_{in}$$
$$= \bar{A}BC_{in} + A\bar{B}C_{in} + ABC_{in} + AB\overline{C_{in}}$$
$$= \bar{A}BC_{in} + A\bar{B}C_{in} + AB \quad (\because C + \bar{C} = 1)$$

$$= (\bar{A}B + A\bar{B})C_{in} + AB$$
$$= (A \oplus B)C_{in} + AB \qquad \ldots(ii)$$

From equation (i) and (ii) we can draw the logic diagram of a full adder using two half adders as shown in Fig. 6.7.

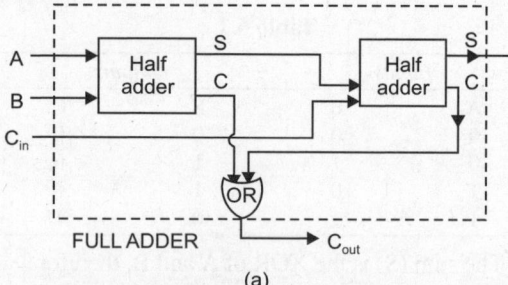

(a)

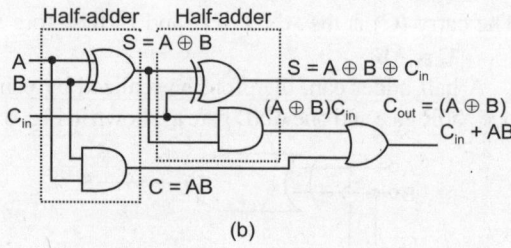

(b)

Fig. 6.7: Logic diagram of a full adder using two half adders (a) Half adder-full adder relations (b) logic diagram

It is to be noted here that even though a full adder can be constructed using two half adders, the disadvantage is that the bits must propagate through several gates in succession which make the total propagation delay greater than that of the full-adder circuit using AOI logic shown in Fig. 6.8.

The full adder can be realized using universal logic, i.e. either only NAND gates or only NOR gates as discussed below.

NAND logic

$$A \oplus B = A\bar{B} + \bar{A}B = A\bar{B} + A\bar{A} + \bar{A}B + B\bar{B}$$
$$= A(\bar{A} + \bar{B}) + B(\bar{A} + \bar{B})$$
$$= A \cdot \overline{AB} + B \cdot \overline{AB}$$
$$= [A \cdot \overline{AB} \cdot B \cdot B.AB] \equiv X$$

Then

$$S = A \oplus B \oplus C_{in} = X \oplus C_{in}$$
$$= [X \cdot \overline{XC_{in}} \cdot C_{in} \cdot \overline{XC_{in}}]'$$

Note: Dash outside the bracket denotes complement or inverse of the expression inside the bracket

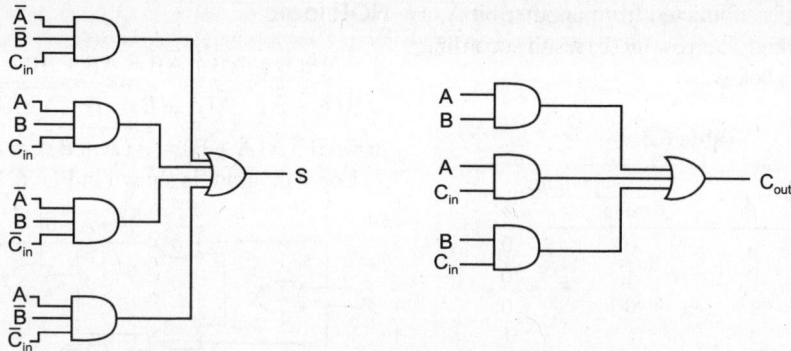

Fig. 6.8: Sum and carry bits of a full adder using And-Or-Invert logic

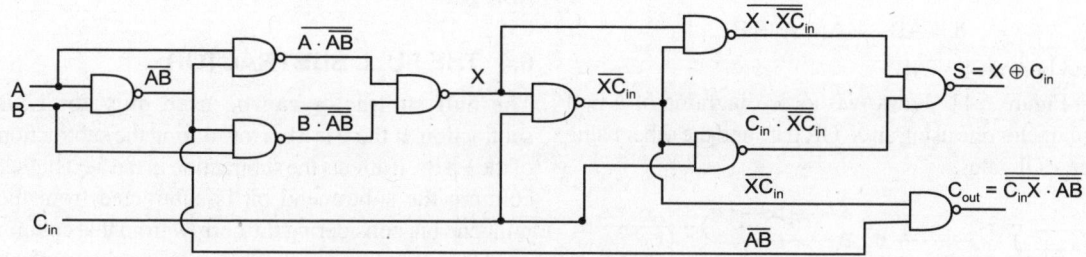

Fig. 6.9: Logic diagram of a full adder using only two input NAND gates

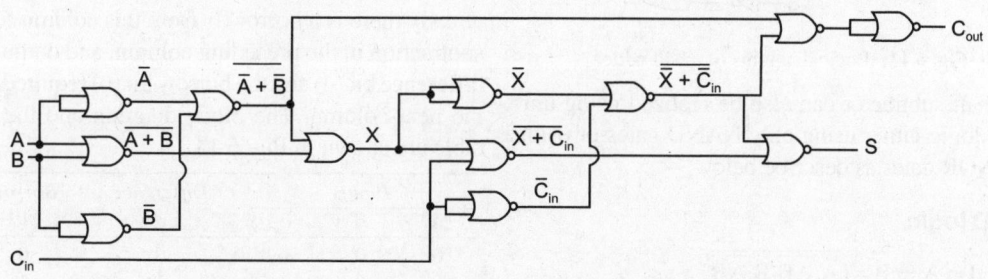

Fig. 6.10: Logic diagram of a full adder using only 2-input NOR gates

$$C_{out} = C_{in}(A \oplus B) + AB = \overline{\overline{C_{in}(A \oplus B) + AB}}$$

$$= [\overline{C_{in}(A \oplus B) \cdot \overline{AB}}]'$$

NOR Logic

Let $A \oplus B = [(\overline{A + B}) + (\overline{\overline{A} + \overline{B}})]' \equiv X$

Then

$$S = A \oplus B \oplus C_{in} = X \oplus C_{in}$$

$$= [\overline{X + C_{in}} + \overline{\overline{X} + \overline{C_{in}}})]'$$

$$C_{out} = AB + C_{in}(A \oplus B) = \overline{\overline{A} + \overline{B}} + \overline{\overline{C_{in}} + \overline{A \oplus B}}$$

$$(\because AB = \overline{\overline{A} + \overline{B}} \text{ see fig. 4.8})$$

or $\quad C_{out} = [(\overline{A} + \overline{B}) + (\overline{C_{in}} + \overline{X})]'$

6.3 THE HALF SUBTRACTOR

A half subtractor is an arithmetic circuit that subtracts one bit from the other. It is used to subtract the LSB of the subtrahend from the LSB of the minuend when one binary number is subtracted from the other.

When a bit B is subtracted from another bit A, a difference bit (d) and a borrow bit (b) result according to the rules given below:

Table 6.3

Inputs		Outputs	
A	B	d	b
0	0	0	0
1	0	1	0
1	1	0	0
0	1	1	1

A circuit working according to the above table gives difference d described by

$$d = A\overline{B} + B\overline{A} = A \oplus B$$

and borrow $b = \overline{A}B$

Figure 6.11 shows two logic diagrams of a half subtractor one using an X-OR gate and the other using the AOI gates.

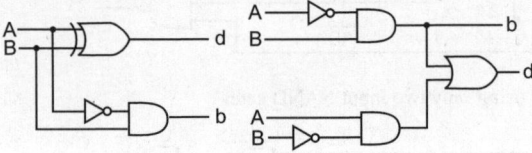

Fig. 6.11: Logic diagrams of a half adder

A half subtractor can also be realized using universal logic-either using only NAND gates or using only NOR gates as describe below:

NAND logic

$$d = A \oplus B = [\overline{A.\overline{AB}}.\overline{B.\overline{AB}}]'$$

$$b = \overline{A}B = B(\overline{A} + \overline{B}) = B\overline{(AB)} = [\overline{B.\overline{AB}}]'$$

Logic diagram is shown in Fig. 6.12

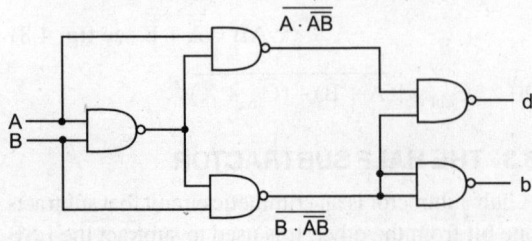

Fig. 6.12: Logic diagram of a half subtractor using only NAND gates

NOR logic

$$d = A \oplus B = A\overline{B} + \overline{A}B = A\overline{B} + B\overline{B} + \overline{A}B + A\overline{A}$$

$$= \overline{B}(A + B) + \overline{A}(A + B) = \overline{B + \overline{A + B}} + \overline{A + \overline{A + B}}$$

$$b = \overline{A}B = \overline{A}(A + B) = [\overline{\overline{A}(A + B)}]' = \overline{A + (\overline{A + B})}]'$$

Logic diagram is shown in Fig. 6.13.

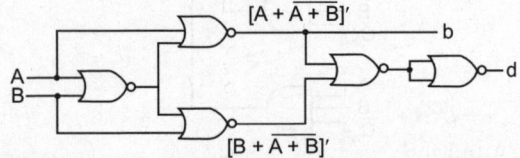

Fig. 6.13: Logic diagram of a half subtractor using only NOR gates

6.4 THE FULL SUBTRACTOR

The half subtractor can be used only for LSB subtraction. If there is a borrow during the subtraction of the LSBs, it affects the subtraction in the next higher column; the subtrahend bit is subtracted from the minuend bit, considering the borrow from that column used for subtraction in the proceeding column. Such a subtraction is performed by a full-subtractor. It subtracts one bit (B) from another bit (A), when already there is a borrow b_i from this column for the subtraction in the preceding column, and outputs the difference bit (d) and the borrow bit (b) required from the next column. The block diagram and the truth table are shown in Fig. 6.14.

Inputs			Difference	Borrow
A	B	b_i	d	b
0	0	0	0	0
0	0	1	1	1
0	1	0	1	1
0	1	1	0	1
1	0	0	1	0
1	0	1	0	0
1	1	0	0	0
1	1	1	1	1

(a) Truth Table

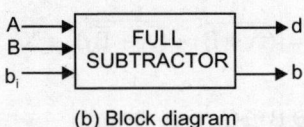

(b) Block diagram

Fig. 6.14: (a) Truth table and (b) block diagram of a full subtractor

From the truth table, a circuit that will produce the correct difference and borrow bits in response to every possible combination of A, B and b_i is described by

$$d = \overline{A}\overline{B}b_i + \overline{A}B\overline{b}_i + A\overline{B}\overline{b}_i + ABb_i$$

$$= b_i(AB + \overline{A}\overline{B}) + b_i(A\overline{B} + \overline{A}B)$$

$$= b_i(\overline{A \oplus B}) + \overline{b}_i(A \oplus B)$$

$$= A \oplus B \oplus b_i$$

$$b = \overline{A}\overline{B}b_i + \overline{A}B\overline{b}_i + \overline{A}Bb_i + ABb_i$$

$$= \overline{A}B + (\overline{A \oplus B})b_i$$

A full subtractor can, therefore, be realized using X-OR gates and AOI (AND-OR-INVERT) gates as shown in Fig. 6.15.

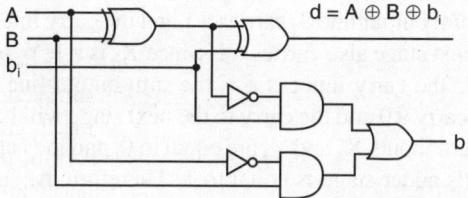

Fig. 6.15: Logic diagram of a full-subtractor

NAND logic

Since $A \oplus B = [\overline{A.\overline{AB}.B.\overline{AB}}]'$

Therefore

$$d = A \oplus B \oplus b_i$$

$$= [\overline{(A \oplus B) \cdot \overline{(A \oplus B)b_i}} \cdot \overline{b_i \overline{(A \oplus B)b_i}}]'$$

$$b = \overline{A}B + \overline{(A \oplus B)}b_i$$

$$= [\overline{\overline{A}B + b_i \cdot \overline{(A \oplus B)}}]'$$

$$= [\overline{(\overline{A}B)} \cdot \overline{b_i(A \oplus B)}]' \qquad (\because \overline{A + B} = \overline{A} \cdot \overline{B})$$

$$= \{\overline{B(\overline{A} + \overline{B})} \cdot \overline{b_i[\overline{b}_i + (A \oplus B)]}\}'$$

$$(\because \overline{AB} = [(\overline{A} + \overline{B})B]'$$

$$= \{\overline{B.\overline{AB}} \cdot \overline{b_i[\overline{b}_i(A \oplus B)]}\}' \quad (\because \overline{AB} = \overline{A} + \overline{B})$$

NOR logic

$$d = A \oplus B \oplus b_i = \overline{(A \oplus B)b_i + (\overline{A \oplus B})\overline{b}_i}$$

$$(\because A \oplus C = \overline{(AC + \overline{A}\overline{C})})$$

$$= \{[\overline{(A \oplus B) + (\overline{A \oplus B})\overline{b}_i][b_i + (\overline{A \oplus B})\overline{b}_i}]\}'$$

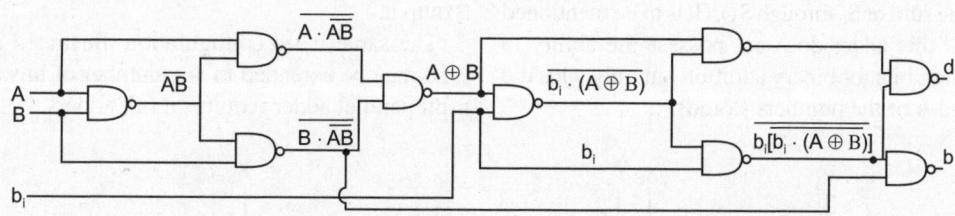

Fig. 6.16: Logic diagram of a full subtractor using 2-input NAND gates

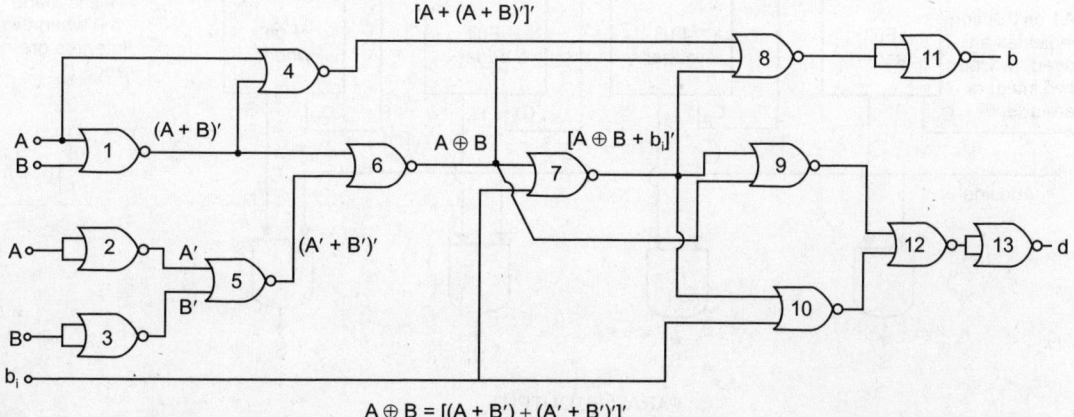

$$A \oplus B = [(A + B') + (A' + B')']'$$

Fig. 6.17

$$= \overline{(A \oplus B) + \overline{(A \oplus B) + b_i} + \overline{b_i (A \oplus B) + b_i}}$$

$$(\because \overline{AB} = \overline{A} + \overline{B})$$

$$= \overline{[(A \oplus B) + \overline{(A \oplus B) + b_i} + \overline{b_i + (A \oplus B) + b_i}]'}$$

(Double negation)

$$b = \overline{AB} + b_i \overline{(A \oplus B)}$$

$$= \overline{A}(A + B) + \overline{(A \oplus B)} + [\overline{(A \oplus B) + b_i}]$$

$$= [A + \overline{(AB)}]' + \overline{(A \oplus B)} + [\overline{(A \oplus B) + b_i}]'$$

(using $= \overline{A.B} = \overline{A} + \overline{B}$)

$$= \{\overline{[A + (A + B)]' + \overline{(A \oplus B)} + [\overline{(A \oplus B) + b_i}]'}\}'$$

6.5 A PARALLEL BINARY ADDER

A parallel binary adder is a digital circuit that performs arithmetic sum of two binary numbers in parallel. A 4-bit parallel binary adder is illustrated in Fig. 6.18. This adder adds two 4-bit binary integers. The addend inputs are named X_0 through X_3 and the augend bits are represented by Y_0 through Y_3 (these inputs would normally be from flip-flop registers X and Y and the adder would add the number in X to the number in Y, giving the sum or S_0 through S_3). (It is to be mentioned here that this adder does not possess the ability to handle sign bits for binary addition but only adds the magnitudes of the numbers stored).

Consider the addition of the following two 4-bit binary numbers.

0111   where $X_3 = 0$, $X_2 = 1$, $X_1 = 1$ and $X_0 = 1$

0011 $Y_3 = 0$, $Y_2 = 0$, $Y_1 = 1$ and $Y_0 = 1$

Sum = 1010

The sum should therefore be $S_3 = 1$, $S_2 = 0$, $S_1 = 1$ and $S_0 = 0$. The operation of the adder may be checked as follows.

Since X_0 and Y_0 are the least significant bits, they cannot receive a carry from a previous stage. In the above example, X_0 and Y_0 are both 1s, their sum is therefore 0 and a carry is generated and added into the full adder for bits X_1 and Y_1. Bits X_1 and Y_1 are both 1s as is the carry input to this stage. Therefore the sum output line S_1 carries a 1 and the carry line to the next stage also carries a 1. Since X_2 is a 1, Y_2 is a 0 and the carry input is a 1, the sum output line S_2 will carry a 0 and the carry to the next stage will be a 1. Both inputs X_3 and Y_3 are equal to 0, and the carry to this adder stage is equal to 1. Therefore the sum output line S_3 will represent a 1, and the carry output line (designated as overflow in the figure) will have a 0 output.

The same basic configuration illustrated in Fig. 6.18 may be extended to any number of bits and an n-bit parallel adder requires n full adders.

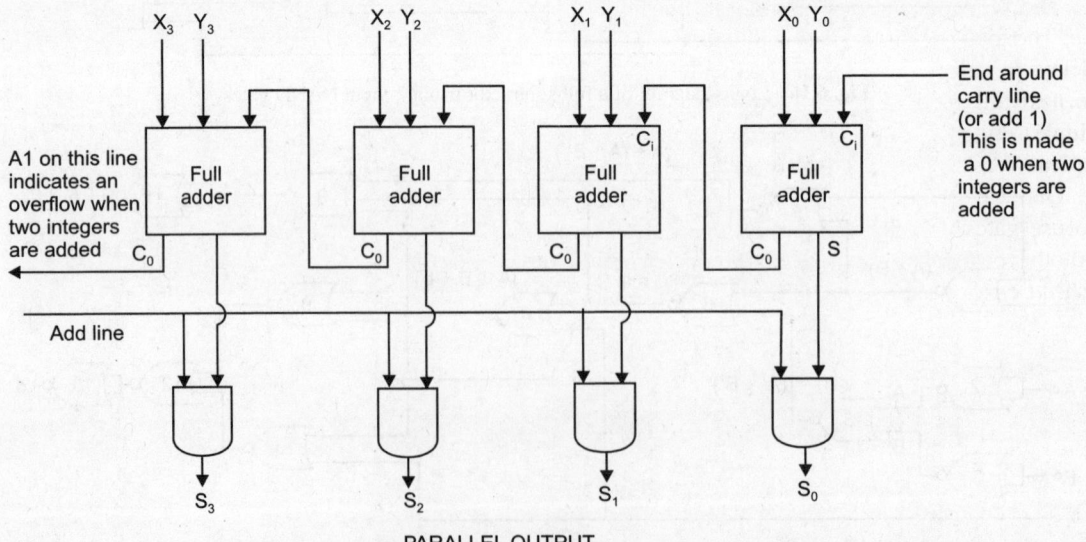

Fig. 6.18: A 4-bit parallel binary adder

The AND gates connected to the S output lines from the four adders are used to gate the sum into the correct register.

If the n-bit adder is implemented using the above scheme of Fig. 6.18, the carry has to ripple down the line of cascaded adders from the LSB to MSB position. This decreases the operating speed of the adder. The time required for addition operation to be completed is limited by the amount of time required to complete the ripple carry operation. A technique known as the look-ahead carry generation process is used for increasing the speed of operation and many high speed adders are available in IC form which utilize the look-ahead, carry for reducing overall propagation delays.

6.6 ADDER WITH LOOK-AHEAD, CARRY

In the half adder and full adder circuits, even if the augend, addend and carry inputs are present simultaneously, the sum and carry outputs will be delayed due to propagation delays of gates through which the signals are passing. For an n-bit parallel adder, the S_o output is the LSB of the final result and it is not required to be passed through other gates so it does not get delayed. The carry output C_0 acts as carry input of the next full adder [FA1] whose outputs S_1 and C_1 will reach steady state only after arrival of C_0 and propagation delay introduced by this FA1, i.e. The total delay upto FA1 will be the sum of the delays introduced by FA0 and FA1. Similarly going to next full adders towards MSB, the carry has to ripple through all FAs, thereby reducing the speed of operation as the number of adder stages are increased.

One way to reduce the propagation delay time is to use gates with lower propagation delay times. Another better approach is to use the concept of *look-a head, carry*. Although it requires additional circuitry,

but the speed of the adder becomes independent of the number of bits which does not hold true for the case of using only gates with lower propagation delay times.

Let us consider a full adder circuit (Fig. 6.19) and its XOR realization.

Here

$$P_i = A_i \oplus B_i \qquad \qquad ...(1)$$
$$G_i = A_i B_i \qquad \qquad ...(2)$$
$$S_i = P_i \oplus C_{i-1} = A_i \oplus B_i \oplus C_{i-1} \qquad ...(3)$$
$$C_i = G_i + P_i C_{i-1} \qquad \qquad ...(4)$$

The output G_i of the first half adder is 1 if A_i and B_i are 1 and a carry is generated. The variable P_i is known as a *carry generate*. Its value is independent of the input carry. The variable P_i is known as a *carry propagate* because this is the term associated with the propagation of the carry from C_{i-1} to C_i. Using eqn. (4) we write the Boolean expression for carry output of each stage:

$$C_{i+1} = G_{i+1} + P_{i+1} \cdot C_i$$

Substituting the value of C_i from eqn. (4) in the above eqn. we get

$$C_{i+1} = G_{i+1} + P_{i+1} G_i + P_{i+1} P_i C_{i-1} \qquad ...(5)$$

Similarly, for an n stage adder, the final carry C_{n-1} can be determined. For a clear understanding of the advantage of this approach, consider a 4 bit adder and formulate Boolean expressions for the carry outputs C_0, C_1, C_2 and C_3. We obtain

$$P_0 = A_0 \oplus B_0 \text{ and } G_0 = A_0 B_0 \qquad ...(6)$$
$$C_0 = G_0 + P_0 C_{-1} \qquad \qquad ...(7)$$
$$C_1 = G_1 + P_1 G_0 + P_1 P_0 C_{-1} \qquad ...(8)$$
$$C_2 = G_2 + P_2 G_1 + P_2 P_1 G_0 + P_2 P_1 P_0 C_{-1} \quad ...(9)$$
$$C_3 = G_3 + P_3 G_2 + P_3 P_2 G_1 + P_3 P_2 P_1 G_0$$
$$+ P_3 P_2 P_1 P_0 C_{-1} \qquad \qquad ...(10)$$

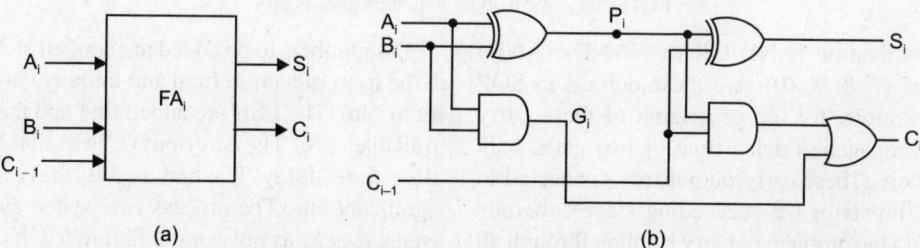

Fig. 6.19: (a) Block diagram of ith state of a full adder (b) X-OR implementation of a full adder

Now let us observe the logic variables involved in eqns (7), (8), (9) and (10). These are

$$G_0, G_1, G_2, G_3, P_0, P_1, P_2, P_3 \text{ and } C_{-1}.$$

The G variables can be generated directly from A and B inputs using AND gates (eqn. 2). The P variables are obtained again directly from A and B inputs using XOR gates (eqn 1). C_{-1} is the carry input. If G_s, P_s and C_{-1} are available simultaneously, the carry outputs C_0, C_1, C_2 and C_3 are produced by using

Logic circuit of a look-ahead carry generator can be prepared using eqns. (7 to 10). A 4-bit adder with look ahead carry is shown in Fig. 6.20.

6.7 SERIAL ADDER

In serial adder, addition is done serially. The basic serial adder is shown in Fig. 6.21. The circuit differs from the configuration of parallel adder by inclusion of a time-delay. The serial adder takes the two input numbers A_n and B_n and delivers the output waveform 'S_n'

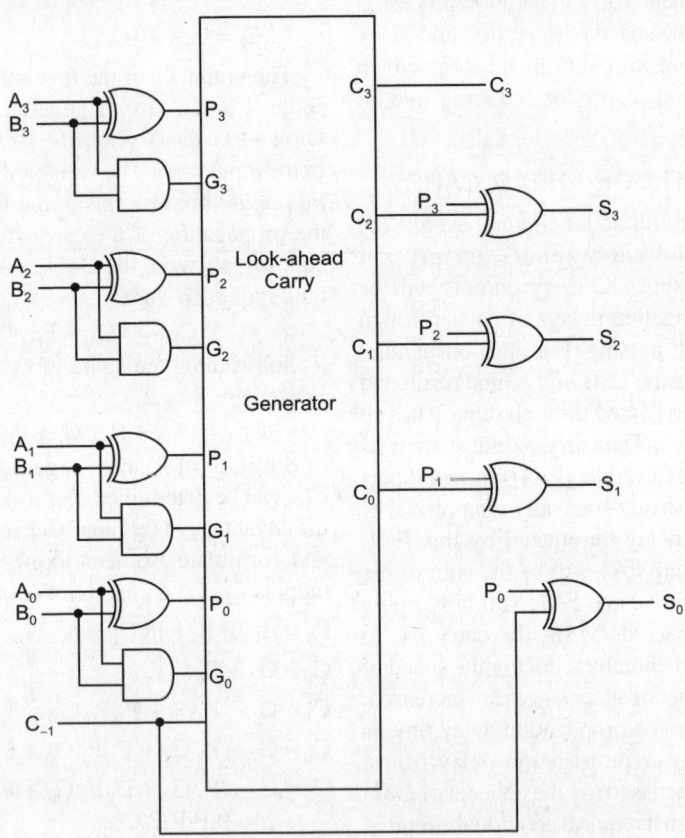

Fig. 6.20: A 4-bit adder with look ahead carry

2-level realization (AND-OR or NAND-NAND) since eqns (7, 8, 9, 10) give these outputs in SOP form. Therefore, for the generation of these carry outputs, propagation delay time of two gates only will be there. These carry outputs are connected to the carry inputs of the succeeding stages, thereby eliminating the problem of carry rippling through all the stages.

The numbers to be added are applied at A_n and B_n in the form of a pulse train and initially carry C_{n-1} is set to zero. The LSB are added first and the result is available at S_n. The carry out C_n is available as C_{n-1} after time delay T when A_n, B_n have the next significant bits. The process is repeated and sum is available at S_n as pulse train. Figure 6.21b illustrates the addition of 1001 and 1000.

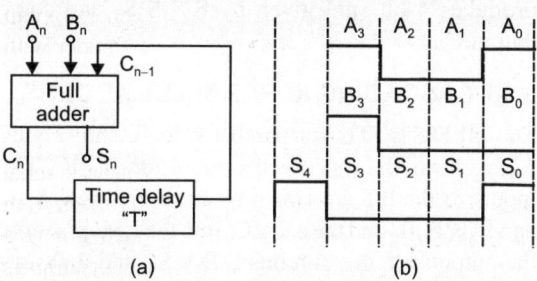

Fig. 6.21: Serial full adder (a) block diagram and (b) pulse train showing addition

The sum S_n is pulse train 10001 as in Fig. 6.21b. A comparison of serial and parallel adders indicates that parallel addition is fast as compared to serial addition, because in parallel addition, all bits are added simultaneously. However, only one full adder is needed to implement serial adder whereas in case of parallel adder, the number of full adders required is equal to total number of bits to be added simultaneously.

6.8 BCD ADDER

BCD adder can add BCD digits in parallel and produce sum digit in BCD. In BCD, addition is performed as follows:

(i) Add two BCD code groups for each decimal digit position using ordinary (binary) addition method.

(ii) If sum is less than or equal to 9, no correction is needed. If it is greater than 9, a correction of 0110 should be added to that sum to produce the proper BCD result. This will produce a carry to be added to the next decimal position.

If two BCD code groups represented by $A_3A_2A_1A_0$ and $B_3B_2B_1B_0$ are applied to a four bit parallel adder, the adder will perform following operation

$$
\begin{aligned}
A_3A_2A_1A_0 &\quad \leftarrow \quad \text{Code 1} \\
+ B_3B_2B_1B_0 &\quad \leftarrow \quad \text{Code 2} \\
\hline
S_4S_3S_2S_1S_0 &\quad \leftarrow \quad \text{Binary sum}
\end{aligned}
$$

S_4 is the carry out of the MSB. The sum output can lie anywhere between 00000 to 10010, when both BCD numbers are 9. Thus BCD adder must include the logic required to detect whenever the sum is greater than 9 (01001), so that the correction can be added. Table 6.4 shows the cases when sum is greater than 9.

Table 6.4: Cases when sum is greater than 9

S_4	S_3	S_2	S_1	S_0	
0	1	0	1	0	10
0	1	0	1	1	11
0	1	1	0	0	12
0	1	1	0	1	13
0	1	1	1	0	14
0	1	1	1	1	15
1	0	0	0	0	16
1	0	0	0	1	17
1	0	0	1	0	18

It is obvious from the table that correction is needed when binary sum has output carry, i.e. $S_4 = 1$ and whenever $S_3 = 1$, and either S_2 or S_1 or both are 1.

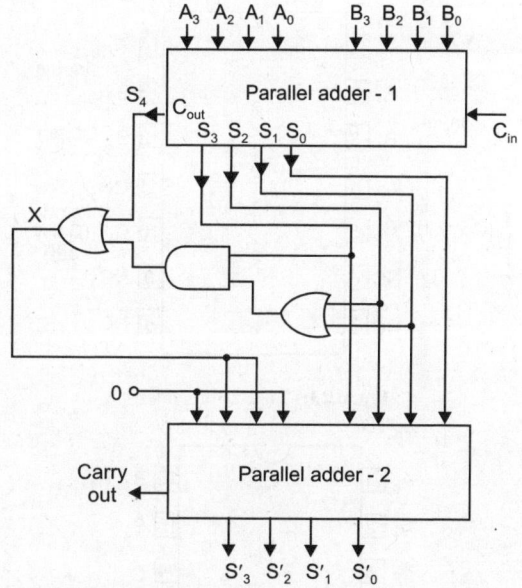

Fig. 6.22: Block diagram of a BCD adder

This can be expressed logically as

$$X = S_4 + S_3 (S_2 + S_1)$$

Whenever $X = 1$, it is necessary to add the correction 0110 to the sum bits and to generate a carry. Figure 6.22 shows complete circuit block diagram for BCD adder.

6.9 IC ADDERS

Adders are available in ICs as 1 bit, 2 bit and 4 bit full adders each in one package. Some of the popular adder ICs are given in Table 6.5. The pin out for 2-bit full adder 7482 is shown in Fig. 6.23. All other adders are 4-bit full adders and are functionally equivalent; however their pin-layout differs. They will add two 4-bit binary words plus one incoming carry. The binary sum appears on sum outputs with output carry. The pin-out for 7483 4-bit parallel adder is shown in Fig. 6.24.

Table 6.5: Adder ICs

Device	Family	Description
7482	TTL	2 bit full adder
7483	TTL	4 bit full adder, Fast carry
74HC283	CMOS	4 bit full adder, Fast carry
4008	CMOS	4 bit full adder, Fast carry

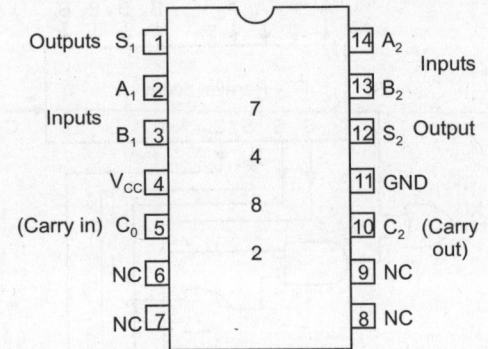

Fig. 6.23: 7482 2- bit adder

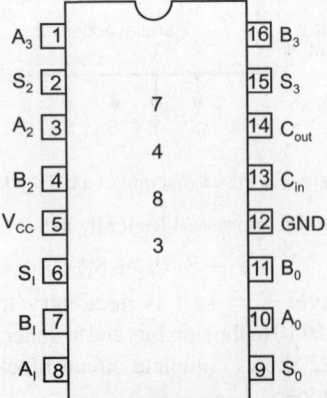

Fig. 6.24: 7483 4-bit adder

All pins are self explanatory. Here 4-bit of the A inputs are connected to the A_0, A_1, A_2, A_3. The 4-bit of B inputs are similarly connected. The adder produces 4-bit sum given by $S_0 S_1 S_2 S_3$ and carry output C_{out}.

6.10 CASCADING IC PARALLEL ADDERS

The 74LS83 is TTL 4 bit parallel adder IC chip. Figure 6.25a shows the functional symbol for this IC. The inputs to this IC are two 4-bit numbers $A_3 A_2 A_1 A_0$ and $B_3 B_2 B_1 B_0$ and the carry C_0 into the LSB position, the outputs are the sum bits $S_3 S_2 S_1 S_0$, and the carry C_4 out of the MSB position.

The addition of large binary numbers can be accomplished by cascading two or more parallel adder chips. Figure 6.25b shows the cascading of two 74LS83 chips to add two 8-bit numbers. The adder on the right adds the 4 LSBs of the numbers.

The C_4 output of this right adder is connected as the input carry to the first position of the second adder which adds the 4 MSBs of the numbers. The eight sum outputs represent the resultant sum of the two eight bit numbers. The carry out C_8 of last position MSB can be used as an overflow bit or as a carry into another higher stage if large binary numbers are to be handled.

6.11 TWO'S COMPLEMENT ADDITION AND SUBTRACTION USING PARALLEL ADDERS

Most modern computers use the 2's complement system to represent negative numbers and to perform subtraction. Both the addition and subtraction operations of signed numbers can be performed using only the addition operation if we use the 2's complement form to represent negative numbers.

Addition

Positive and negative numbers including the sign bits can be added together in the basic parallel adder circuit when the negative numbers are in the 2's complement form. This is illustrated in Fig. 6.26 for the addition of –4 and +7. The –4 is represented in its 2's complement form as 1100, where the left most bit 1 is the sign bit, the +7 is represented as 0111 with the left most bit 0 as the sign bit. These numbers are stored in their corresponding registers. The 4 bit parallel adder produces the sum output 0011 which represents +3. The carry out C_4 is a 1, but is discarded in the 2's complement method.

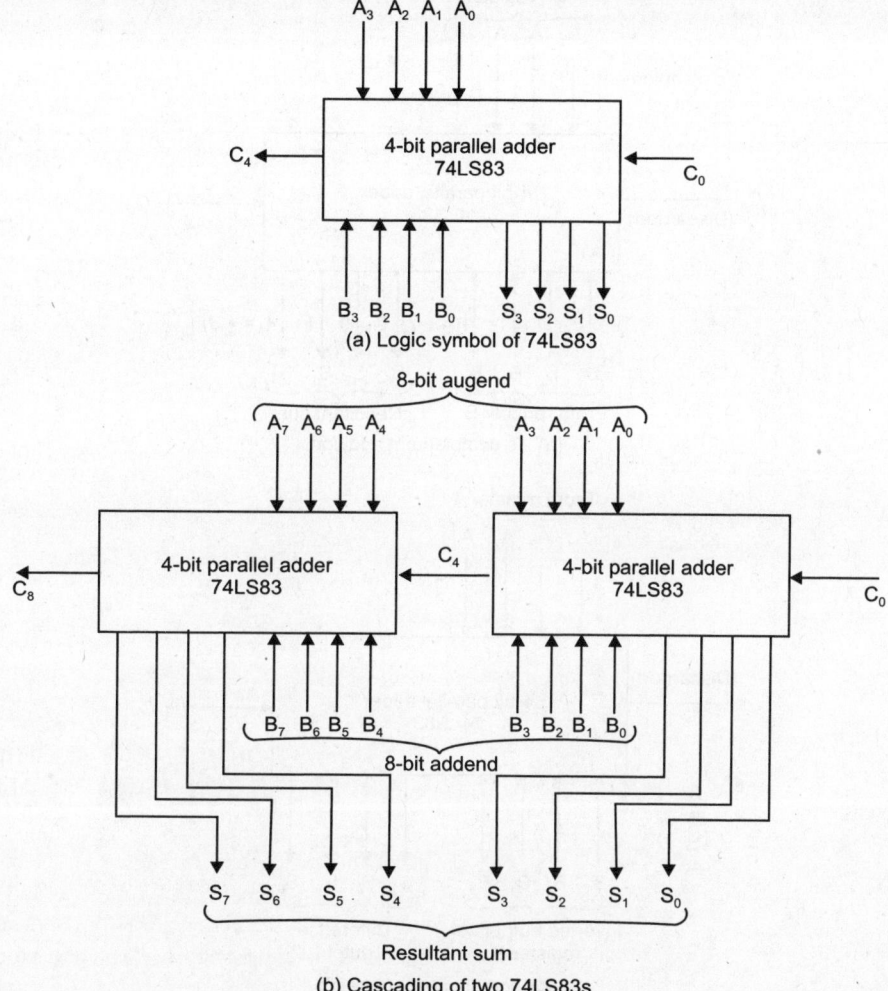

Fig. 6.25: Cascading of two IC parallel adders

Subtraction

In 2's complement method, the subtrahend is changed to its 2's complement form and then added to the minuend. The sum outputs of the adder circuit represent the difference between the minuend and subtrahend.

Figure 6.26b shows an arrangement to subtract the number $B_3 B_2 B_1 B_0$ from the number $A_3 A_2 A_1 A_0$. These numbers are first stored in registers A and B respectively. Since the 2's complement of a number is obtained by complementing its each bit and then adding 1 to the LSB, the complemented number $\overline{B}_3 \overline{B}_2 \overline{B}_1 \overline{B}_0$ is fed to the adder along with $A_3 A_2 A_1 A_0$. The C_0 is made a 1 and added to the LSB of the adder; this accomplishes the same effect as adding a 1 to the LSB of $\overline{B}_3 \overline{B}_2 \overline{B}_1 \overline{B}_0$ for forming its 2's complement. The output $S_3 S_2 S_1 S_0$ represents the result of the subtraction operation S_3 is though the sign bit of the result and indicates whether the result is negative or positive. The carryout C_4 is disregarded.

6.12 DIGITAL COMPARATOR

It is sometimes necessary to know whether a binary number A is greater than, equal to or less than another

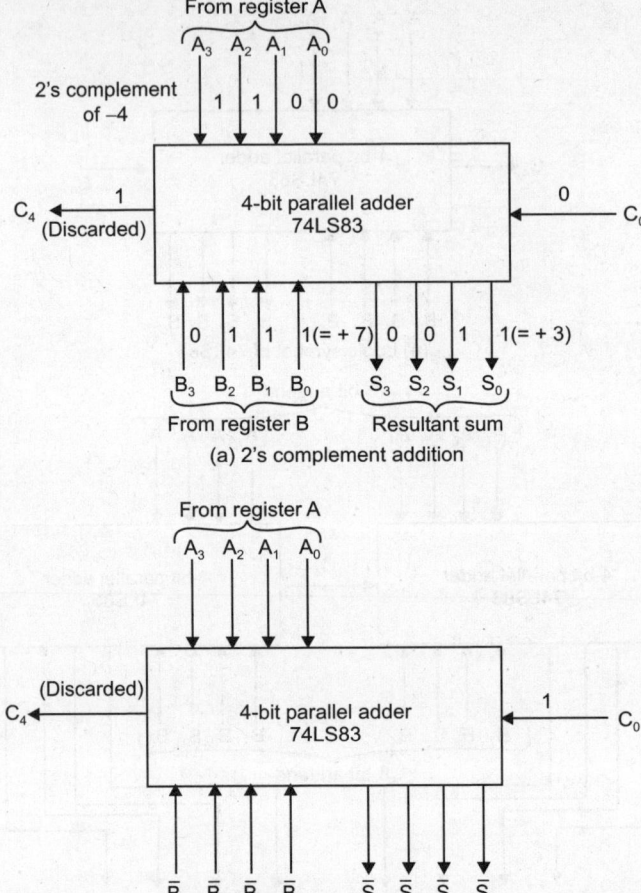

From register A

A_3 A_2 A_1 A_0

2's complement of -4

1 1 0 0

1

C_4 (Discarded)

4-bit parallel adder 74LS83

0 C_0

0 1 1 1(= + 7)   0 0 1 1(= + 3)

B_3 B_2 B_1 B_0

From register B

S_3 S_2 S_1 S_0

Resultant sum

(a) 2's complement addition

From register A

A_3 A_2 A_1 A_0

(Discarded)

C_4

4-bit parallel adder 74LS83

1 C_0

$\bar{B}_3$ $\bar{B}_2$ $\bar{B}_1$ $\bar{B}_0$

Inverted outputs of register B

$\bar{S}_3$ $\bar{S}_2$ $\bar{S}_1$ $\bar{S}_0$

Difference output

(b) 2's complement subtraction

Fig. 6.26: Two's complement addition and subtraction using parallel adders

number B. The system for making this determination is called a magnitude digital (or binary) comparator.

Consider first single bit numbers. Now, the exclusive-nor (XNOR) gate is an equality detector because

$$E = \overline{A\bar{B} + \bar{A}B} = \begin{cases} 1 & A = B \\ 0, & A \neq B \end{cases}$$

The condition $A > B$ is given by

$$C = A\bar{B} = 1$$

because if $A > B$, then $A = 1$ and $B = 0$ so that $C = 1$. On the other hand, if $A = B$ or $A < B$, $(A = 0, B = 1)$ then $C = 0$.

Similarly, the restriction $A < B$ is determined from

$$D = D = \bar{A}B = 1.$$

The logic block diagram for the nth bit has been given in Fig. 6.27. It has all three desired outputs C_n, D_n and E_n.

6.13 PARITY CHECKER/GENERATOR

An arithmetic operation that is often required in a digital system is that of determining whether the sum of the binary bits in a word is odd (called odd parity) or even (designated even parity).

The output of an XOR gate is 1 if and only if one input is 1 and the other is 0. In other words, the output

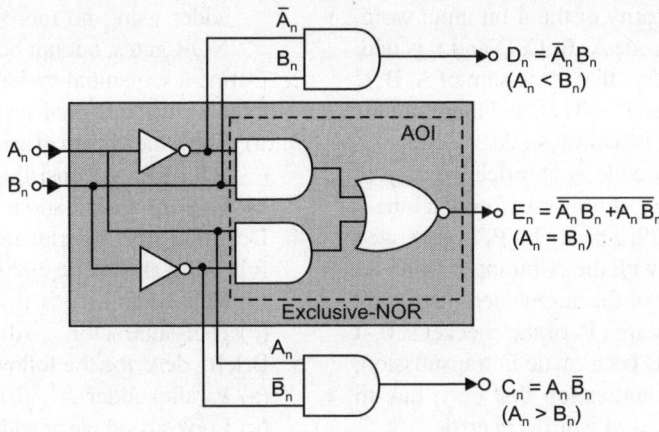

Fig. 6.27: A 1-bit digital comparator

is 1 if the sum of the digits is 1. An extension of this concept to the EXCLUSIVE-OR tree of Fig. 6.28 leads to the conclusion that $Z = 1$ (or $Y = 0$) if the sum of the input bits A,B,C and D is odd. Hence if, the input P' is grounded ($P' = 0$), then $P = 0$ for odd parity and $P = 1$ for even parity.

The system of Fig. 6.28 is not only a parity checker but it may also be used to generate a parity bit P.

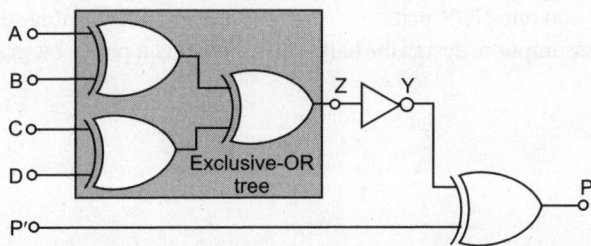

Fig. 6.28: An odd-parity checker or parity-bit generator system for a 4-bit input word. Assume $P' = 0$ and then $P = 0(1)$ represents odd (even) parity

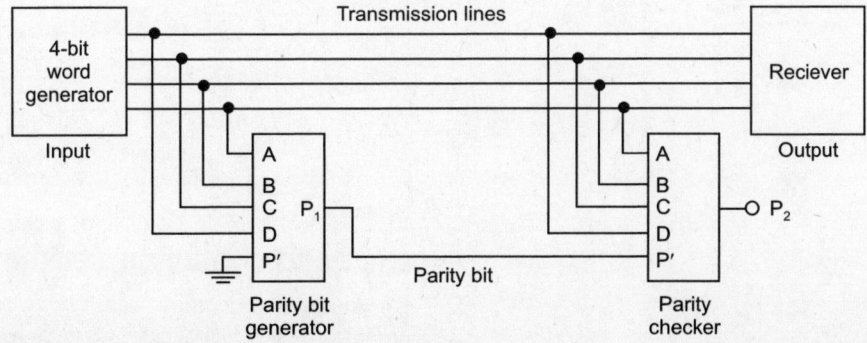

Fig. 6.29: Testing reliability of binary data transmission by generating a parity bit at the input to a line and checking the parity of the transmitted bits plus the generated bit at the receiving end (*)

---

* After Jacob Millman and Christos C. Halkas, McGraw Hill (1985).

Independently of the parity of the 4-bit input word, the parity of the 5-bit code A, B, C, D and P is odd. This follows from the fact that if the sum of A, B, C and D is odd (even) then P is 0 (1), and therefore the sum of A, B, C, D and P is always odd.

The use of a parity code is an effective way of increasing the reliability of transmission of a binary information. In Fig. 6.29, a parity bit P_1 is generated and transmitted along with the N-bit input word. At the receiver, the parity of the augmented (N +1)-bit signal is tested. If the output P_2 of the checker is 0, it means that no error has been made in transmission, whereas $P_2 = 1$ is an indication that (say, due to random noise), the received word is in error.

Note that only errors in an odd number of digits can be detected with a single parity check.

EXERCISE

1. (a) Assuming that only uncomplemented inputs are available, show an implementation of the half adder which requires only three two-input diode gates and one NOR gate.
   (b) Under the above assumption, design the half adder using no more than five NAND or NOR gates, but not both.
2. Define a sequential system. How does it differ from a combinational system?
3. (a) Show the system of a three bit parallel binary full adder as consisting of half adders
   (b) Explain its operation.
4. Describe the operation performed by the following arithmetic circuits:
   (a) Half-adder   (b) Full-adder
   (c) Half-subtractor   (d) Full-subtractor
5. Briefly describe the following:
   (a) Parallel adder   (b) Serial adder
   (c) Look-ahead carry adder
   (d) BCD adder
6. How does the look-ahead carry adder speed up the addition process?
7. When is a carry generated and when is a carry propagated?
8. How do you compare serial and parallel adders?
9. What do you mean by cascading of parallel adders? Why is it required?
10. What is a parity bit generator?

7

Flip-Flops

Gates are decision-making elements. We know that they can perform binary addition and subtraction, as we have seen in the preceding chapter. However, decision making elements are not enough. We also need memory elements in a computer; devices that can store a binary digit. The memory elements are called flip-flops, which we will discuss in this chapter.

7.1 RS LATCHES

A *flip-flop* is a device with two stable states:

It remains in one of these states until triggered into the other. The RS Latch (discussed in this section) is one of the simplest flip-flops.

A transistor-latch is shown in Fig. 7.1a.

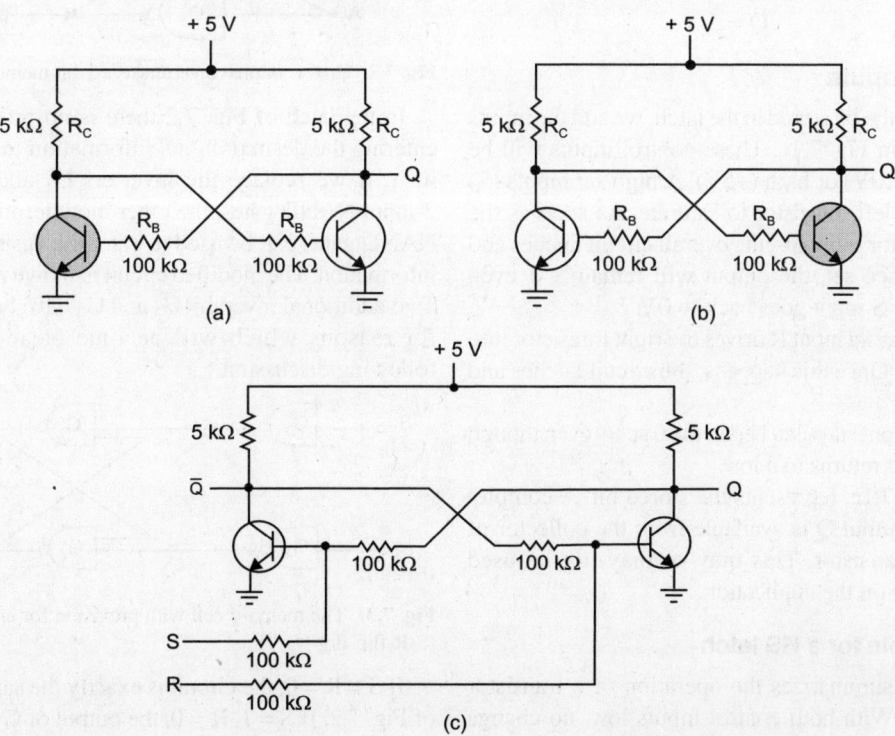

Fig. 7.1: (a) Latched state (b) alternative state (c) trigger inputs

Here each collector drives the opposite base through a 100kΩ resistor. In this circuit, one of the transistors is saturated and the other is cut off. For instance, if the right transistor is saturated, its collector voltage is approximately 0V. This means that there is no base drive for the left transistor, so it cuts off and its collector voltage approaches +5V. This high voltage produces enough base current in the right transistor to sustain its saturation. The overall circuit is *latched* with the left transistor cut off (dark shaded), and the right transistor saturated. Q is approximately 0V.

Similarly, if the left transistor is saturated, the right transistor is cut off, Fig. 7.1b illustrates this other state. Q is approximately 5V for this condition.

Output Q can be low or high (binary 0 or 1). If latched as shown in Fig. 7.1a, the circuit is storing a binary 0 because

$$Q = 0$$

In the other hand, when latched as shown in Fig. 7.1b, the circuit stores a binary 1 because

$$Q = 1$$

Control Inputs

To control the bit stored in the latch, we add the inputs as shown in Fig. 7.1c. These control inputs will be either low (0V) or high (+5V). A high *set* inputs (S) forces the left transistor to saturate. As soon as the left transistor saturates, the overall circuit latches and Q = 1. Once set, the output will remain a 1 even though the S input goes back to 0V.

A high *reset* input R drives the right transistor into saturation. Once this happens, the circuit latches and Q = 0.

The output stays latched in the 0 state, even though the R input returns to a low.

In Fig. 7.1c, represents the stored bit. A complementary output Q is available from the collector of the left transistor. This may or may not be used depending on the application.

Truth Table for a RS latch

Table 7.1 summarizes the operation of a transistor RS latch. With both control inputs low, no change can occur in the output and the circuit remains latched in its last state. This condition is called the inactive state because nothing changes.

Table 7.1: Truth table for a transistor latch

R	S	Q	Comments
0	0	NC	No change
0	1	1	Set
1	0	0	Reset
1	1	*	Race

Thus, if the circuit is in 1 state (*or set state*), it continues to remain in this state and similarly, if it is in 0 state (*Reset state*) it continues to remain in this state. This property of the circuit is referred to as *memory*, i.e. it can store 1-bit of digital information. Since this information is locked or latched in this circuit, this circuit is also known as a *latch*. A 1-bit memory cell using NAND gates is depicted in Fig. 7.2.

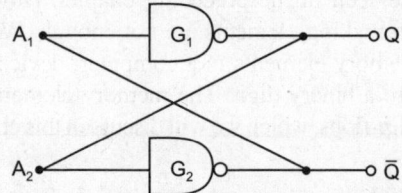

Fig. 7.2: Cross-coupled inverters as a 1-bit memory element

In the latch of Fig. 7.2 there is no provision for entering the desired digital information to be stored in it. If we replace the inverters G_1 and G_2 with 2-input NAND gates, the other input terminal of the NAND gates can be used to enter the desired digital information. The modified circuit is shown in Fig. 7.3. Two additional inverters G_3 and G_4 have been added for reasons which will become clear from the following discussion.

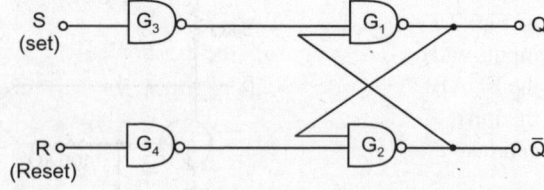

Fig. 7.3: The memory cell with provision for entering data (S-R flip-flop)

If S = R = 0, the circuit is exactly the same as that of Fig. 7.2. If S = 1, R = 0, the output of G_3 will be 0 and the output of G_4 will be 1. Since one of the inputs of G_1 is 0, its output will be certainly 1. Consequently, both the inputs of G_2 will be 1 giving an output $\overline{Q} = 0$.

Hence, for this input condition Q = 1 and $\overline{Q} = 0$. Similarly if S = 0, R = 1 then the outputs will be Q = 0 and $\overline{Q} = 1$. The first of these two input conditions (S = 1, R = 0) makes Q = 1 which is referred to as the *set state* whereas the second input condition (S = 0, R = 1) makes Q = 0 which is referred to as the *reset state* or *clear state*. This provides us the means for entering the desired bit in the latch.

Now we see what happens if the input conditions are changed from S = 1, R = 0 to S = R = 0 or from S = 0, R = 1 to S = R = 0.

Suppose initially Q = 1 and $\overline{Q} = 0$. If now S = 0 and R = 0, these make G_1 inputs (1, 0) and G_2 inputs (1, 1). So, outputs of G_1 and G_2 are Q = 1 and $\overline{Q} = 0$ i.e. the output remains unaltered.

If Q = 0, $\overline{Q} = 1$, and then inputs are S = 0, R = 0. Input to G_1 is (1, 1) and input to G_2 is (0, 1). Consequently, the outputs of G_1 and G_2 are

$$Q = 0, \overline{Q} = 1.$$

Thus output remains unaltered for S = 0 and R = 0. For Q = 1, $\overline{Q} = 0$ and S = 1, R = 0, inputs at G_1 are (0, 0) and at G_2 are (1, 1). This makes Q = 1, $\overline{Q} = 0$, i.e. S makes the circuit in *set state*.

For Q = 1, $\overline{Q} = 0$, S = 0, R = 1. Inputs at G_1 are (1, 0) and at G_2 are (0, 1) outputs at G_1 and G_2 are Q = 0 and $\overline{Q} = 0$. R makes the circuit in *reset state*.

If S = R = 1, both the outputs Q and $\overline{Q}$ will try to become 1, this input condition is prohibited.

7.2 CLOCKED SR FLIP-FLOP

In the latches described above, the output can change state any time the input conditions are changed so they are called *asynchronous* latches. A gated SR latch of Fig. 7.3 requires an ENABLE input. Its S and R inputs will control the state of the flip-flop only when the ENABLE is 'high'. When the ENABLE is 'low', the inputs become ineffective and no change of state can take place. The ENABLE input may be a clock.

So, a gated SR latch with a 'clock' is called *clocked S-R flip-flop* or a *synchronous* SR latch.

In a clocked SR flip-flop the memory cell can be set or reset in synchronism with a train of pulses from a clock (CLK). Such a circuit is shown in Fig. 7.4. In this circuit, if a clock pulse is present (CLK = 1), its operation is exactly the same as that of Fig. 7.3. On the other hand, when the clock pulse is not present (CLK = 0), the gates G_3 and G_4 are inhibited, i.e. their outputs are 1 irrespective of the values of S or R.

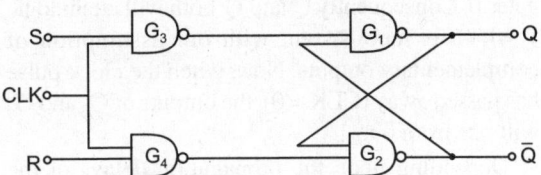

Fig. 7.4: A clocked SR flip-flop

The circuit responds to the inputs S and R only when the clock pulse is present. Assuming that the inputs do not change during the presence of the clock pulse, the operation of the flip-flop of Fig. 7.4 is given in Table 7.2.

Table 7.2: Truth table of S-R flip-flop

Inputs		Output
S_n	R_n	Q_{n+1}
0	0	Q_n
1	0	1
0	1	0
1	1	?

Here S_n and R_n denote the inputs and Q_n the output during the bit time n (Fig. 7.5).

Q_{n+1} denotes the output Q after the pulse passes, i.e. in the bit time n + 1.

If $S_n = R_n = 0$, and the clock pulse is applied, the output at the end of clock pulse is same as the output before the clock pulse, i.e. $Q_{n+1} = Q_n$. This is indicated

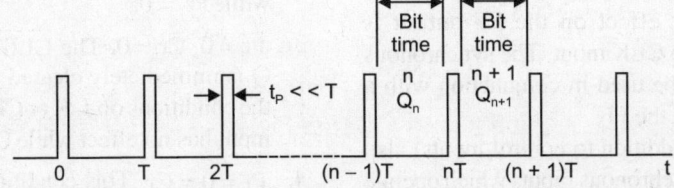

Fig. 7.5: A train of pulses generated by a system clock

in the first row of the truth table. If $S_n = 1$ and $R_n = 0$, the output at the end of the clock pulse will be 1, whereas if $S_n = 0$ and $R_n = 1$ then $Q_{n+1} = 0$. These are indicated in the second and third rows of the truth table.

In the circuit of Fig. 7.5 it was mentioned that $S = R = 1$ is not allowed. Let us see what happens in the SR flip-flop of Fig. 7.4 if $S_n = R_n = 1$. When the clock is present, the outputs of gates G_3 and G_4 are both 0, making one of the inputs of G_1 and G_2 NAND gates 0. Consequently Q and $\bar{Q}$ both will attain logic 1 which is inconsistent with our assumption of complementary outputs. Now, when the clock pulse has passed away (CLK = 0), the outputs of G_3 and G_4 will rise from 0 to 1.

Depending upon the propagation delays of the gates either the stable state $Q_{n+1} = 1$, ($\bar{Q}_{n+1} = 0$) or $Q_{n+1} = 0$ ($\bar{Q}_{n+1} = 1$) will result. That means the state of the circuit is undefined or ambiguous or indeterminate and therefore is indicated by a question mark (fourth row of the truth table). Thus the condition $S_n = R_n = 1$ is forbidden and it must not be allowed to occur. The logic symbol of clocked S-R flip-flop is given in Fig. 7.6.

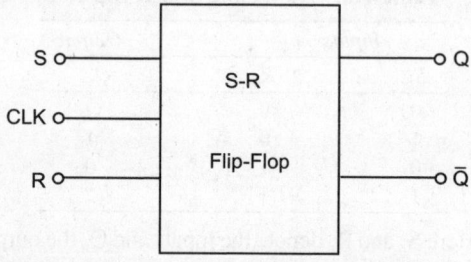

Fig. 7.6: Logic symbol of clocked S-R flip-flop

7.3 THE EFFECT OF ASYNCHRONOUS INPUTS 'PRESET' AND 'CLEAR'

For the clocked FFs, the S, R or J, K inputs are referred to as *control* inputs. These are also called *synchronous inputs* because their effect on the FF output is synchronized with the CLK input. The synchronous control inputs are to be used in conjunction with a clock signal to trigger the FF.

Clocked FFs, (in addition to control inputs) also have one or more asynchronous inputs which operate independent of the synchronous and clock inputs.

These asynchronous inputs can be used to set the FF to the 1 state or clear the FF to the 0 state at any time regardless of the conditions at the other inputs. In other words, the asynchronous inputs are used to over ride all the other inputs, to place the FF in one state or the other.

Figure 7.7 shows a J-K flip-flop with two asynchronous inputs designated as $\bar{Pr}$ and $\bar{Cr}$. These are active LOW inputs (as indicated by the bubbles on the FF symbol). The given truth table summarizes how these affect the FF output.

$\bar{Pr}$	$\bar{Cr}$	FF response
1	1	clocked operation*
0	1	Q = 1
1	0	Q = 0
0	0	not used

* Q will respond to J, K and CLK.

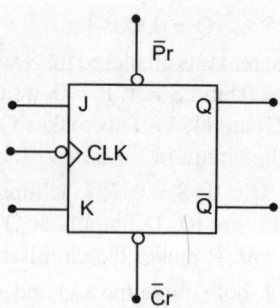

Fig. 7.7: clocked J-K FF with asynchronous inputs $\overline{Preset}$ and $\overline{Clear}$. Let us consider the various cases

1. $\bar{Pr} = \bar{Cr} = 1 =$ the asynchronous inputs are inactive and the FF is free to respond to the J, K and CLK inputs, i.e. the clocked operation can take place.

2. $\bar{Pr} = 0$, $\bar{Cr} = 1$. The $\overline{PRESET}$ is activated and Q is immediately set to 1 no matter what conditions are present at the J, K and CLK inputs. The CLK input can not affect the FF while $\bar{Pr} = 0$.

3. $\bar{Pr} = 0$, $\bar{Cr} = 0$. The $\overline{CLEAR}$ is activated and Q is immediately cleared to 0 independent of the conditions on J, K or CLK inputs. The CLK input has no effect while $\bar{Cr} = 0$.

4. $\bar{Pr} = 0 = \bar{Cr}$. This condition is to be avoided, since it can result in an ambiguous response.

It should be noted here that these asynchronous inputs respond to dc levels. For a constant 0 held on the $\overline{Pr}$ input, the FF will remain in the Q = 1 state regardless of what is occurring at the other inputs, similarly a constant 0 on the $\overline{Cr}$ input holds the FF in the Q = 0 state. Thus the asynchronous inputs can hold the FF in a particular state for any desired interval. Further, these are used to set or clear the FF to the desired state by application of a momentary pulse.

Sometimes FFs have asynchronous inputs that are active-high rather than active-low. For these FFs, the flip-flop symbol would not have a bubble on the asynchronous inputs.

We will consider the working of a J-K FF in section 7.5.

7.4 PRESET AND CLEAR IN CLOCKED S-R FLIP-FLOP

In the flip-flop of Fig. 7.8 when the power is switched on, the state of the circuit is uncertain. It may come

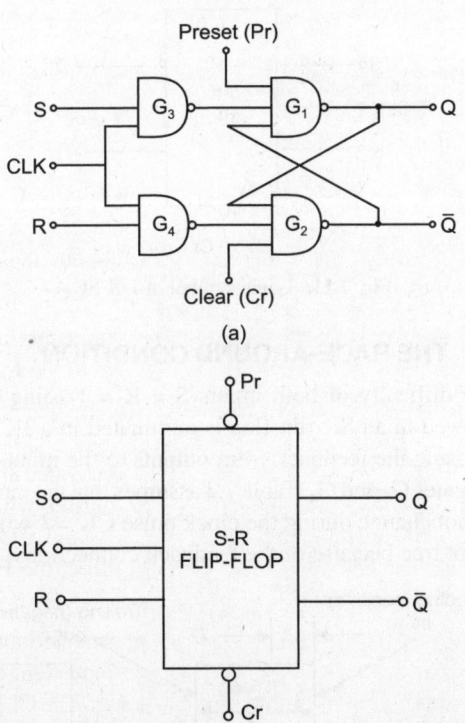

(a)

(b)

Fig. 7.8: (a) An S-R flip-flop with preset and clear (b) Its logic symbol

to set (Q = 1) or reset (Q = 0) state. In many applications it is desired to initially set or reset the flip-flop, i.e. the initial state of the flip-flop is to be assigned. This is accomplished by using the direct or *asynchronous* inputs referred to as *preset* (Pr) and clear (Cr) inputs. These inputs may be applied at any time between clock pulses and are not in synchronism with the clock. An S-R flip-flop with preset and clear is shown in Fig. 7.8.

If Pr = Cr = 1, the circuit operates in accordance with the truth table of S-R flip-flop given in Table 7.2. If Pr = 0 and Cr = 1, the output of G_1 (i.e. Q) will certainly be 1 consequently, all the three inputs to G_2 will be 1 which will make $\overline{Q}$ = 0. Hence making Pr = 0 sets the flip-flop.

Similarly, if Pr = 1 and Cr = 0, the flip-flop is reset. Once the state of the flip-flop is established asynchronously, the asynchronous inputs Pr and Cr must be connected to logic 1 before the next clock is applied.

The condition Pr = Cr = 0 must not be used, since this leads to an uncertain state.

In the logic symbol of Fig. 7.8b, bubbles are used with Pr and Cr inputs, which means these are active low, i.e. the intended function is performed when the signal applied to Pr or Cr is LOW. The operation of Fig. 7.8 is summarized in Table 7.3.

Table 7.3: Operation of S-R flip-flop with preset and clear inputs

Inputs			Output	Operation performed
CLK	*Cr*	*Pr*	*Q*	
1	0	0	Q_{n+1} (Table 7.2)	Normal flip-flop
1	0	1	0	Clear
1	1	0	1	Preset

7.5 J-K FLIP-FLOP

The uncertainty in the state of an S-R flip-flop for $S_n = R_n = 1$ (fourth row of the truth table) can be eliminated by converting it into a J-K flip-flop. The data inputs are J and K which are ANDed with $\overline{Q}$ and Q respectively, to obtain respectively S and R inputs

$$S = J \cdot \overline{Q}$$
$$R = K \cdot Q$$

A J-K flip-flop thus obtained is shown in Fig. 7.9.

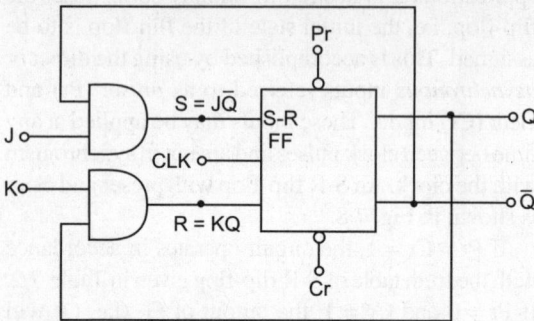

Fig. 7.9: An S-R flip-flop converted into a J-K flip-flop

Its truth table is given in Table 7.4a which is reduced to Table 7.4b for convenience. Table 7.4a has been prepared for all the possible combinations of J and K inputs, and for each combination both the states of the output have been considered. The reader can verify this.

Table 7.4: (a) Truth table for Fig. 7.9

Data inputs		Output		Inputs to S-R FF.		Output
J_n	K_n	Q_n	$\bar{Q}_n$	S_n	R_n	Q_{n+1}
0	0	0	1	0	0	1 $\Big\} = 1$
0	0	1	0	0	0	1
1	0	0	1	1	0	0 $\Big\} = 0$
1	0	1	0	0	0	0
0	1	0	1	0	0	1 $\Big\} = \bar{Q}_n$
0	1	1	0	0	1	0
1	1	0	1	1	0	0 $\Big\} = Q_n$
1	1	1	0	0	1	1

Table 7.4: (b) Truth table of J-K flip-flop

Inputs		Output
J_n	K_n	Q_{n+1}
0	0	Q_n
1	0	1
0	1	0
1	1	$\bar{Q}_n$

It is not necessary to use the AND gates of Fig. 7.9 since the same function can be performed by

adding an extra input terminal to each NAND gate G_3 and G_4 of Fig. 7.8. With this modification incorporated in Fig. 7.8, we obtain the J-K flip-flop using NAND gates as shown in Fig. 7.10. The logic symbol of J-K flip-flop is given in Fig. 7.11.

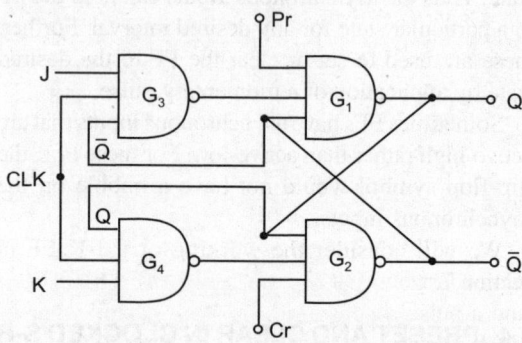

Fig. 7.10: A J-K FF using NAND gates

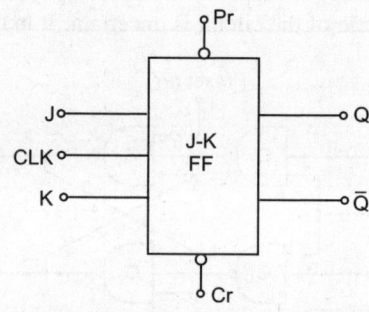

Fig. 7.11: Logic symbol of J-K FF

7.6 THE RACE-AROUND CONDITION

The difficulty of both inputs S = R = 1 being not allowed in an SR flip-flop is eliminated in a JK FF by using the feedback from outputs to the inputs of the gates G_3 and G_4. Table 7.4 assumes that the inputs do not change during the clock pulse CK = 1 which is not true because of the feedback connections.

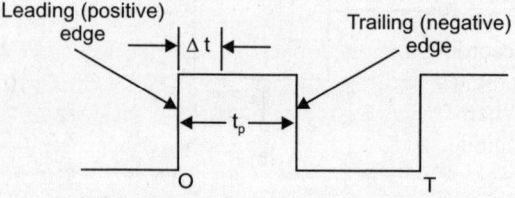

Fig. 7.12: A clock pulse

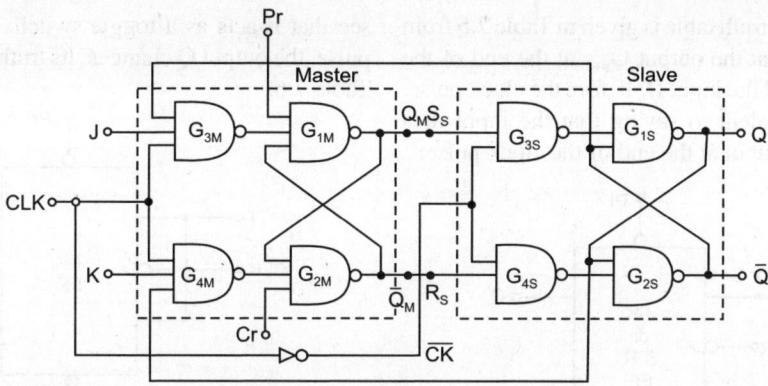

Fig. 7.13: A master-slave J-K flip-flop

Consider now that the inputs are J = K = 1 and Q = 0 and a pulse as shown in Fig. 7.12 is applied at the clock input.

After a time interval Δt equal to the propagation delay through two NAND gates in series, the output will change to Q = 1. (fourth row of Table 7.4b). Now we have J = K = 1 and Q = 1and after another time interval of Dt, the output will change back to Q = 0. Hence we conclude that for the duration t_p of the clock pulse, the output will oscillate back and forth between 0 and 1. At the end of the clock pulse, the value of Q is uncertain. This situation is referred to as the *race-around condition*. This can be avoided if $t_p < Δt < T$. However it may be difficult to satisfy because of very small propagation delays in ICs. A more practical way is to use the master-slave (M-S) configuration.

7.7 THE MASTER-SLAVE J-K FLIP FLOP

A master-slave JK FF is a cascade of two SR flip-flops with feedback from the outputs of the second to the inputs of the first (Fig. 7.13).

Positive clock pulses are applied to the 1st FF and the clock pulses are inverted before these are applied to the 2nd FF.

When CLK = 1, the first FF is enabled and the outputs QM and $\overline{\text{QM}}$ respond to the inputs J and K according to Table 7.4. At this time, the second FF is inhibited because its clock is "LOW" ($\overline{\text{CLK}}$ = 0). When CK goes 'low' ($\overline{\text{CLK}}$ = 1), the first FF is inhibited and the second FF is enabled, because, now, its clock is 'High' ($\overline{\text{CLK}}$ = 1). Therefore the outputs Q and $\overline{\text{Q}}$ follow the outputs QM and $\overline{\text{QM}}$ respectively (2nd and 3rd rows of Table 7.4b). Since the

2nd FF, simply follows the 1st one, it is referred to as the *slave* and the first one as the *master.* Hence the name master-slave (M-S) flip-flop.

In this circuit, the inputs to the gates G₃M and G₄M do not change during the clock pulse, therefore the race-around condition does not exist. The state of the master-slave FF, changes at the negative transition (trailing edge) of the clock pulse. The logic symbol of a M-S flip-flop is given in Fig. 7.14. At the clock input terminal, the symbol > is used to illustrate that the output changes when the clock makes a transition and the accompanying bubble signifies negative transition (change in CLK from 0 to 1).

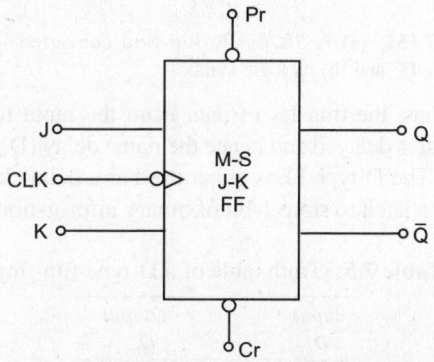

Fig. 7.14: Logic symbol of a master-slave J-K FF

7.8 D-TYPE FLIP-FLOP

If we use only the middle two rows of the truth table of the SR or JK (Tables 7.2 and 7.4b respectively) flip-flop, we obtain a D-type flip-flop as shown in Fig. 7.15. It has only one input, referred to as D-input or data-input. Its truth table is given in Table 7.5 from which we see that the output Q_{n+1} at the end of the clock pulse equal the input D_n before the clock pulse.

This is equivalent to saying that the input data appears at the output at the end of the clock pulse.

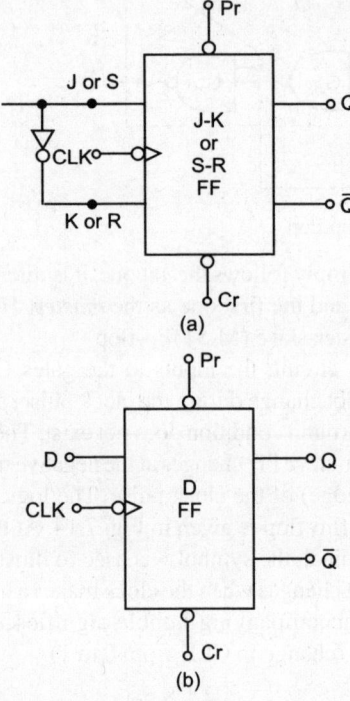

Fig. 7.15: (a) A SR or JK flip-flop converted into a D-type FF and (b) its logic symbol

Thus, the transfer of data from the input to the output is delayed and hence the name delay (D) flip-flop. The D-type FF is either used as a delay device or as a latch to store 1-bit of binary information.

Table 7.5: Truth table of a D-type flip-flop

Input D_n	Output Q_{n+1}
0	0
1	1

7.9 T-TYPE FLIP-FLOP

In a J-K flip-flop, if $J = K$, the resulting FF is referred to as a T-type flip-flop and is shown in Fig. 7.16. It has only one input, referred to as T input. Its truth table is given in Table 7.6 from which we can

see that it acts as a toggle switch. For every clock pulse, the output Q changes. Its truth table is given in Table 7.6.

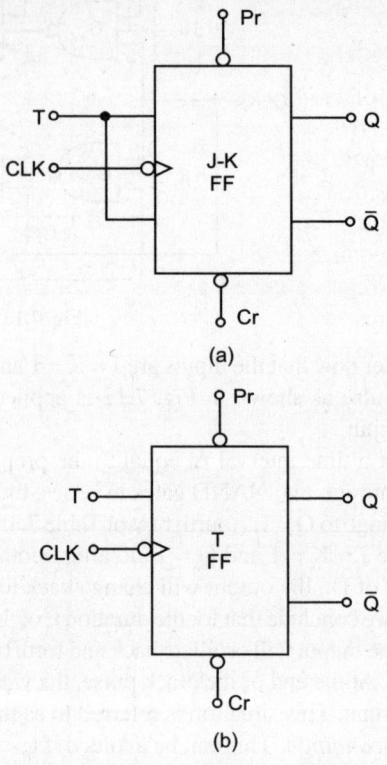

Fig. 7.16: (a) A J-K FF converted into a T-type flip-flop (b) Its logic symbol

Table 7.6: Truth table of a T-type flip-flop

Input T_n	Output Q_{n+1}
0	Q_n
1	$\overline{Q}_n$

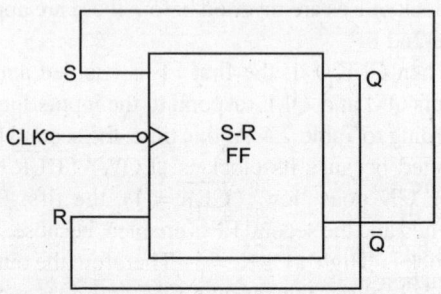

Fig. 7.17: An S-R flip-flop as a toggle switch

An SR FF cannot be converted into a T-type flip-flop since S = R = 1 is not allowed. However, the output of Fig. 7.17 acts as a toggle switch, i.e. the output Q changes with every clock pulse.

7.10 EXCITATION TABLE OF FLIP-FLOPS

The truth table of a flip-flop is also known as the *characteristic table* and specifies the operational characteristic of the flip-flop.

In the design of sequential circuits, we may be knowing the *present state* and the *next state* and it may be required to find the input conditions that must prevail to cause the desired transition of state. Here, by the present state and the next state we mean the state of the circuit prior to, and after the clock pulse respectively. For example, the output of an S-R flip-flop before the clock pulse in $Q_n = 0$ and it is desired that the output does not change when the clock pulse is applied. What input conditions (S_n and R_n values), then must exist to achieve this?

From the characteristic table of an S-R FF, we obtain the following conditions:

1. $S_n = R_n = 0$ (first row of the Table 7.2)
2. $S_n = 0$, $R_n = 1$ (third row of the Table 7.2)

From the above conditions, we conclude that the S_n input must be 0, whereas the R_n input may be either 0 or 1 (don't care). Similarly, input conditions can be found for all possible situations. A tabulation of these conditions is known as the *excitation table*. It forms a useful design aid for sequential circuits. Table 7.7 gives the excitation table of S-R, J-K, T and D flip-flops. This has been derived from the characteristic tables of the flip-flops.

Table 7.7: Excitation table of flip-flops

Present state	Next state	S-R FF S_n	R_n	J-K FF J_n	K_n	T FF T_n	D FF D_n
0	0	0	×	0	×	0	0
0	1	1	0	1	×	1	1
1	0	0	1	×	1	1	0
1	1	×	0	×	0	0	1

There are four possible transitions from present state to the next state. The required input conditions for each of the four transitions are derived from the information available in the characteristic tables of the S-R, J-K, T and D FFs. The symbol × in the excitation table represents a don't care condition, i.e. it does not matter whether the input is 1 or 0.

7.11 EFFECT OF CLOCK SIGNALS IN CLOCKED FLIP-FLOPS

We know that digital systems can operate either asynchronously or synchronously. In an asynchronous system, the outputs of logic circuits can change state any time one or more of the inputs change. In synchronous systems, the exact times at which any output can change states are determined by a signal called the *clock*. The clock signal is distributed to all parts of the system and most (if not all) of the system outputs can change state only when the clock makes a transition. When the clock changes from a 0 to a 1, this is called a positive going or leading edge transition. When the clock goes from 1 to 0, this is the negative going or trailing edge transition. (We use the abbreviations PGT and NGT respectively for these transitions).

The synchronous action of the clock signals is accomplished through the use of clocked Flip-flops. These FF's are designed to change states on one or the other of the clock's transitions.

The following points are to be noted in regard to the operation of clocked Flip-flops.

1. In most clocked FFs, the clock input CK is *edge-triggered*, which means that it is activated by a signal transition, this is indicated by the presence of a small triangle on the clock input. This contrasts with the latches which are level triggered. (In level-triggered latches, the outputs respond to the inputs as long as the clock is present).

2. Clocked FFs also have one or more *control inputs* (e.g. S and R in S-R FF, J and K in J-K FF) The control inputs will have no effect on Q until the active clock transition occurs. In other words, their effect is *synchronous* with the signal applied to clock CK. For this reason they are called synchronous control inputs.

3. In summary, we can say that the control inputs get the FF outputs ready to change, while the active transition at the CK input actually triggers the change. The control inputs control the WHAT (i.e. what state the output will go to) while the clock input CLK determines the WHEN.

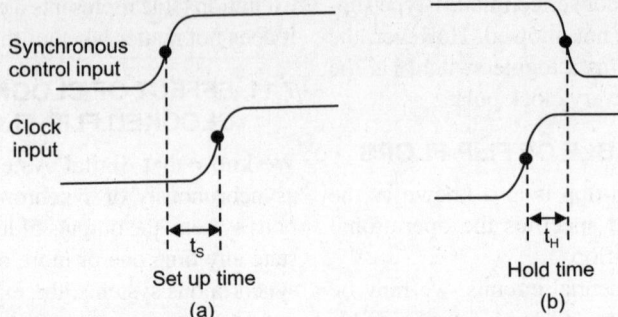

Fig. 7.18: Control inputs have to be held stable for (a) time t_s prior to active clock transition and for (b) a time t_H, after the active clock transition for PGT

7.11.1 Set up and Hold Times

Two timing requirements must be met if a clocked FF is to respond reliably to its control inputs when the active CLK transition occurs. These are illustrated in Fig. 7.18 for a FF that triggers on a PGT.

Set up time: It is the time required for the input data to settle in before the triggering edge of the clock. In other words it is the time interval immediately preceding the active transition of the CLK signal during which the synchronous input has to be maintained at the proper level. IC manufacturers usually specify the minimum allowable set up time. If this time requirement is not met, the FF may not respond reliably when the clock edge occurs.

The *hold time* is the time interval immediately following the active transition of the CLK signal during which the synchronous input has to be maintained at the proper level. IC manufacturers usually specify the minimum acceptable value of hold time. If this requirement is not met, the FF will not trigger reliably.

IC flip-flops have minimum allowable t_S and t_H values in the nanosecond range. Set up times are usually in the range 5 to 50 ns while hold times are generally from 0 to 10 ns. Note that these times are measured between the 50 percent point on the transitions.

7.12 WAVE FORMS IN CLOCKED S-R FLIP-FLOP

Figure 7.19 shows the logic symbol for a clocked S-R flip-flop that is triggered by the positive going edge of the clock signal. The FF can change state only

when a clock signal makes a transition from 0 to 1. The up arrow (–) in the truth table indicates that a PGT is required at CLK and Q_0 indicates the level at Q prior to the PGT. The waveforms in Fig. 7.19c illustrate the operation of the clocked S-R FF. If the set up and hold time requirements are being met in all cases, we can analyze the waveforms as follows:

1. Initially all inputs are 0 and the output Q is assumed to b 0.
2. When the PGT of the first clock pulse occurs (point a), the S and R inputs are both 0, so the FF is not affected and remains in the Q = 0 state (i.e. $Q = Q_0$)
3. At the occurrence of the PGT of the second clock pulse (point c), the S input is now high, with R still low. Thus the FF sets to the 1 state at the rising edge of this clock pulse.
4. When the third clock pulse makes its positive transition (point e), it finds that S = 0 and R = 1, which causes the FF to clear to the 0 state.
5. The fourth pulse sets the FF once again to the Q = 1 state (point g) because S = 1 and R = 0 when the positive edge occurs.
6. The fifth pulse also finds that S = 1 and R = 0 when it makes its positive going transition. However Q is already high, so it remains in that state.
7. The S = R = 1 condition should be avoided, because it results in an ambiguous condition.

7.13 WAVE FORMS IN CLOCKED J-K FLIP-FLOP

Figure 7.20 shows a clocked J-K, that is triggered by the positive going edge of the clock signal. The J and K

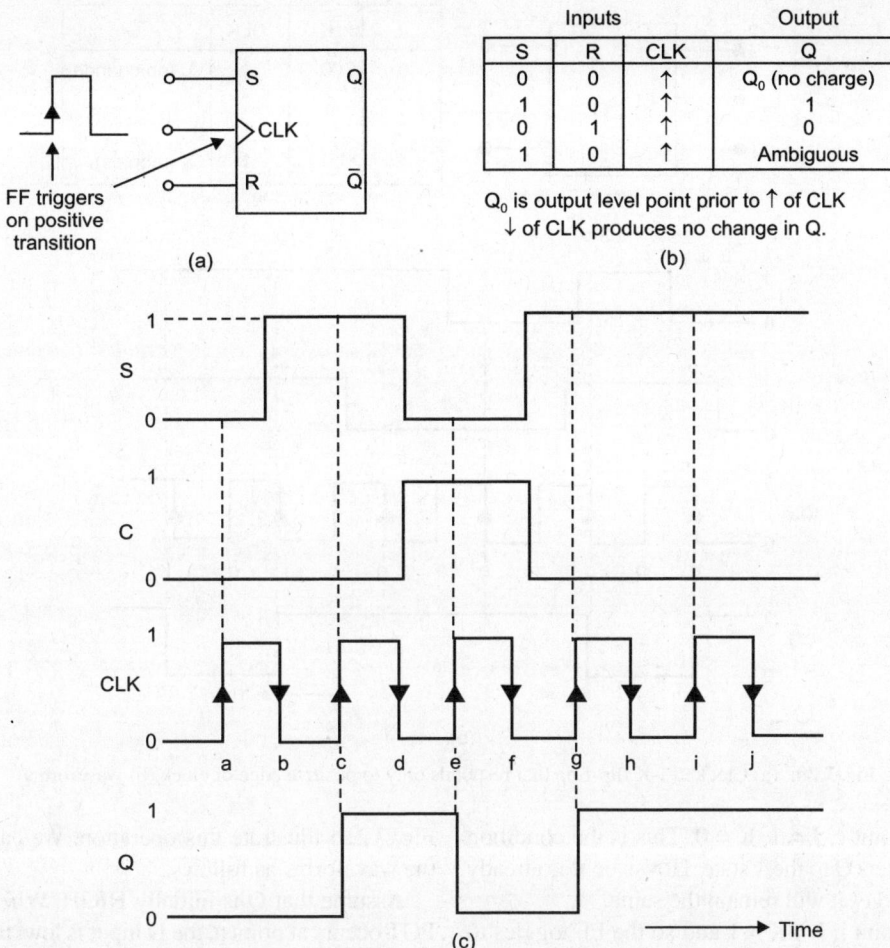

Inputs			Output
S	R	CLK	Q
0	0	↑	Q_0 (no charge)
1	0	↑	1
0	1	↑	0
1	0	↑	Ambiguous

Q_0 is output level point prior to ↑ of CLK
↓ of CLK produces no change in Q.

(b)

(c)

Fig. 7. 19: (a) Clocked S-R FF, that responds only to the positive going edge of a clock pulse. (b) truth table (c) typical waveforms

inputs control the state of the FF in the same way as S and R inputs in clocked S-R FF, except for one major difference; the J = K = 1 condition does not result in an ambiguous output. For this condition, the FF will always go to its opposite state upto the positive transition of the clolck signal. This is called the *toggle* mode of operation. In this mode, if both J and K are left HIGH, the will change state (toggle) for each positive going transition (PGT) of the clock.

The operation of this FF is illustrated by the waveforms in Fig. 7.20b. We assume that the set up and hold time requirements are satisfied.

We can analyze the waveforms as follows:

1. Initially all inputs are 0 and the Q output is assumed to be 0, i.e. Q = 0.
2. When the positive-going edge of the first clock pulse occurs (point a) the J = 0, K = 1 condition exists. Thus the FF will be cleared to the Q = 0 state.
3. The second clock pulse finds J = K = 1 when it makes its positive transition (point c). This causes the FF to toggle to its opposite state Q = 1.
4. At point e on the clock waveform, J and K are both 0, so that the FF does not change states on this transition.

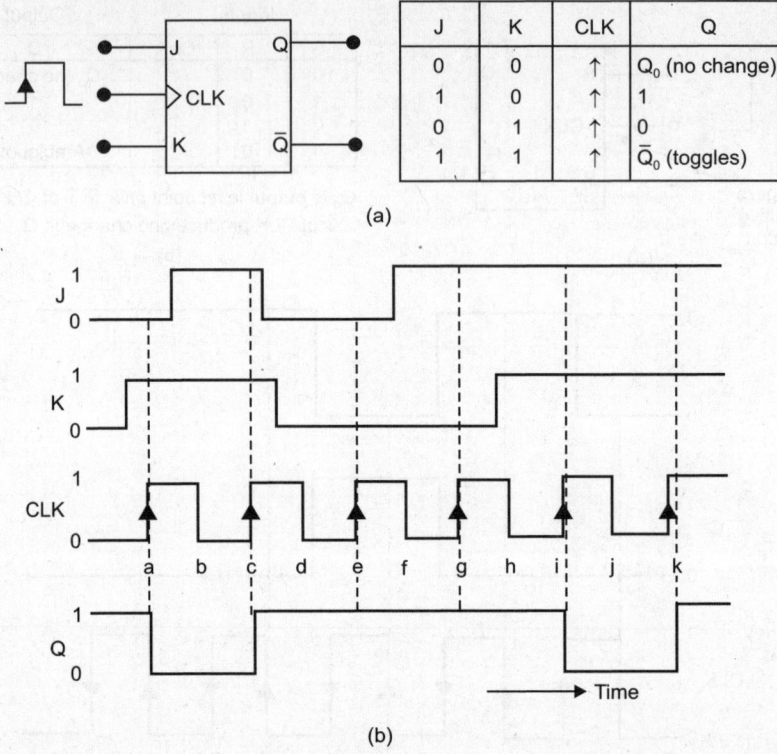

Fig. 7.20: (a) Clocked J-K flip-flop that responds only to positive edge of clock (b) waveforms

5. At point g, J = 1, K = 0. This is the condition that sets Q to the 1 state. However, it is already 1, and e_0 it will remain the same.

6. At point i, J = K = 1 and so the FF toggles to its opposite state. The same thing occurs at point K.

It should be noted here that the FF is not affected by the negative going edge of the clock pulses. Also, the J and K inputs by themselves cannot cause the FF to change states.

7.14 WAVE FORMS IN CLOCKED D-FLIP-FLOP

Figure 7.21a shows the symbol and truth table for an edge triggered D FF, that triggers off a P.G.T. Unlike the SR FF this FF has only one synchronous control input D (which stands for *data*). The operation of the D flip-flop is very simple; Q will go to the same state that is present on the D input. When a PGT occurs at CK, i.e. the level present at D will be stored in the FF at the instant the PGT occurs. The waveforms in

Fig. 7.21b illustrate this operation. We can analyze the waveforms as follows.

Assume that Q is initially HIGH. When the first PGT occurs at point a, the D input is low; thus Q will go to the 0 state. Even though the D input level changes between points a and b, it has no effect on Q. Q is storing the LOW that was on D at point a. When the PGT at b occurs, Q goes HIGH since D is HIGH at that time. Q stores this HIGH until the PGT at point c causes Q to go LOW, since D is low at that time. In a similar manner, the Q output takes on the levels present at D when the PGTs occur at points d, f and g. Note that Q stays HIGH at point e because D is still HIGH.

It is important to note that Q can change only when a PGT occurs. The D input has no effect between PGTs.

7.15 GENERAL MODEL FOR CONVERSION FROM ONE TYPE OF FF TO ANOTHER TYPE

In previous sections we have discussed conversion from SR to JK, SR (or JK) to D-type, and JK to

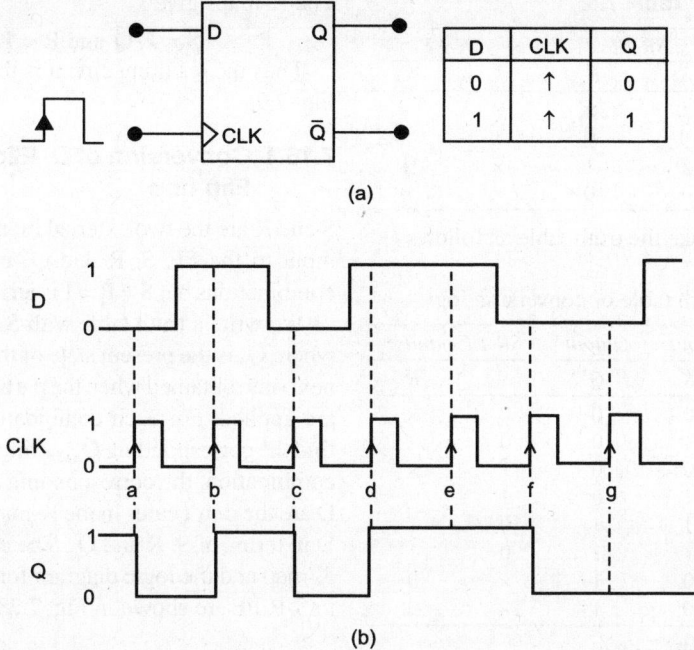

(a)

(b)

Fig. 7.21: (a) Flip-flop that triggers only on positive going transitions (PGTs) (b) waveforms

T-type FFs. Now, we shall effect the conversion from one type of FF to another type by giving a formal technique which is useful in the design of clocked sequential circuits.

Consider Fig. 7.22 in this, we are required to design the "conversion logic" for converting new input definitions into input codes which will cause the given FF to perform as desired.

To design the conversion logic, we need to combine the excitation tables for both flip-flops and make a truth table with data input (s) and Q as the inputs and the inputs of the given FF as the output (s). The combinational logic design then follows as usual.

The above method will become clear by considering following example-

Example 7.1: Convert an SR-flip-flop to a JK- flip-flop.

Solution: The excitation tables (as given in Table 7.8) for SR and JK FLIPS are as follows

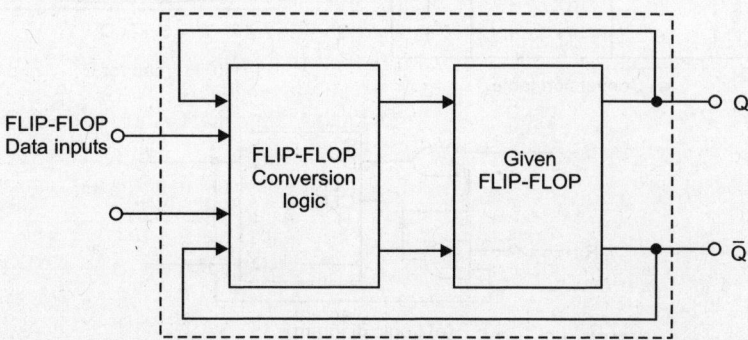

Fig. 7.22: The general model used to convert one type of FF to another

Table 7.8

Present state	Next state	SR-FF		JK-FF	
		S_n	R_n	J_n	K_n
0	0	0	×	0	×
0	1	1	0	1	×
1	0	0	1	×	1
1	1	×	0	×	0

From this, we make the truth table as follows

Table 7.9: Truth table of conversion logic

Row	FF data input		Output	SR-FF inputs	
	J	K	Q	S	R
1	0	0	0	0	×
2	0	1	0	0	×
3	1	0	0	1	×
4	1	1	0	1	×
5	0	1	1	0	1
6	1	1	1	0	1
7	0	0	1	×	0
8	1	0	1	×	0

The K-map are as follows

JK\ Q	00	01	11	10
0	0	0	1	1
1	X	0	0	X

For S

JK\ Q	00	01	11	10
0	X	X	0	0
1	0	1	1	0

For R

The K-maps give

$$S = J \cdot \overline{Q} \text{ and } R = K \cdot Q$$

Thus the resulting circuit is the same as shown in Fig. 7.9.

7.16.1 Conversion of D- Flip-flop to S-R Flip-flop

S and R are the two external inputs and D, the actual input to the FF. S. R and Q_n make eight possible combinations but S = R = 1 is an invalid combination.

We write a truth table with S, R, Q_n, Q_{n+1} and D, where Q_n is the present state of the FF and Q_{n+1} is the next state obtained when the particular S and R inputs are applied. For each combination of S, R and Q_n, find the corresponding Q_{n+1}. Since S = R = 1 is invalid combination, the corresponding entries for Q_{n+1} and D are the don't cares in the K-map. Then, we express D in terms of S, R and Q_n. The conversion table, the K-map and the logic diagram for conversion from D to S-R FF are shown in Fig. 7.23.

7.16.2 Conversion of D- Flip-flop to J-K Flip-flop

J and K are the external inputs and D is the actual input to the flip-flop. J, K and Q_n make eight combinations (Fig. 7.24a). Express D in terms of J, K and

S - R Inputs		Outputs		D
S	R	Q_n	Q_{n+1}	inputs
0	0	0	0	0
0	0	1	1	1
0	1	0	0	0
0	1	1	0	0
1	0	0	1	1
1	0	1	1	1

(a) Conversion table

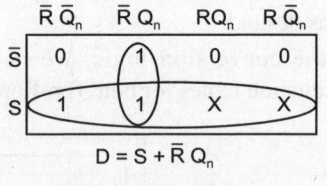

$$D = S + \overline{R} Q_n$$

(b) k - map for D

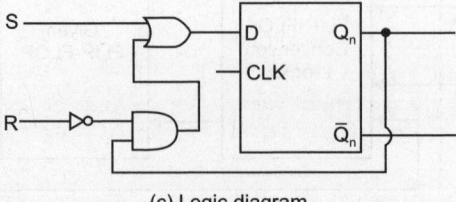

(c) Logic diagram

Fig. 7.23: Conversion of a D flip-flop to S-R FF

Q_n. The K-map for D in terms of J, K and Q_n and the logic diagram showing the conversion from D to J-K FF are shown respectively in Figs. 7.24b and c.

J - K Input		Outputs		D Input
J	K	Q_n	Q_{n+1}	
0	0	0	0	0
0	0	1	1	1
0	1	0	0	0
0	1	1	0	0
1	0	0	1	1
1	0	1	1	1
1	1	0	1	1
1	1	1	0	0

(a) Conversion table

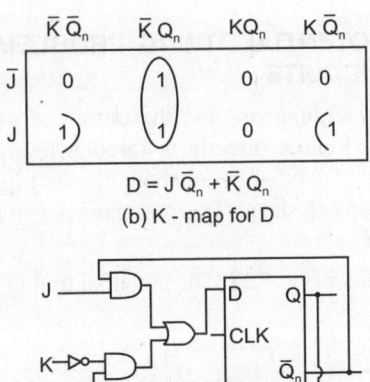

$$D = J \bar{Q}_n + \bar{K} Q_n$$

(b) K - map for D

(c) Logic diagram

Fig. 7.24: Conversion of a D fllip-flop to J-K FF

7.17 FLIP-FLOP TIMING CONSIDERATIONS

Manufactures of IC FFs specify several important timing parameters that must be considered before a FF is used in any circuit application. Most important of these are described below:

Set up and Hold Times: These have already been described earlier and these represent requirements that must be met for reliable FF triggering. The manufactures IC data sheet specify the minimum values of t_S and t_H.

Propagation Delays: Whenever a signal is to change the state of a FF's output, there is a delay from the time the signal is applied to the time when the output makes its change. Figure 7.25 illustrate the propa-

gation delays that occur in response to a positive transition on the CLK input. Note that these delays are measured between the 50% points on the input and output waveforms. The same types of delays occur in response to signals on a flip-flop's asynchronous inputs (preset and clear). The manufactures data sheet usually specify the maximum values for propagation delays.

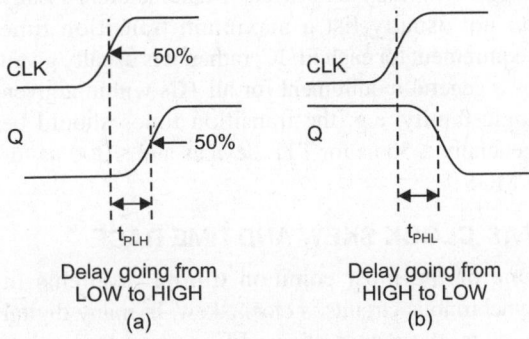

Fig. 7.25: Flip-flop propagation delays

Maximum Clocking Frequency, f_{Max}: This is the highest frequency that may be applied to the CK input of a FF and still have it trigger reliably.

Clock Pulse HIGH and LOW Times: The manufacturer also specify the *minimum* time duration that the CK signal must remain LOW before it goes HIGH sometimes called t_W (L) and the *minimum* time that CK must be kept HIGH before it returns LOW called t_W (H). Fig. 7.26a. Failure to meet these can result in unreliable triggering. These time values are measured between the halfway points on the signal transitions.

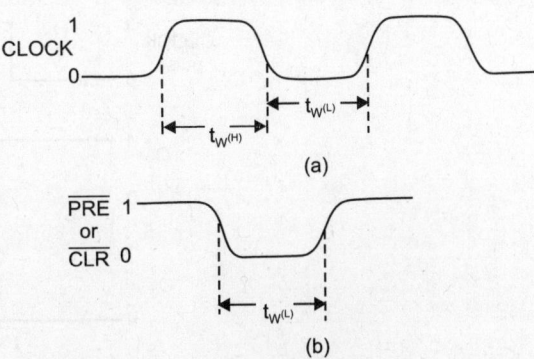

Fig. 7.26: (a) Clock LOW and HIGH times (b) asynchronous pulse width

Asynchronous Active Pulse Width: The manufacturer also specifies the minimum time duration that a preset or clear input has to be kept in its active state in order to reliably set or clear the FF. Figure 7.26b shows t_W (L) for active-LOW asynchronous inputs.

Clock Transitions Times: For reliable triggering, the clock waveform transition times (rise and fall times) should be kept very short (otherwise, the FF may trigger erratically or not at all). Manufacturers usually do not usually list a maximum transition time requirement for each FF IC, rather, it is usually given as a general requirement for all 1Cs within a given logic family, e.g. the transition times should be generally ≤ 50 ns for TTL devices and ≤ 200 ns for CMOS.

7.18 CLOCK SKEW AND TIME RACE

One of the most common timing problems in synchronous circuits is clock skew. In many digital circuits, the output of one FF is connected either directly or through logic gates to the input of another FF and both flip-flops are triggered by the same clock signal. The propagation delay of a FF and/or the delays of the connecting gates make it difficult to predict precisely when the changing state of one FF will be experienced at the input of another.

The clock signal which is applied simultaneously to all FFs in a synchronous system, may undergo varying degrees of delay caused by wiring between components and arrive at the CLK inputs of different FFs at different times. This is referred to as *clock skew*. If the clock skew is minimal a FF may get clocked before it receives a new input (from the output of another clocked FF). On the other hand, if the clock pulse is delayed significantly, the inputs to a FF may have changed before clock pulse arrives. In these situations, we have a kind of race between the two competing signals that are attempting to accomplish opposite effects. This is known as *time race*. It is clear that reliable system operation is not possible when the responses of a FF depend on the outcome of a race.

7.19 POTENTIAL TIMING PROBLEM IN FF CIRCUITS

In many digital circuits, the output of one FF is connected either directly or through logic gates to the input of another FF and both FFs are triggered by the same clock signal. This presents a potential timing problem.

Consider Fig. 7.27. The potential timing problem is this:

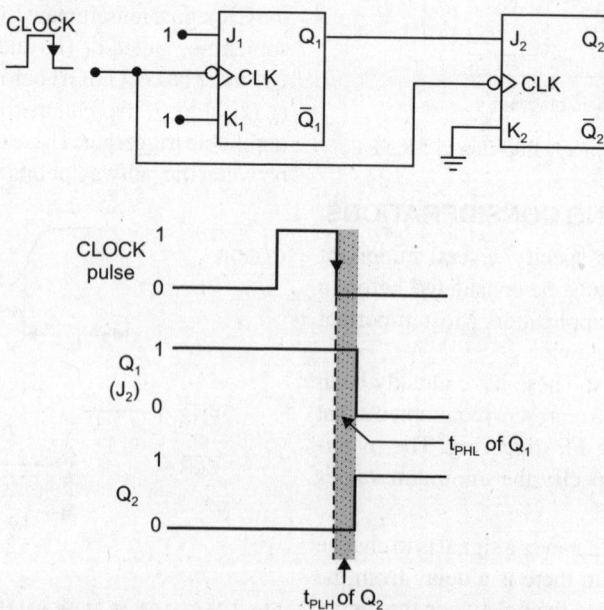

Fig. 7.27: Pertaining to timing problem in FF circuits

Since Q_1 will change or the NGT of the clock pulse, the J_2 input of Q_2 will be changing as it receives the same NGT. This could lead to an unpredictable response at Q_2.

Let us assume that initially $Q_1 = 1$ and $Q_2 = 0$. Thus the Q_1 FF has $J_1 = K_1 = 1$ and Q_2 FF has $J_2 = Q_1 = 1 = K_2 = 0$ prior to the NGT of the clock pulse.

When the NGT occurs, Q_1 toggles to the LOW state but it will not actually go LOW until after its propagation delay, t_{PHL}. The same NGT will reliably clock Q_2 to the HIGH state provided t_{PHL} is greater than Q_2's hold time requirement t_H. If this condition is not met, the response of Q_2 will be unpredictable. Fortunately, all modern edge-triggered FFs satisfy this condition. Thus, for all FFs it may be assumed that FF's hold time requirement is short enough to respond reliably according to the following rule:

The FF output will go to a state determined by the logic levels present at its synchronous control inputs just prior to the active clock transition.

Thus, for Fig. 7.27 this rule says that Q_2 will go to a state determined by the $J_2 = 1$, $K_2 = 0$ condition that is present just prior to the NGT of the clock pulse. The fact that J_2 is changing in response to the same NGT has no effect. Here Q_2's hold time requirement t_H is less than Q_1's propagation delay.

7.20 APPLICATIONS OF FLIP-FLOPS

There are a large number of applications of FFs. Some important applications are

1. Parallel data storage
2. Serial data storage
3. Transfer of data
4. Frequency division
5. Counting
6. Parallel to serial conversion
7. Serial to parallel conversion
8. Synchronizing the effect of asynchronous data
9. Detection of an input sequence

These are discussed in detail as follows.

Parallel data storage

A group of flip-flops is called a register. To store a data of N bits, N flip-flops are required. Since the data is available in parallel form, i.e. all bits are present at a time, these bits may be made available at the D input terminals of the FFs and when a clock pulse is applied to all the FFs simultaneously, these bits will be transferred to the Q outputs of the flip-flops and the FFs (register) then stores the data.

Serial data storage

To store a data of N bits available in serial form. N number of D FFs are connected in cascade. The clock signal is connected to all the FFs The serial data is applied to the D input terminal of the first FF. Each clock pulse transfers the D input to its Q output. Thus, after N clock pulses, the register (group of FFs) contains the data and then stores it.

Transfer of data

Data stored in flip-flops may be transferred out in a serial manner, i.e. bit-by-bit from the output of one FF or may be transferred out in parallel form i.e. all bits at a time from the Q outputs of each of the flip-flops.

Serial to parallel conversion

To convert the data available in serial form into parallel form, the serial data are first entered and stored in a serial-in parallel-out shift register (a group of FFs connected in cascade) and then, since the data are available simultaneously at the outputs of the FFs, the data may be taken out in parallel form. To convert an N-bit serial data to parallel form, N FFs are required. N clock pulses are required to enter the data in serial form and one clock pulse is required to shift the data out in parallel form.

Parallel to serial conversion

To convert the data available in parallel form into serial form, the parallel data are first entered in the parallel-in-serial-out shift register in parallel form i.e. all bits simultaneously and then that data is shifted out of the register serially, i.e. bit by bit by the application of clock pulses. To convert an N-bit parallel data to serial form, N flip-flops are required. One clock pulse is required to shift the parallel data into the register and N clock pulses are required to shift the data out of the register serially.

Counting

A number of FFs may be connected in a particular fashion to count the pulses electronically. One FF can count upto two pulses, two FFs, upto $2^2 = 4$ pulses. In general N flip-flops can count upto 2^N pulses. In a simple counter, all the FFs are connected in a toggle mode. The clock pulses are applied to the first FF and the clock terminal of each subsequent FF is connected to the Q output of the previous FF, Feedback may be provided if the maximum count required is not 2^N.

Frequency division

Flip-flops may be used to divide the input signal frequency by 2. Two FFs may be used to divide the input frequency by 4. In general N FFs may be used to divide the input frequency by 2^N. If N flip-flops are connected as a ripple counter (i.e. a counter in which the external signal is applied to the clock signal of the first FF and Q output of each FF is connected to the clock input of the next FF) and if the input signal of frequency f is fed to the first FF, the output of this FF wil be of frequency f/2, the output of the 2nd FF will be of frequency f/4 and so on and the output of Nth flip-flop will be of frequency $f/2^N$.

7.21 RANDOM ACCESS MEMORY

In computers, digital control systems etc, it is necessary to *store* digital data and *retrieve* the data as desired. For this purpose, earlier, magnetic memory devices were possible, however, nowadays, it has become possible to make memory devices using semiconductors, semiconductor memories have become very popular because of their small size (available in ICs) and convenience to use. Chapter 11 deals with various memories.

Flip-flops can be used for making memories in which data can be stored for any desired length of time and then read out whenever required. In such a memory, data can be put into (*writing* into the memory) or retrieved from (*reading* from the memory) the memory in a random faction and is known as random-access memory (RAM).

A 1-bit read/write memory is shown in Fig. 7.28 which is the basic memory element for circuit using ICs.

In this memory cell, a level D FF is used which has Q output that follows the D input as long as CK terminal is at logic 1. The moment the CK input changes to logic 0, the Q output does not change and it retains the D input level that existed just before the transition from 1 to 0 at input CK. This input is used to select the memory cell. In the 1-bit cell shown, there are 3 inputs-D_i (data input). A_n (address select) and R/$\overline{W}$ (read/write control) and one output D_0 (data output).

$A_n = 1$ enables the cell for reading or writing operation, R/$\overline{W}$ at logic 1 is for reading from the cell and logic 0 for writing into the cell. As long as $A_n = 0$, all input and output activities are blocked, and the cell is in the hold mode where its stored output is protected. The complete function of this cell can be understood from the function-table (Table 7.10). The

Table 7.10: Function table of 1-bit R/$\overline{W}$ memory cell

An	R/$\overline{W}$	D_i	Mode
	Inputs		
0	×	×	Hold, $D_0 = 0$
1	0	0	write 0 into memory $D_0 = 0$
1	0	1	write 1 into memory $D_0 = 0$
1	1	×	Read, D_0 = stored D_i bit.

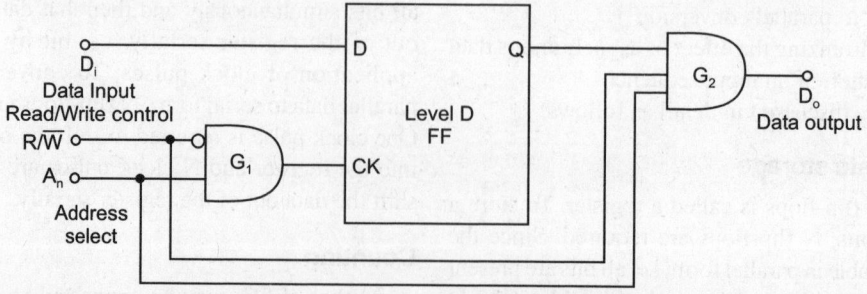

Fig. 7.28: A 1-bit read/write memory cell

read operation is non destructive, i.e. the stored bit can be read out any number of times without disturbing it. The stored bit will be protected as long as power is on. Therefore, this type of memory is known as *volatile memory*. For writing into the cell, it is not required to be cleared before entering the new bit.

Whenever the new bit is entered, the earlier one gets destroyed automatically.

7.22 EDGE-TRIGGERED FLIP-FLOPS

We know that FFs are synchronous bistable devices. The term synchronous means that the output changes state only at a specified point on a triggering input (clock, designated as a control input C), i.e. changes in the output occur in synchronization with the clock.

The term edge-triggered means that the FF changes state either at the positive edge (leading edge) or at the negative edge (trailing edge) of the clock pulse and is sensitive to its inputs only at this transition of the clock. Figure 7.29 shows three basic types of edge-triggered FFs and their truth tables.

Notice that each type can be either positive edge triggered (no bubble at C input) or negative edge triggered (bubble at C input). The key to identifying an edge-triggered FF by its logic symbol is the small 'cap' at the clock (C) input.

7.23 PULSE-TRIGGERED (MASTER-SLAVE) FLIP-FLOPS

The term pulse-triggered means that data are entered into the flip-flop on the leading edge of the clock pulse, but the output does not reflect the input state until the trailing edge of the clock pulse. The inputs must be set up prior to the clock pulse's leading edge, but the output is postponed until the trailing edge of the clock.

A major restriction of the pulse triggered FF is that the data inputs must not change while the clock pulse is HIGH because the FF is sensitive to any change of input levels during this time.

As with edge-triggered FFs, there are three basic types of pulse-triggered FFs: S-R, D and J-K. These are shown in Fig. 7.30 along with their truth tables.

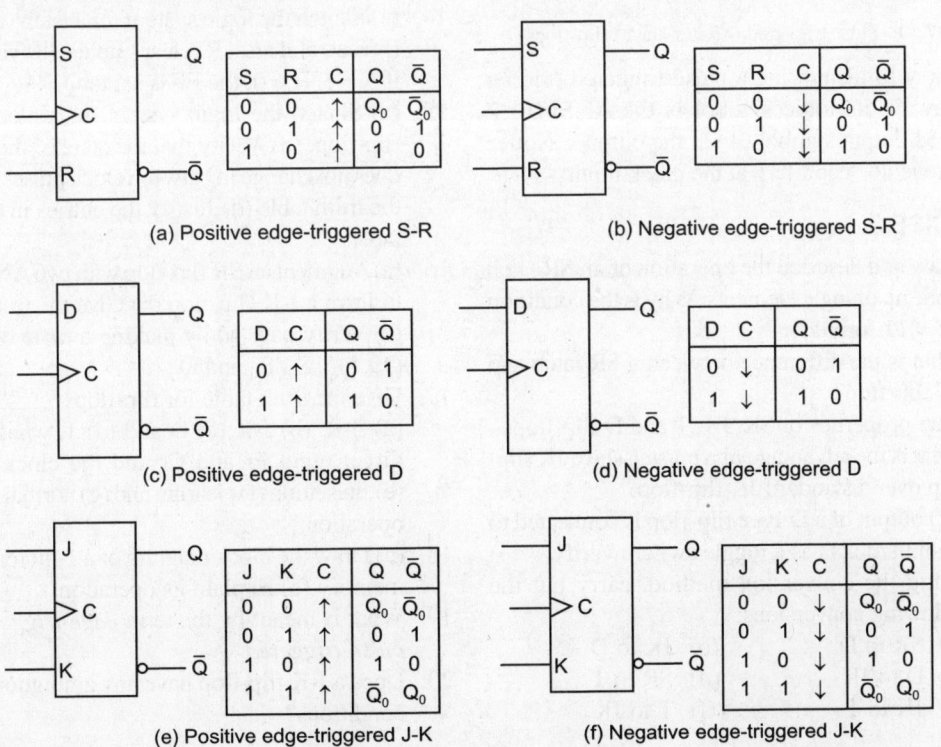

(a) Positive edge-triggered S-R

S	R	C	Q	$\bar{Q}$
0	0	↑	Q_0	$\bar{Q}_0$
0	1	↑	0	1
1	0	↑	1	0

(b) Negative edge-triggered S-R

S	R	C	Q	$\bar{Q}$
0	0	↓	Q_0	$\bar{Q}_0$
0	1	↓	0	1
1	0	↓	1	0

(c) Positive edge-triggered D

D	C	Q	$\bar{Q}$
0	↑	0	1
1	↑	1	0

(d) Negative edge-triggered D

D	C	Q	$\bar{Q}$
0	↓	0	1
1	↓	1	0

(e) Positive edge-triggered J-K

J	K	C	Q	$\bar{Q}$
0	0	↑	Q_0	$\bar{Q}_0$
0	1	↑	0	1
1	0	↑	1	0
1	1	↑	$\bar{Q}_0$	Q_0

(f) Negative edge-triggered J-K

J	K	C	Q	$\bar{Q}$
0	0	↓	Q_0	$\bar{Q}_0$
0	1	↓	0	1
1	0	↓	1	0
1	1	↓	$\bar{Q}_0$	Q_0

Fig 7.29: Edge-triggered flip-flops

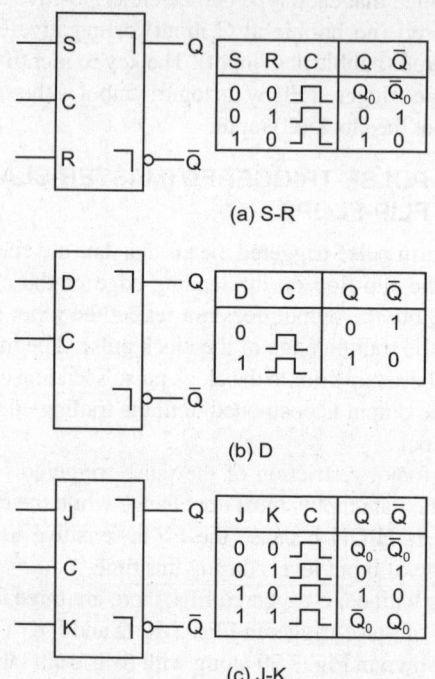

S	R	C	Q	Q̄
0	0	⎍	Q_0	$\bar{Q}_0$
0	1	⎍	0	1
1	0	⎍	1	0

(a) S-R

D	C	Q	Q̄
0	⎍	0	1
1	⎍	1	0

(b) D

J	K	C	Q	Q̄
0	0	⎍	Q_0	$\bar{Q}_0$
0	1	⎍	0	1
1	0	⎍	1	0
1	1	⎍	$\bar{Q}_0$	Q_0

(c) J-K

Fig. 7.30: Pulse triggered (master-slave) flip-flops

The key to identifying a pulse-triggered (master slave) by FF its logic symbol is the ANSI/IEEE postponed output symbol (⎤) at the outputs. Notice that there is no "cap" (▷) at the clock input.

EXERCISE

1. Draw and describe the operation of an SR latch made up of logic elements. Why is the condition SR = 11 forbidden?
2. What is the difference between a SR latch and SR flip-flop?
3. State properties of SR, JK, T and D flip-flops.
4. What is the advantage of a master-slave JK flip-flop over a standard JK flip-flop?
5. If Q output of a D-type flip-flop is connected to D input, it acts as a toggle switch. Verify.
6. Using the conversion method, carry out the following conversions:
   (a) SR to D   (b) JK to D
   (c) D to JK   (d) SR to T
   (e) JK to T   (f) T to JK

7. How does a J-K FF differ from an S-R FF in its basic operation? What is its advantage over an S-R FF.
8. What is a master-slave flip-flop? Discuss its working.
9. What do you mean by (a) clock skew and (b) time race?
10. Define the following terms as applied to flip-flops:
    (a) Set up time
    (b) Hold time
    (c) Propagation delay
    (d) Maximum clock frequency
    (e) Power dissipation
11. Explain what is meant by a *race-around* condition in connection with a flip-flop.
12. (a) Draw a master-slave J-K flip-flop system.
    (b) Explain its operation and show that the race-around condition is eliminated.
13. (a) Define a latch (b) Show how to construct this unit from NOT gates. (c) has two stable states.
14. (a) Sketch the logic system for a latch with set S (preset) and reset R (clear) inputs. (b) Verify that if S = 1, R = 0, the FF is set to Q = 1.
15. (a) Sketch the logic system for a clocked SR flip-flop. (b) Verify that the state of the system does not change in between clock pulses (c) Give the truth table (d) Justify the entries in the truth table.
16. (a) Augment an SR flip-flop with two AND gates to form a J-K flip-flop (b) Give the truth table. (c) Verify part (b) by making a table of J_n, K_n, Q_n, $\bar{Q}_n$, S_n, R_n and Q_{n+1}.
17. Give the truth table for flip-flops (a) S-R, (b) J-K (c) D and (d) T, what are the direct input Pr and Cr and the clock Ck for (e) presetting (f) clearing and (g) normal clocked operation.
18. (a) Draw the block diagram of a 1-bit read/write memory (b) Explain its operation.
19. What is meant by the term *edge-triggered* and *pulse-triggered*.
20. Does a J-K flip-flop have any ambiguous input conditions?

8

Shift Registers

8.1 INTRODUCTION

A shift register is a very important digital building block. These are often used to momentarily store binary information appearing at the output of an encoding matrix. Similarly these are also often used to momentarily store binary data at the output of a decoder. The registers as a result form an important link between the main digital system and the input-output channels.

A binary register also forms the basis for some very important arithmetic operations like complementation, multiplication and division. A shift register can also be connected to form a number of different types of counters which often offer some very distinct advantages.

8.1.1 What is a Shift Register?

A register is simply a group of flip-flops that can be used to store a binary number. There must be one FF for each bit in the binary number. For example, a register used to store an 8-bit binary number must have 8 flip-flops. Obviously the FFs must be connected such that the binary number can be entered into the register and possibly shifted out. A group of flip-flops connected to provide either or both of these functions is called a *shift register*.

8.2 FUNCTIONS

Shift registers are useful in applications involving the *storage* and *transfer* of data in a digital system.

The basic difference between a register and a counter (discussed in next chapter) is that a register

has no specified sequence of states except in certain very specialized applications. A register in general, is used solely for *storing* and *shifting* data (1s and 0s) entered into it from an external source and possesses no characteristic internal sequence of states.

As already described, a register is a group of FFs connected in a sequence. Figure 8.1 describes how a FF acts as a storage element. It illustrates how a 1 or a 0 is stored in a flip-flop.

A 1 is applied to the input as shown, and a clock pulse is applied that stores the 1 by setting the FF. When the 1 on the input is removed, the FF remains in the SET state, thereby storing the 1. Similarly bit 0 is also stored as illustrated in Fig. 8.1.

Registers are used for *temporary storage* of data in a digital system. The *shift* capability of a register permits the movement of data from one stage to next within a register or into or out of the register upon application of clock pulses. Figure 8.2 shows symbolically the types of data movement in shift register operations. In Fig. 8.2 the block represents any arbitrary 4-bit register and the arrow indicates the direction and type of data movement.

8.3 TYPES OF REGISTERS

The bits in a binary number (data) can be moved from one place to another in either of the following two ways:(1) shifting the data one bit at a time in a serial fashion (beginning with either the MSB or the LSB). This is known as *serial shifting*. (2) shifting all data bits simultaneously and referred to parallel shifting. Consequently, there are two ways to shift data into a

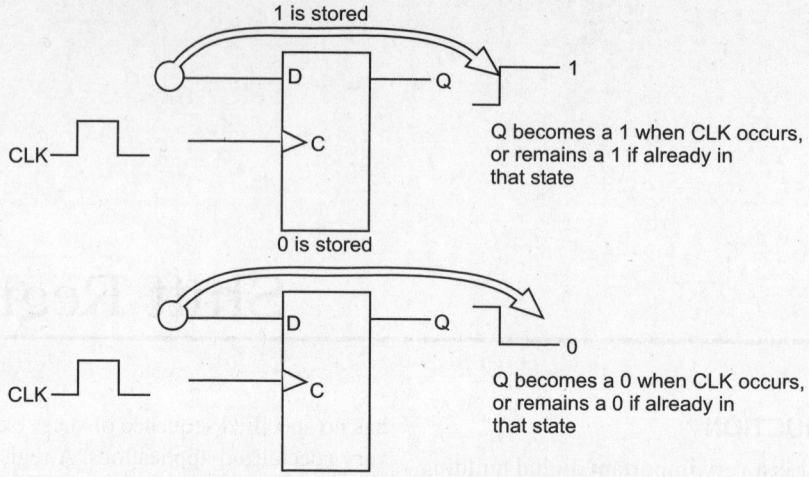

Fig. 8.1: The flip-flop as a storage element

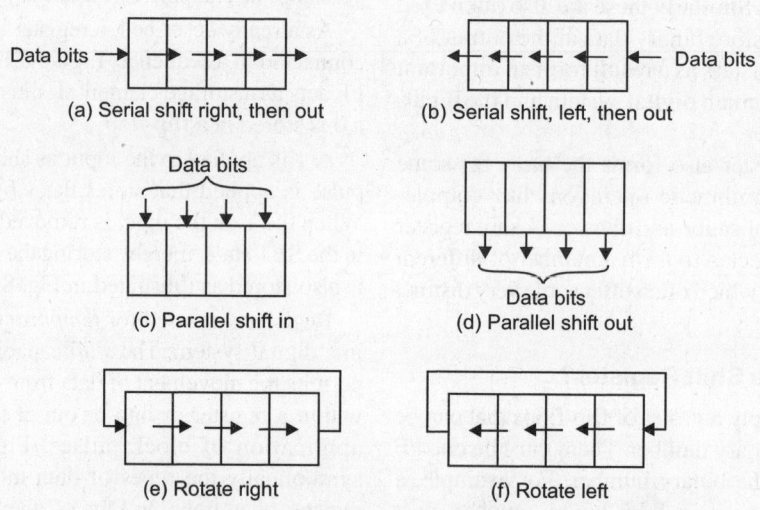

(a) Serial shift right, then out

(b) Serial shift, left, then out

(c) Parallel shift in

(d) Parallel shift out

(e) Rotate right

(f) Rotate left

Fig. 8.2: Basic types of data movement in registers

register (serial or parallel). Similarly there will be two ways to shift data out of the register. This leads to the construction of four basic register types, viz. (i) serial-in, serial-out (ii) serial-in, parallel-out (iii) parallel-in, serial-out and (iv) parallel-in, parallel-out.

8.3.1 Serial-in-Serial Out Shift Register

This type of shift register accepts data serially, i.e. one bit at a time on a single line. It produces the stored information on its output also in serial form. Let us first consider the serial entry of data into a typical

shift register with Fig. 8.3 which shows a four bit register implemented with D filip-flops.

With four stages, this register can store upto four bits of data. Suppose, 4-bit binary number 1010 has to be entered into the register beginning with the LSB 0. The 0 is put onto the data input line making $D = 0$ for FFA. When the first clock pulse is applied, FFA is RESET, thus "storing" the 0. Next the 1 is applied to the data input making $D = 1$ for FFA and $D = 0$ for FFB, because D of FFB is connected to the Q_A output.

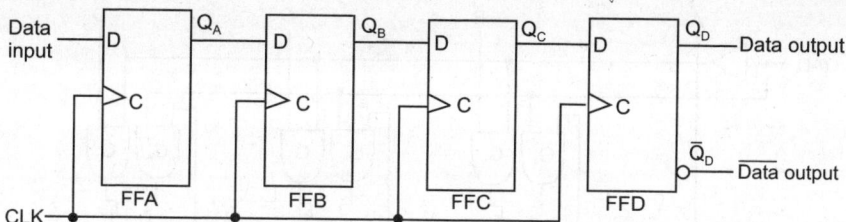

Fig. 8.3: Serial-in, serial-out shift register

When the second clock pulse is applied, the 1 on the data input is "shifted" into FFA because FFA SETS and the 0 that was in FFA is shifted into FFB. The next 0 in the binary number is now put onto the data-input line and a clock pulse is applied. The 0 is entered into FFA, the 1 stored in FFA is shifted into FFB and the 0 stored in FFB is shifted into FFC. The last bit in the binary number, a 1 is now applied to the data input and a clock pulse is applied, and 1 is entered into FFA, the 0 stored in FFA is shifted into FFB, the 1 stored in FFB is shifted into FFC and the 0 stored in FFC is shifted into FFD. This completes the serial entry of the given binary number into the shift register, where it can be stored any length of time.

If we want to get the data out of the register, they must be shifted out serially and taken off the Q_D output.

After CLK_4 in the data entry operation described above, the rightmost 0 in the number appears on the Q_D output. When clock pulse CLK_5 is applied, the second bit appears on the Q_D output. CLK_6 shifts the 3rd bit to the output and CLK_7 shifts the 4th bit to the output. Note that while the original four bits are being shifted out, a new four bit number can be shifted in.

8.3.2 Serial-in, Parallel-Out Shift Register

Data bit are entered into this type of register in the same manner as serial in - serial out shift register.

The difference is in the way, the data bits are taken out of the register. In the parallel output register, the output of each stage is available. Once the data bits are stored, each bit appears on its respective output line and all bits are available simultaneously rather than on a bit-by-bit basis as with the serial output. Figure 8.4 shows a four-bit serial-in parallel-out shift register.

8.3.3 Parallel-in, Serial-Out Shift Register

For a register with parallel data inputs, the bits are entered simultaneously into their respective stages on parallel lines rather than bit-by-bit on one line as with serial data inputs. The serial output is executed as in serial in-serial out shift register.

Figure 8.5 illustrates a four-bit parallel in-serial out register. Note that there are four data input lines A, B, C and D and a SHIFT/$\overline{LOAD}$ input that allows four bits of data to be entered in parallel into the register. When SHIFT/$\overline{LOAD}$ is LOW, gates G_1 through G_3 are enabled, allowing each data bit to be applied to the D input of its respective FF. When a clock pulse is applied, the FF with D = 1 will SET and these with D = 0 will RESET, thereby storing all four bits simultaneously.

When SHIFT/$\overline{LOAD}$ is HIGH, gates G_1 through G_3 are disabled and gates G_4 through G_6 are enabled,

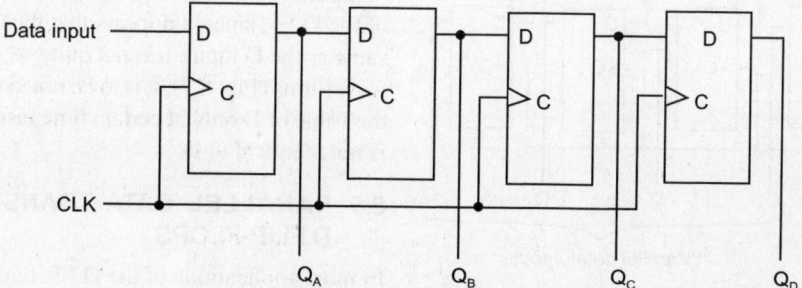

Fig. 8.4: Serial-in, parallel-out shift register

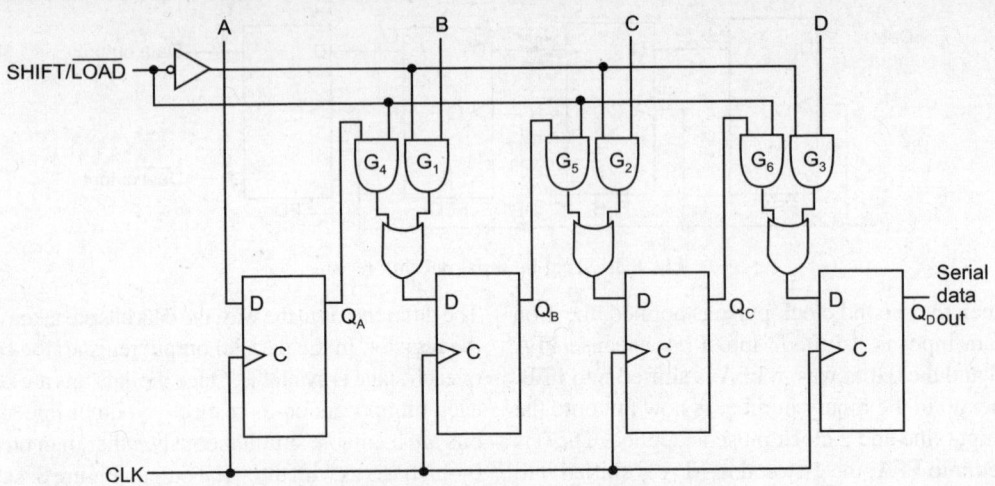

Fig. 8.5: A four-bit parallel-in, serial-out shift register

allowing the data bits to shift right from one stage to the next.

The OR gates allow either the normal shifting operation or the parallel data entry operation, depending on which AND gates are enabled by the level on the $\overline{\text{SHIFT/LOAD}}$ input.

8.3.4 Parallel-in Parallel-Out Shift Register

We described parallel entry of data in section 8.3.3 and parallel output of data in section 8.3.2. The parallel in-parallel out register employs both of these methods, immediately after the simultaneous entry of all data bits, the bits appear on the parallel outputs. This register is shown in Fig. 8.6.

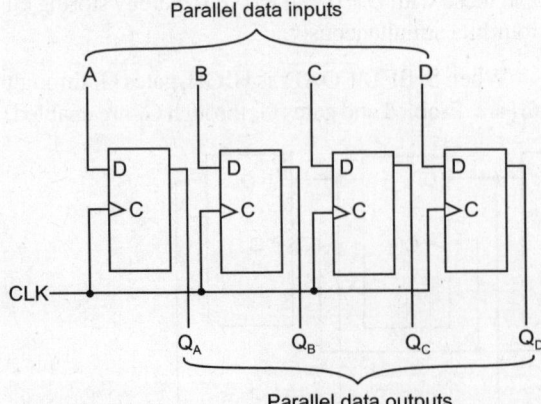

Fig. 8.6: A parallel-in, parallel-out register

8.4 BUFFER REGISTER

Some registers do nothing more than storing a binary word. The buffer register is the simplest of registers. It simply stores a binary word. Most of the buffer registers use D flip-flops. Figure 8.7 shows a 4-bit buffer register. The binary word to be stored is applied to the data terminals D_1, D_2, D_3 and D_4. On the application of clock pulse, the output word becomes the same as the word applied at the input terminals, i.e. the input word is loaded into the register by the application of the clock pulse.

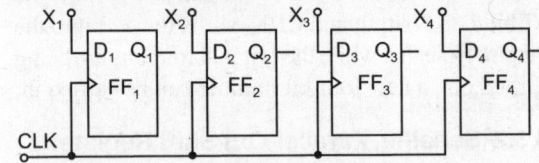

Fig. 8.7: Logic diagram of a 4-bit buffer register

At this point, one may wonder about the usefulness of the D FF, since it appears that the Q output is the same as the D input. It is not quite so, from the D FF waveforms (Fig. 8.7), it is to be noted that Q takes on the value of D only at certain time instances and so it is not identical to D.

8.5 PARALLEL DATA TRANSFER USING D FLIP-FLOPS

In most applications of the D FF, the Q output must take on the value at its D input only at precisely

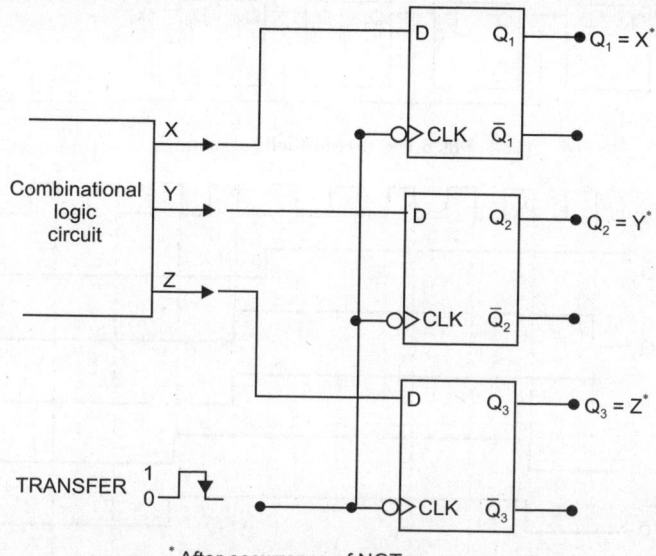

* After occurrence of NGT

Fig. 8.8: Parallel transfer of binary data using D flip-flops

defined times. An example of this is illustrated in Fig. 8.8. Logic circuit outputs X, Y, Z are to be transferred to FFs Q_1, Q_2 and Q_3 for storage. Using the D FFs, the levels present at X, Y and Z will be transferred to Q_1, Q_2 and Q_3 respectively, upon application of a *transfer pulse* to the common CLK inputs. The flip-flops can store these values for subsequent processing. This is an example of *parallel transfer* of binary data; the three bits X, Y and Z are all transferred simultaneously.

8.6 SERIAL SHIFT REGISTER

A (serial) shift register is capable of shifting its binary information either to the right or to the left. The simplest possible shift register is one that uses flip-flops connected in cascade, with the output of one connected to the input of the next, as shown in Fig. 8.9. The Q output of a given FF is connected to the D

input of the FF at its right. Each clock pulse shifts the contents of the register one bit position to the right. The serial input determines what goes into the leftmost FF during the shift. The serial output is taken from the output of the right-most FF prior to the application of a pulse.

Such a unidirectional shift register can function either as shift right or as a shift-left register. The register in Fig. 8.9 shifts its contents with every clock pulse during the NGT (indicated by the small circle associated with the clock input in all FFS). The shift can easily be controlled by means of an external AND gate as shown in the subsequent section.

Example 8.1: Shift Left Register:
Figure 8.1a is a shift-left register. As shown, D_{in} sets up the right FF, Q_0 sets up the second FF, Q_1 the third and so on. When the next positive clock edge strikes, therefore, the stored bits move one position to the left.

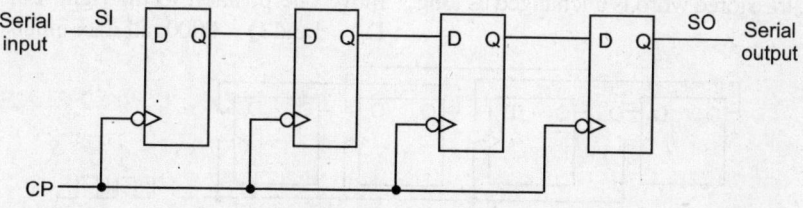

Fig. 8.9: Shift register

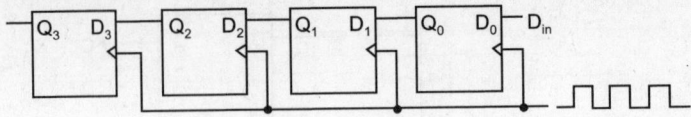

Fig. 8.10: (a) Shift-left register

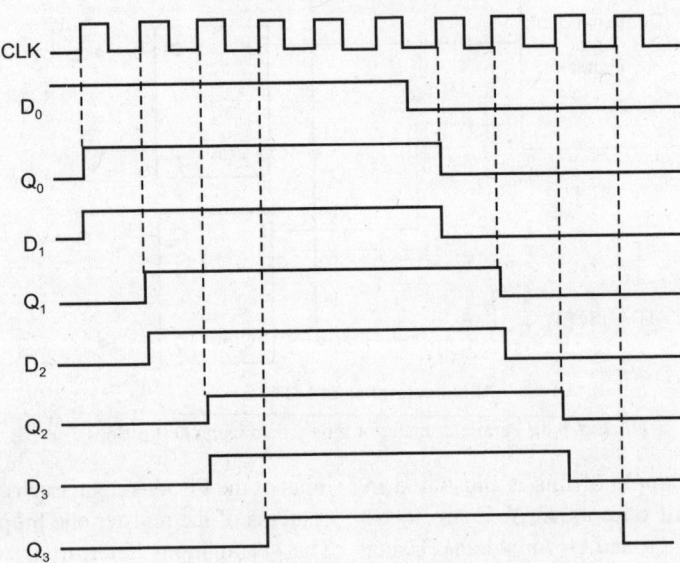

Fig. 8.10: (b) Shift-left timing diagram

As an example, here is what happens with $D_{in} = 1$ and Q = 0000. All data inputs except the one on the right are 0s. The arrival of the first rising clock edge sets the right FF and the stored word becomes

$$Q = 0001$$

This new word means D_1 now equals 1, as well as D_0.

When the next positive clock edge hits, the Q_1 FF sets and the register contents become

$$Q = 0011$$

The 3rd PGT results in

$$Q = 0111$$

and the fourth PGT gives

$$Q = 1111$$

Henceforth, the stored word is unchanged as long as $D_{in} = 1$.

Suppose D_{in} is now changed to 0. Then, successive clock pulse produce these register contents:

$$Q = 1110$$
$$Q = 1100$$
$$Q = 1000$$
$$Q = 0000$$

As long as $D_{in} = 0$, subsequent clock pulse have no further effect. The timing diagram of Fig. 8.10b summarizes the above discussion.

Example 8.2: Shift Right Register:
Figure 8.11 shows a shift right register. As shown, each Q output sets up the D input of the preceding FF. When the rising clock edge arrives the stored bits move one position to the right. For example, with $D_{in} = 1$ and Q = 0000, all data inputs except the one

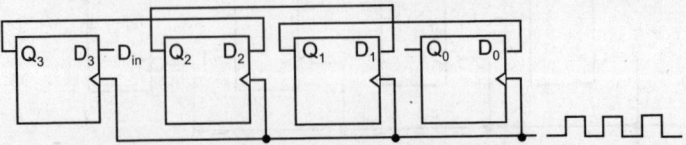

Fig. 8.11: Shift right register

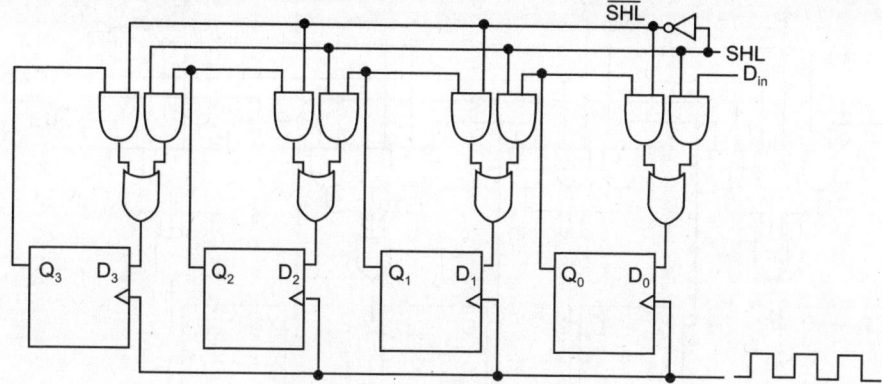

Fig. 8.12: Controlled shift register

on the left are 0s. The first PGT sets the left FF and the stored word becomes

$$Q = 1000$$

With the appearance of this word, D_3 and D_2 are 1s. The second rising clock edge gives

$$Q = 1100$$

The 3rd clock pulse gives

$$Q = 1110$$

and the 4th clock pulse gives $Q = 1111$.

8.7 CONTROLLED SHIFT REGISTER

A *controlled shift register* has control inputs that determine what it does on the next clock pulse.

SHL Control: Figure 8.12 shows how the shift-left operation can be controlled. SHL is the control signal.

When SHL is LOW, the inverted signal $\overline{SHL}$ is HIGH. This forces each FF output to feed back to its data input. Thus the data is retained in each FF as the clock pulses arrive. In this way, a digital word can be stored indefinitely.

When SHL goes HIGH, D_{in} sets up the right FF, Q_0 sets up the second FF, Q_1 the third FF and so on. In this mode, the circuit acts like a shift-left register. Each PGT shift the stored bits one position to the left.

Serial Loading: Serial loading means storing a word in the shift register by entering 1 bit per clock pulse. To store a 4-bit word, we need four clock pulses. for example, here is how to serially store the word

$$X = 1010$$

With SHL HIGH in Fig. 8.12, make $D_{in} = 1$ for the first clock pulse, $D_{in} = 0$ for the second clock

pulse $D_{in} = 1$ for the third clock pulse and $D_{in} = 0$ for the fourth clock pulse. If the register is clear before the 1st clock pulse, the successive register contents will be like this:

$Q = 0001$ ($D_{in} = 1$, 1st CLK pulse)
$Q = 0010$ ($D_{in} = 0$, 2nd CLK pulse)
$Q = 0101$ ($D_{in} = 1$, 3rd CLK pulse)
$Q = 1010$ ($D_{in} = 0$, 4th CLK pulse)

In this way data is entered serially into the right end of the register and shifted left until all four bits have been stored. After the last bit has entered, SHL is taken LOW to freeze the register contents.

Parallel Loading

Figure 8.13 shows a shift register with parallel-loading and the circuit can load X bits directly into the flip-flops, like a buffer register. This requires only one clock pulse to store a digital word.

If LOAD and SHL are LOW, the output of the NOR gate is HIGH and FF outputs return to their data inputs. This forces the data to be retained in each FF as the PGT edges of clock arrive. In other words, the register is inactive when LOAD and SHL are LOW, and the contents are stored indefinitely.

When LOAD is LOW and SHL is HIGH, the circuit acts like a shift-left register (as described previously). On the other hand, when LOAD is HIGH and SHL is LOW, the circuit acts like a buffer register because the X bits set up the FFs for parallel loading. (**Note** that having LOAD and SHL simultaneously HIGH is forbidden, since it is impossible to do both operations on a single clock edge).

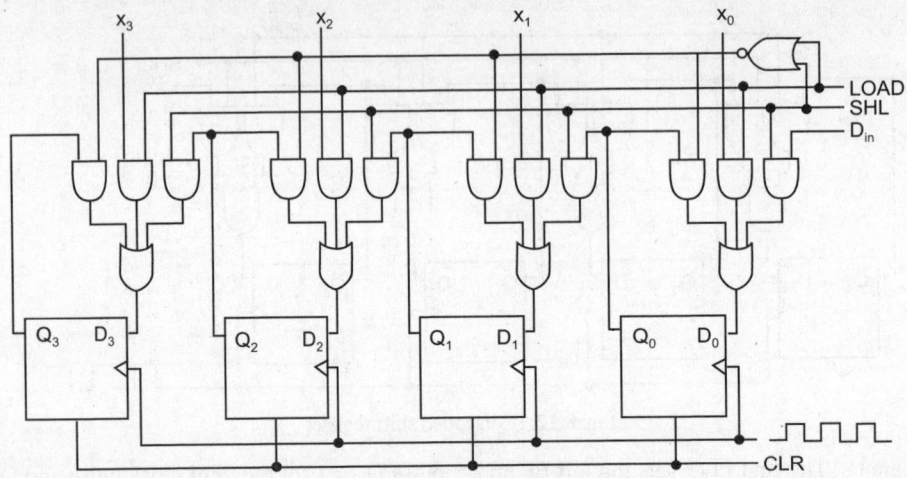

Fig. 8.13: Shift register with parallel loading

8.8 BIDIRECTIONAL SHIFT REGISTER

There are applications in which shifting data to the right or to the left is required. For this purpose, bidirectional shift register shown in Fig. 8.14 can be used.

When $M = 1$, all the A AND gates are enabled and the data at D_R is shifted to the right when clock pulses are applied. When $M = 0$, the A gate are inhibited

and B gates are enabled making the data at D_L to be shifted to the left. M should be changed only when $CK = 0$, otherwise the data stored in the register may be altered.

8.9 SERIAL DATA TRANSFER

A digital system is said to operate in a serial mode when information is transferred and manipulated one

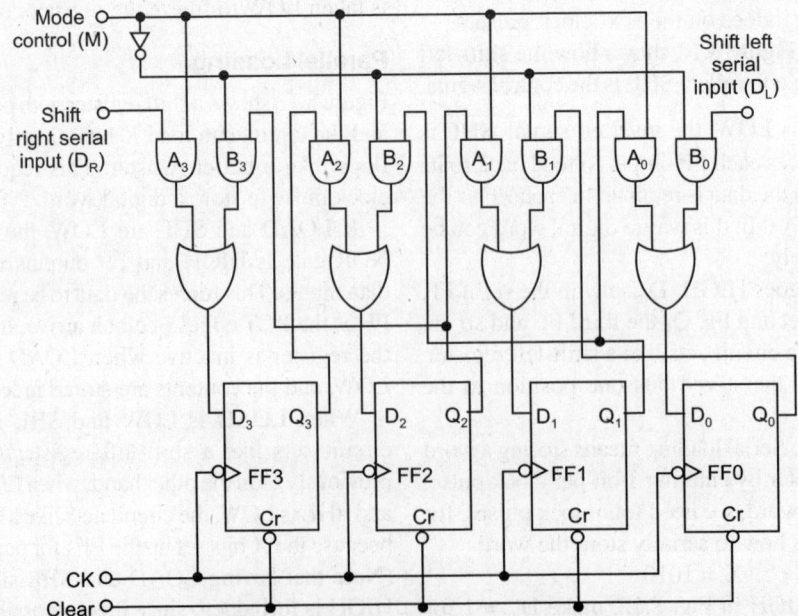

Fig. 8.14: A 4-bit bidirectional shift register

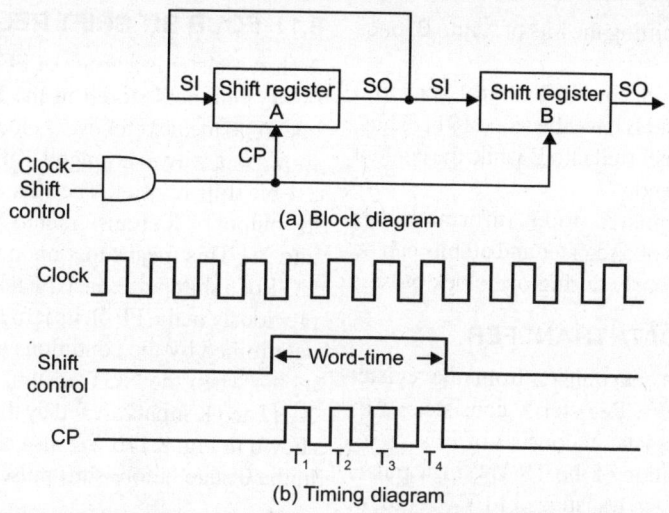

Fig. 8.15: Serial transfer from register A to register B

bit at a time. The content of one register is transferred to another by shifting the bits (one bit at a time) to the other.

The serial transfer of information from register A to register B is done with shift registers as shown in block diagram of Fig. 8.15a. The serial output (SO) of register A goes to the serial input (SI) of register B. To prevent the loss of information stored in the source register A, the A register is made to circulate its information by connecting the serial output to its serial input terminal.

The initial content of register B is shifted out through its serial output and is lost unless transferred to a third register. The shift control input determines when and by how many times the registers are shifted. This is done by the AND gate that allows clock pulses to pass into the CP terminals only when the shift control is 1.

Suppose the shift registers have four bits each. The shift control units must be designed in such a way

that it enables through the shift control signal for a fixed time duration equal to four clock pulses. This is shown in timing diagram of Fig. 8.15b. The shift control signal is synchronized with the clock and changes value just after the NGT of a clock pulse. The next four clock pulses find the shift-control signal in the 1 state and the output of the AND gate connected to the CP terminals produces the four pulses T_1, T_2, T_3 and T_4. The fourth pulse changes the shift control to 0 and the shift registers are disabled.

Now suppose that the binary content of A before the shift is 1011 and that of B 0010. The serial transfer from A to B will occur in four steps as shown in Table 8.1.

After the first pulse T_1, the rightmost bit of A is shifted into the leftmost bit of B and at the same time, this bit is circulated into the leftmost position of A. The other bits of A and B are shifted once to the right. The previous serial output from B is lost and its value changes from 0 to 1. The next three pulses perform

Table 8.1: Serial transfer of 1011 from register A to register B

Timing pulse	Shift register A				Shift register B				Serial output of B
Initial value	1	0	1	1	0	0	1	0	0
After T_1	1	1	0	1	1	0	0	1	1
After T_2	1	1	1	0	1	1	0	0	0
After T_3	0	1	1	1	0	1	1	0	0
After T_4	1	0	1	1	1	0	1	1	1

identical operations shifting the bits of A into B, once at a time.

After the fourth shift, the shift control goes to 0 and both registers A and B have the value 1011. Thus, the content of A is transferred into B while the content of A remains unchanged.

In the parallel transfer node, information is available from all bits of a register and all bits can be transferred simultaneously during one clock pulse.

8.10 PARALLEL DATA TRANSFER

Figure 8.16 illustrates data transfer from one register to another using D FFs. Register X consists of FFs X_1, X_2 and X_3 and register Y consists of FFs Y_1, Y_2 and Y_3. Upon application of the TRANSFER pulse, the level stored in X_1 is transferred to Y_1, X_2 to Y_2 and X_3 to Y_3. The transfer of the contents of the X register to the Y register is a synchronous transfer. It is also referred to as a *parallel transfer*, since the contents of X_1, X_2, X_3 are transferred simultaneously into Y_1, Y_2 and Y_3.

It is to be noted here that parallel transfer does not change the contents of the register that is the source of data. For example, in Fig. 8.5 if $X_1 X_2 X_3 = 101$ and $Y_1 Y_2 Y_3 = 011$ prior to the occurrence of the TRANSFER pulse, then both registers will be holding 101 after the transfer pulse.

8.11 FOUR BIT SHIFT REGISTER

A shift register is a group of FFs arranged so that the binary numbers stored in the FFs are shifted from one FF to the next for every clock pulse. Figure 8.17a shows one way to arrange J-K flip flops to operate as a 4-bit shift register. The FFs are connected so that the output of X_3 transfers into X_2, X_2 into X_1 and X_1 into X_0. This means that upon the occurrence of the NGT of a shift pulse, each FF takes on the value stored previously in the FF on its left. FF X_3 takes on a value determined by the conditions present on its J and K inputs when the NGT occurs. We assume here that X_3's J and K inputs are fed by the DATA IN waveform shown in Fig. 8.17b. We also assume that all FFs are in the 0 state before shift pulse is applied.

The waveforms (Fig. 8.17b) show how the input data are shifted from left to right from FF to FF as shift pulses are applied.

When the first NGT occurs at T_1, each of the FF's X_2, X_1 and X_0 will have the J = 0, K = 1 condition present at its input because of the state of the FF on its left. FF X_3 will have J = 1; K = 0 because of DATA IN. Thus at T_1 only X_3 will go HIGH, while all the other FFs remain LOW. When the second NGT occurs at T_2, FF X_3 will have J = 0, K = 1 because of DATA IN. FF X_2 will have J = 1, K = 0 because of the current

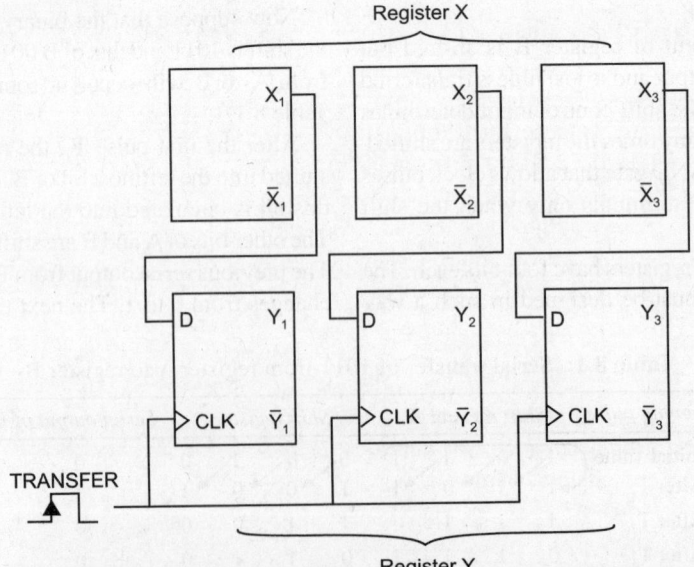

Fig. 8.16: Parallel transfer of contents of register X into register Y

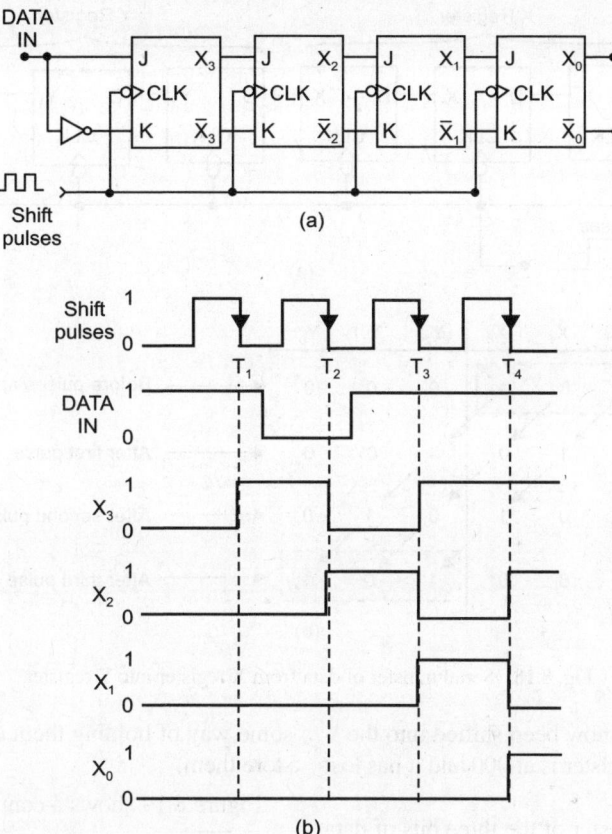

Fig. 8.17: Four bit shift register

HIGH at X_3. FF X_1 and X_0 will still have $J = 0$, $K = 1$. Thus at T_2 only FF X_2 will go HIGH, FF X_3 will go LOW and FFs X_1 and X_0 will remain LOW. Similar reasoning can be used to determine how the waveforms change at T_3 and T_4. It is to be noted that on each NGT of the shift pulse, each FF output takes on the level that was present at the output of the FF on its left prior to the NGT. However X_3 takes on the level that was present at DATA IN prior to the NGT.

8.12 SERIAL TRANSFER FROM ONE REGISTER TO NEXT REGISTER

Though, we have discussed serial transfer earlier, this will become further clear here where we consider two registers X and Y and contents of register X will be serially shifted into register Y. D FFs have been used because this requires fewer connections than JK FFs. Note that X_0, the last FF of register X is

connected to the D input of Y_2, the first FF of register Y. Thus as the shift pulse are applied, the data transfer takes place as follows: $X_2 \rightarrow X_1 \rightarrow X_0 \rightarrow Y_2 \rightarrow Y_1 \rightarrow Y_0$. The X_2 FF will go to a state determined by its D input. For now, D will be held LOW, so that X_2 will go LOW on the first pulse and will remain there.

Consider the case that before any shift pulses are applied, the contents of the X register are 101. (i.e. $X_2 = 1$, $X_1 = 0$, $X_0 = 1$) and the Y register is at 000.

Refer to the Table in Fig. 8.18b which shows how the states of each FF change as shift pulses are applied: The following points are noteworthy:

1. On the NGT of each pulse, FF takes on the value that was stored in the FF on its left prior to the occurrence of the pulse.

2. After three pulses, the 1 that was initially in X_2 is in Y_2, the 0 initially in X_1 is in Y_1 and the 1 initially in X_0 is in Y_0. Thus, the 101 stored in

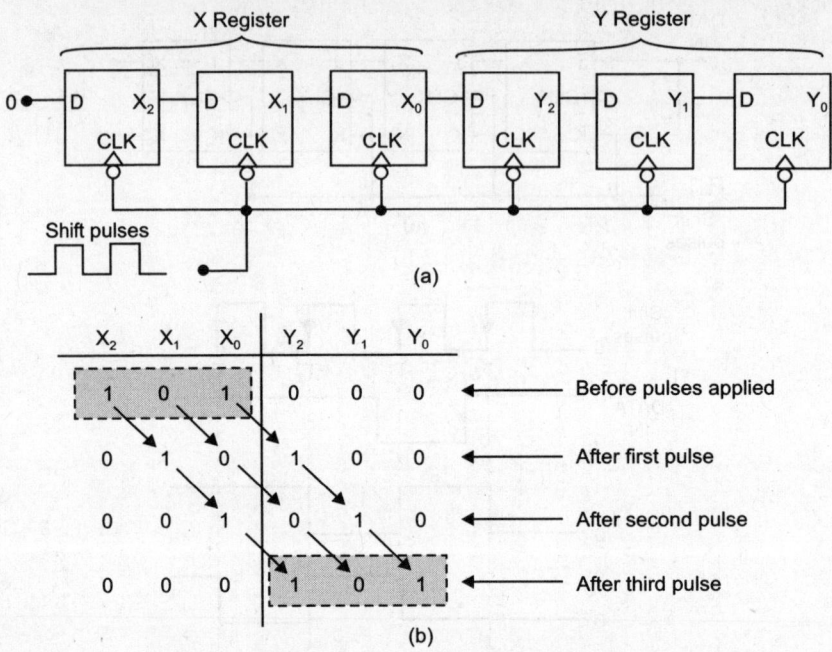

Fig. 8.18: Serial transfer of data from X register into Y register.

the X register has now been shifted into the Y register. The X register is at 000 and it has lost its original data.

3. The complete transfer of the three bits of data requires three shift pulses.

8.13 CONTROLLED BUFFER OUT REGISTER

We discussed buffer register in section 8.4, where the input word could be loaded by the application of a clock pulse. However, that circuit is not of any use. What it needed was some control over the X bits, i.e.

some way of holding them off until we are ready to store them.

Figure 8.19 shows a controlled buffer register.

If $\overline{CLR}$ goes LOW, all the FFs are RESET and the output becomes Q = 0000.

When $\overline{CLR}$ is HIGH, the register is ready for action. LOAD is the control input. When LOAD is HIGH, the data bits X can reach D inputs of FFs. At the PGT of the next clock pulse, the register is loaded, i.e.

$$Q_4\ Q_3\ Q_2\ Q_1 = X_4\ X_3\ X_2\ X_1 \text{ or } Q = X$$

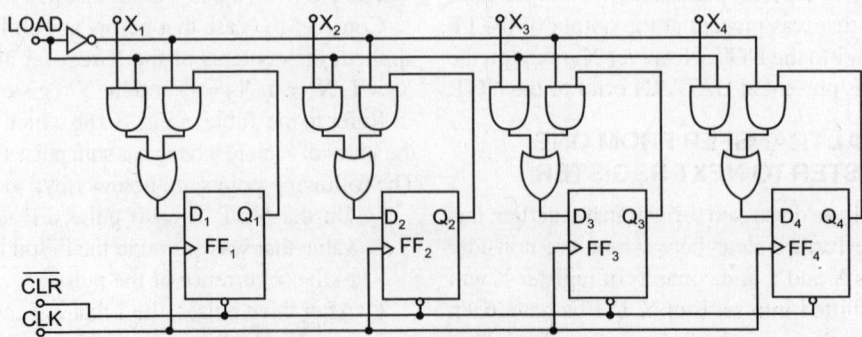

Fig. 8.19: A 4-bit controlled buffer register

when LOAD is LOW, the X bits cannot reach the FFs. At the same time, the inverted signal $\overline{LOAD}$ is HIGH.

This forces each FF output to feedback to its data input. Thus, data is circulated or retained as each clock pulse arrives. In other words, the contents of the register remain unchanged in spite of the clock pulses.

Figure 8.20 shows the symbol for a controlled buffer register X is the word to be loaded and Q is the stored word. When LOAD is LOW, Q is frozen, i.e. Q does not change in spite of changing X bits and arrival of clock pulses. Only when LOAD is HIGH, can the next positive clock edge load X into the register.

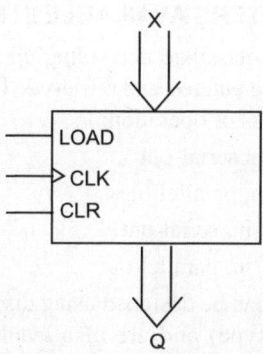

Fig. 8.20: Logic symbol for controlled buffer register

8.14 SERIES-IN, PARALLEL OUT REGISTER

Figure 8.21 shows a 5-bit shift register (Tl 7496). There are 5FFs to store a 5-bit word. To allow the data in the word to be read into the register serially, the output of one FF is connected to the binary inputs of the following one. Each FF is of the S-R (or J-K) master slave type. Note that the stage which is to store the MSB is converted into a D type latch by connecting S and R through an inverter. This register is available (on a single chip) in a 16-pin package (medium-scale integration).

We shall consider the operation by assuming that the serial data 01011 is to be registered. The MSB is 0 and the LSB in 1.

The FFs are cleared by applying 0 to the clear input so that every output (Q_0 to Q_4) is 0. Then Cr is set to 1 and Pr is held constant at 1 (by keeping the preset enable at 0). Then, serial data train and the synchronous clock are applied. The LSB is entered into FF4 when CK changes from a 0 to 1_7 by the action of D-type FF. After the clock pulse, $Q_4 = 1$, while all other outputs remain 0.

At the second clock pulse, the state of Q4 is transferred to the master latch of FF 3. Simultaneously, the next bit (a 1 in the 01011 word) enters the master of FF4. After the second clock pulse, the bit in each master transfers to slave and $Q_4 = 1$, $Q_3 = 1$ and the other outputs remain 0. The readings of the register after each pulse are given in Table 8.2. In this way, the input word may be installed in a n-bit shift register after the nth clock pulse (for an n-bit code), and the clock pulses must stop at the moment the word is registered.

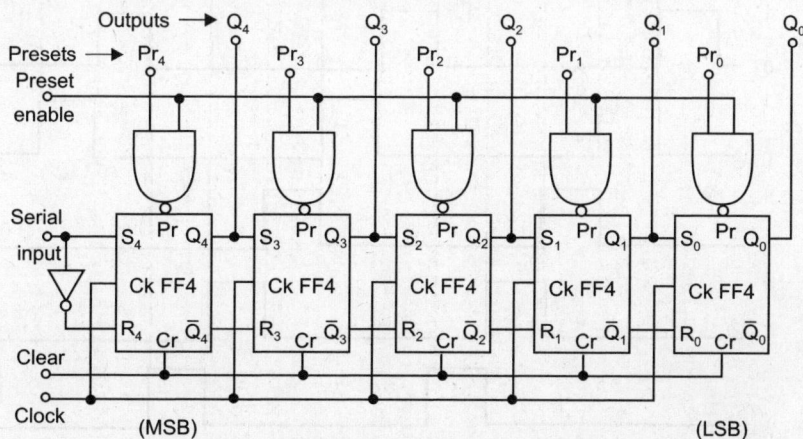

Fig. 8.21: (a) 5-bit series-in, parallel-out shift register

Table 8.2: Reading of register of Fig. 8.21 after each clock pulse

Clock pulse	Word bit	Q_4	Q_3	Q_2	Q_1	Q_0
1	1 → 1	0	0	0	0	
2	1 → 1	1	0	0	0	
3	0 → 0	1	1	0	0	
4	1 → 1	0	1	1	0	
5	0 → 0	1	0	1	1	

Since the each output is available on a separate line, they may be read simultaneously. Further, the data enters the system serially and comes out in parallel, this shift register is a serial-to-parallel converter. A *temporal code* (a time arrangement of bits) has been changed to a *spacial code*. (Information stored in a static memory). It is to be noted here that the use of master-slave flip-flops avoids the race problem between the stages.

8.15 RING COUNTER

If the serial output Q_0 of the shift register of Fig. 8.10a is connected back to the serial input, then an injected pulse will keeps circulating. This circuit is referred to as a *ring-counter*. The pulse is injected by entering 00001 in the parallel form after clearing the FFs.

When the clock pulses are applied, this 1 circulates around the circuit. The waveforms at the Q outputs are shown in Fig. 8.21b.

The outputs are sequential non-overlapping pulses which are useful for control-state counters, for stepper motor (which rotates in steps) which require sequential pulses to rotate it from one position to the next, etc. This circuit can also be used for counting the number of pulses. The number of pulses counted is read by noting which FF is in 1 state. No decoding circuitry is required. Since there is one pulse at the output for each of the N clock pulses, this circuit is referred to as a *divide-by-N counter*.

8.16 REGISTERS AVAILABLE IN TTL SERIES

Registers are classified depending upon the way in which data are entered and retrieved. These are four possible modes of operation:

1. Serial-in, serial-out
2. Serial-in, parallel-out
3. Parallel-in, serial-out
4. Parallel-in, parallel-out

Registers can be designed using discrete FFs (SR or JK as D-type) and are also available as MSI devices. The registers available in 54/74 TTL series are given in Table 8.3.

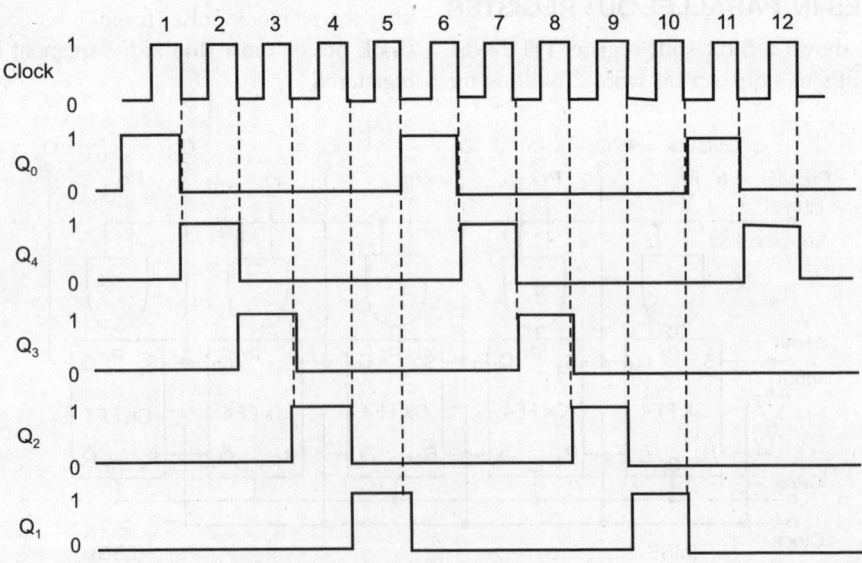

Fig. 8.21: (b) Output waveforms of ring counter

Table 8.3: Shift registers available in TTL series

IC No.	Description
7491, 7491A	8-bit serial-in, serial-out
7494	4-bit parallel-in, serial-out
7495	4-bit serial/parallel-in, parallel-out (right-shift, left-shift)
7496	5-bit parallel-in/ parallel-out, serial-in/serial-out
7499	4-bit bi-directional (universal)
74164	8-bit serial-in, parallel-out
74165	8-bit serial/parallel-in, serial-out
74166	8-bit serial/parallel-in, serial-out
74178, 74179	4-bit bi-directional (universal)
74194	4-bit bi-directional (universal)
74195	4-bit serial/parallel-in, parallel-out
74198	8-bit bi-directional (universal)
74199	8-bit serial/parallel-in, parallel-out
74195A	4-bit TRI-STATE serial/ parallel-in, parallel-out bi-directional
74395	4-bit TRI-STATE cascadable serial/ parallel-in, serial/parallel-out

Note that a register is referred to as a *universal register* if it can be operated in all the four possible modes and also as a bidirectional register (i.e. as a shift-right and shift-left register)

8.17 UNIVERSAL SHIFT REGISTER

We know that a register may operate in any of the following ways: serial-in-serial-out (SISO), serial-in-parallel-out (SIPO), parallel-in-serial-out (PISO) or parallel-in-parallel-out (PIPO) or bidirectional. If a register can be operated in all these five possible ways, it is called a universal shift register. Such a register is available as IC 74194. (Fig. 8.22).

This register has 10 inputs and 4 outputs. The output connections are connected to the normal Q outputs of flip-flops inside the IC. The four inputs A, B, C, D are parallel load inputs.

The data is supplied into the register serially (one bit at a time) to the next two inputs. The input D_{SR} triggers the four FFs on LOW to HIGH transition of the clock pulse. When activated with a LOW, the clear (CLR) input resets each FF to 0. The mode control S_0 and S_1 guide the register through a gating network to shift-right, shift-left, parallel-load or HOLD. In addition to this, TTL IC has a +5V and GND connections for power supply. Parallel loading, which is synchronous with a positive clock, is done by supplying the four bits of data to the parallel inputs and a HIGH to the S_0 and S_1.

Shift right is done synchronously by applying positive clock pulse when S_1 is LOW and S_0 is HIGH. Serial data is entered at the shift right serial input (D_{SR}). When S_1 is HIGH and S_0 is LOW, data bits shift left synchronously with the clock and new data is entered at the shift-left serial input (D_{SL}).

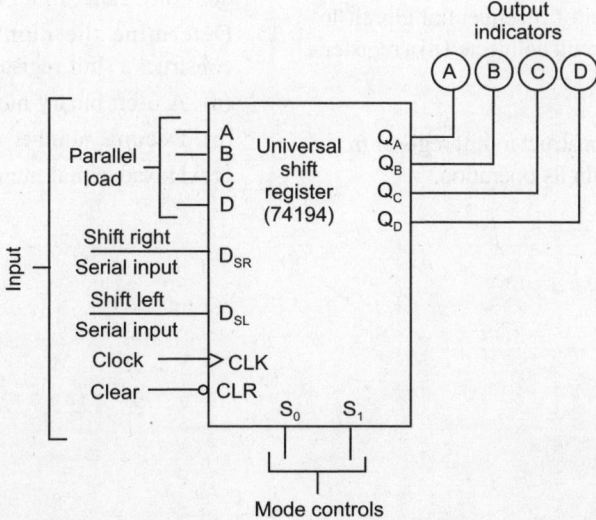

Fig. 8.22: Logic symbol of a TTL 4-bit universal shift register (IC 74194)

8.18 SEQUENTIAL LOGIC IMPLEMENTATION

A clocked sequential circuit consists of a group of flip-flops and combinational gates. Since registers are readily available as MSI circuits, registers can be conveniently employed as part of the sequential circuits. A block diagram of a sequential circuit that uses a register is shown in Fig. 8.23.

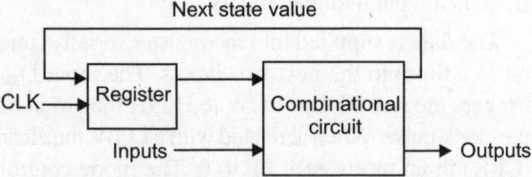

Fig. 8.23: Block diagram of a sequential circuit

The present state of the register and the external inputs determine the next state of the register and the values of external outputs. Part of the combinational circuit determines the next state and the other part generates the outputs. The next state value from the combinational circuit is loaded into the register with a clock pulse. If the register has a load input, it must be set to 1; otherwise, if the register has no load input, the next state value will be transferred automatically at every clock pulse.

The combinational part of the circuit can be implemented by using SSI gates, or ROM or with a programmable logic array (PLA). By using a register, we can reduce the design of a sequential circuit to that of a combinational circuit connected to a register.

EXERCISE

1. Define a *register*. Construct a shift register from SR flip-flops. Explain its operation.

2. Explain how a shift register is used as a converter from
   (a) Serial-to parallel data
   (b) Parallel-to-serial data.

3. Explain how a shift register is used as a ring-counter. Draw the output waveform from each flip-flop of a three stage unit.

4. Explain what do you understand by a register. What is a buffer register?

5. Explain the operation of the bidirectional shift register.

6. Explain what is a universal shift register. Explain its working.

7. Explain the various modes in which the data can be entered or taken out from a register.

8. Explain how a flip-flop can store a data bit.

9. Describe the working of a parallel-in, parallel-out shift register. Explain how number can be shifted in or out from such a register.

10. What is a serial in, serial-out shift register?

11. What is a parallel-in, serial-out shift register?

12. What is a serial-in, parallel-out shift register?

13. What do you understand by
    (a) Buffer register
    (b) Controlled buffer register.

14. Why is a FF called a single-bit register? What does the "state" of a FF indicate?

15. Determine the number of FFs needed to construct a shift register capable of storing
    (a) A 6-bit binary number
    (b) Decimal number upto 32
    (c) Hexadecimal number upto F.

9

Binary Counters
(Asynchronous and Synchronous Counters)

The term asynchronous refers to events that do not occur at the same time. With respect to counter operation, *asynchronous* means that the FFs within the counter are not made to change states exactly at the same time, they do not because the clock pulses are not connected directly to the CLK input of each FF in the counter, i.e. the FFs are not simultaneously triggered. On the other hand, the *synchronous counter* is clocked such that each FF in the counter is triggered at the same time. This is accomplished by connecting the clock-line to CLK input of each FF.

Since both these counters go through a binary sequence, these are also referred to as *binary counters*.

Note that asynchronous counters are commonly referred to as *ripple counters* for the following reason. The effect of the input clock pulse is first "felt" by the first FFA. This effect cannot get to FFB immediately due to the propagation delay through FFA. Then there is the propagation delay through FFB before FFC can be triggered. As a result, the effect of an input clock pulse "ripples" through the counter, taking some time, due to propagation delays to reach the last flip-flop.

We will now first consider asynchronous counters.

9.1 THREE BIT ASYNCHRONOUS BINARY COUNTER

The fundamental purpose of a binary counter is to record the number of occurrences of some input. This is a basic function, that of counting and it is used over and over. Before discussing the counting operation we first need to explain the concept of frequency division.

Frequency Division: Refer to Fig. 9.1a. Each FF has its J and K inputs at the 1 level, so that it will change states (toggle) whenever the signal on its CLK input goes from HIGH to LOW. The clock pulses are applied only to the CLK input of FF X_0. Output X_0 is connected to the CLK input of FF X_1, and output X_1 is connected to the CLK input of FF X_2. The waveforms in Fig. 9.1b show how the FFs change states as the pulses are applied.

The following points are to be noted:

1. FF X_0 toggles on the negative-going transition of each input close. Thus, the X_0 output waveform has a frequency that is exactly one-half of the clock pulse frequency.

2. FF X_1 toggles each time the X_0 output goes from HIGH to LOW. The X_1 waveform has a frequency equal to exactly one-half the frequency of the X_0 output and therefore one-fourth of the clock frequency.

3. FF X_2 toggles each time the X_1 output goes from HIGH to LOW. Therefore the X_2 waveform has one-half the frequency of X_1 and therefore one-eighth of the clock frequency.

4. Each FF output is a square wave.

Thus, in Fig. 9.1, each FF divides the frequency of its input by 2. If we were to add a fourth FF to this chain, it would have a frequency equal to $1/2^4$ and so

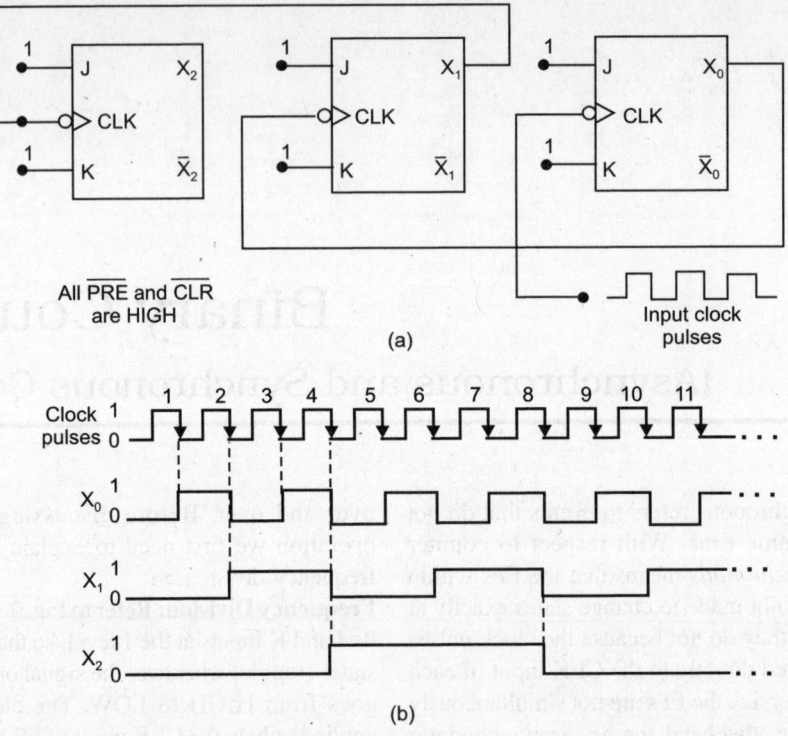

All $\overline{PRE}$ and $\overline{CLR}$ are HIGH

(a)

(b)

Fig. 9.1: J-K FFs wired as a 3-bit binary counter

on. Consequently, using N flip-flops would produce an output frequency from the last FF which is equal to $1/2^N$ of the input frequency.

9.2 COUNTING OPERATION

In addition to functioning as a frequency divider, the circuit of Fig. 9.1 also works as a *binary counter*. This can be understood by examining the sequence of states of the FFs after the occurrence of each clock pulse. The results are given in Table 9.1 known as *state table*.

Let $X_2X_1X_0$ values represent a binary number where X_2 is in the 2^2 position, X_1 is in the 2^1 position and X_0 is in the 2^0 position. The first eight $X_2X_1X_0$ states in the table can be recognized as the binary counting sequence from 000 to 111. After the first NGT, the FFs are in the 001 state ($X_2 = 0$, $X_1 = 0$, $X_0 = 1$) which represents 001_2 (equivalent to decimal 1), after the second NGT the FFs represent 010_2 or 2_{10}; after 3rd pulse; $011_2 = 3_{10}$, after 4 pulses, $100_2 = 4_{10}$ and so on, until after seven pulses, $111_2 = 7_{10}$. On the eighth NGT, the FFs return to the 000 state and

the binary sequence repeats itself for succeeding pulses, i.e. for the first seven input pulses, the circuit functions as a binary counter in which the states of the FFs represent a binary number equivalent to the number of pulses that have occurred. This counter can count upto $111_2 = 7_{10}$ before it returns to 000.

Table 9.1: State table for the circuit of Fig. 9.1

$\dfrac{2^2}{X_2}$	$\dfrac{2^1}{X_1}$	$\dfrac{2^0}{X_0}$	Clock pulse
0	0	0	Before applying clock pulses
0	0	1	After 1st pulse
0	1	0	After 2 pulses
0	1	1	After 3 pulses
1	0	0	After 4 pulses
1	1	0	After 5 pulses
1	1	1	After 6 pulses
0	0	0	After 7 pulses
0	0	1	After 8 pulses (recycles)
.	.	.	After 9 pulses
.	.	.	
.	.	.	

9.2.1 State Transition Diagram

Figure 9.2 shows the state transition diagram. Each circle represents one possible state as indicated by the binary number inside the circle. It shows how the states of the counter FFs change with each applied clock pulse. State transition diagrams help to describe, analyze and design counters and other sequential FF circuits.

9.2.2 MOD Number

The counter of Fig. 9.1 has $2^3 = 8$ different states (000 through 111). It is therefore referred to as a MOD-8 counter, where the MOD number indicates the number of states in the counting sequence. If a fourth FF is added, the sequence of states would count in binary from 0000 to 1111, i.e. total 16 states then, this would be called a MOD-16 counter. In general, N FFs connected in the arrangement of Fig. 9.1 will have 2^N gates and so it will be a MOD-2^N counter. It would be capable of counting upto 2^N-1 before returning to its zero state. In terms of frequency, it could be used to divide the input pulse frequency by a factor of 2^N (the MOD number).

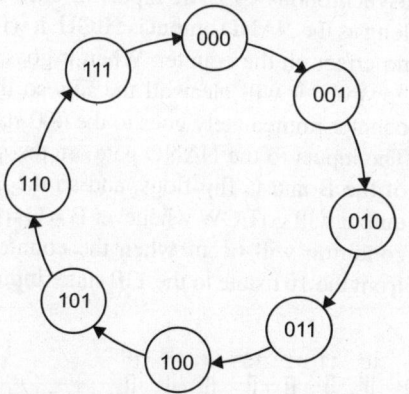

Fig. 9.2: State transition diagram for a 3-bit binary counter. Here each arrow represents the occurrence of a clock pulse

Example 9.1: Given that the MOD-8 counter (Fig. 9.1) is in the 101 state. What will be the state (count) after 13 pulses have been applied?

Solution: Locate the 101 state on the state transition diagram (Fig. 9.2). Proceed around the state diagram through 8 state changes and you are back in the 101 state. Now continue through 5 more state changes (for a total of 8 + 5 = 13) and you end up in the 010 state (Ans.).

9.3 FOUR BIT ASYNCHRONOUS (RIPPLE) COUNTER

Figure 9.3 shows a 4-bit binary counter circuit. There are following important points about its operation.

1. The clock pulses are applied only to the clock input of FF A. Thus flip-flop A will toggle (change to its opposite state) each time the clock pulses make a negative (HIGH to LOW) transition. Note that J = K = 1 for all flip-flops.

2. The normal output of FF A acts as the clock input for FF B and so flip-flop B will toggle each time the A output goes from 1 to 0. Similarly FF C will toggle when B goes from 1 to 0, and FF D will toggle when FF C goes from 1 to 0.

3. FF outputs D, C, B and A represent a 4-bit binary number with D as the MSB. Let us assume that all FFs have been cleared to the 0 state (CLEAR inputs are not shown). Table 9.2 shows that a binary counting sequence from 0000 to 1111 is followed as clock pulses are continuously applied.

4. After the 15th clock pulse has occurred, the counter FFs are in the 1111 condition. On the sixteenth clock pulse, FF A goes from 1 to 0, which causes FF B to go from 1 to 0 and so on until the counter is in the 0000 state. In other words, the counter has gone through one

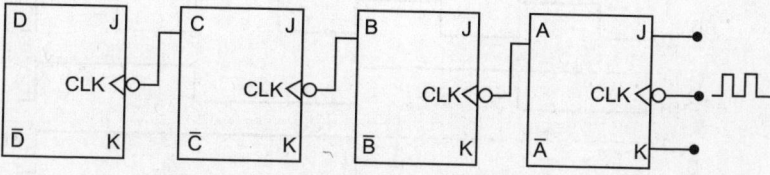

*All J and K inputs assumed to be 1

Fig. 9.3: Four-bit asynchronous (ripple) counter

complete cycle (0000 through 1111) and has recycled back to 0000, from where it will begin a new counting cycle with subsequent clock pulses.

Table 9.2: State-table for 4-bit asynchronous counter

D	C	B	A	Number of Clock pulses
0	0	0	0	0
0	0	0	1	1
0	0	1	0	2
0	0	1	1	3
0	1	0	0	4
0	1	0	1	5
0	1	1	0	6
0	1	1	1	7
1	0	0	0	8
1	0	0	1	9
1	0	1	0	10
1	0	1	1	11
1	1	0	0	12
1	1	0	1	13
1	1	1	0	14
1	1	1	1	15
0	0	0	0	16 (recycles)
0	0	0	1	17
.	.	.	.	.
.	.	.	.	.
.	.	.	.	.

This type of counter where each FF output serves as the clock input signal for the next FF is referred to as an *asynchronous counter*. This is because all the FFs do not change states in exact synchronism with the clock pulses; only FFA responds to the clock pulses, FF B has to wait for FF A to change states before it is toggled, FF C has to wait for FFB, and so on, i.e. there is a delay between the responses of consecutive FFs. Because of the manner in which this type of counter operates, it is also commonly referred to as a *ripple counter*. Thus often we use the terms "asynchronous counter" and "ripple counter" interchangeably.

9.4 COUNTERS WITH MOD NUMBERS < 2^N

A basic ripple counter is limited to MOD number = 2^N where N is the number of flip-flops. This value is actually the maximum MOD number that can be obtained using N FFs. The basic counter can be modified to produce MOD number less than 2^N by allowing the counter to skip states that are normally part of the counting sequence. Most common method for doing this is illustrated in Fig. 9.5 where a 3-bit ripple counter is shown. Disregarding the NAND gate for a moment, it is a MOD-8 binary counter which will count in sequence from 000 to 111. However, the presence of the NAND gate will alter this sequence as follows:

1. The NAND output is connected to the asynchronous CLEAR inputs of each FF. As long as the NAND output is HIGH, it will have no effect on the counter. When it goes LOW, however, it will clear all the FFs so that the counter immediately goes to the 000 state.

2. The inputs to the NAND gate are the outputs of the B and C flip-flops, and so the NAND output will go LOW whenever B = C = 1. This condition will occur when the counter goes from the 101 state to the 110 state (input pulse

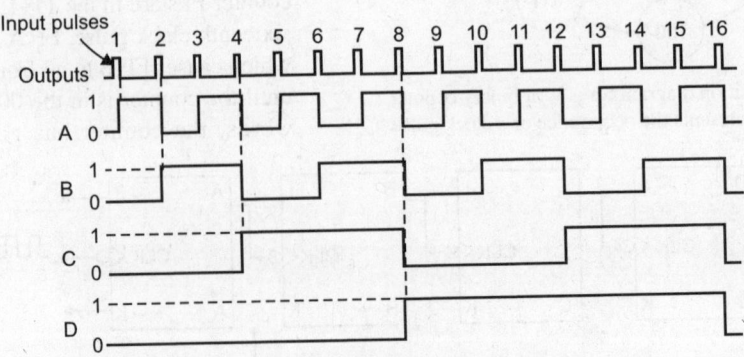

Fig. 9.4

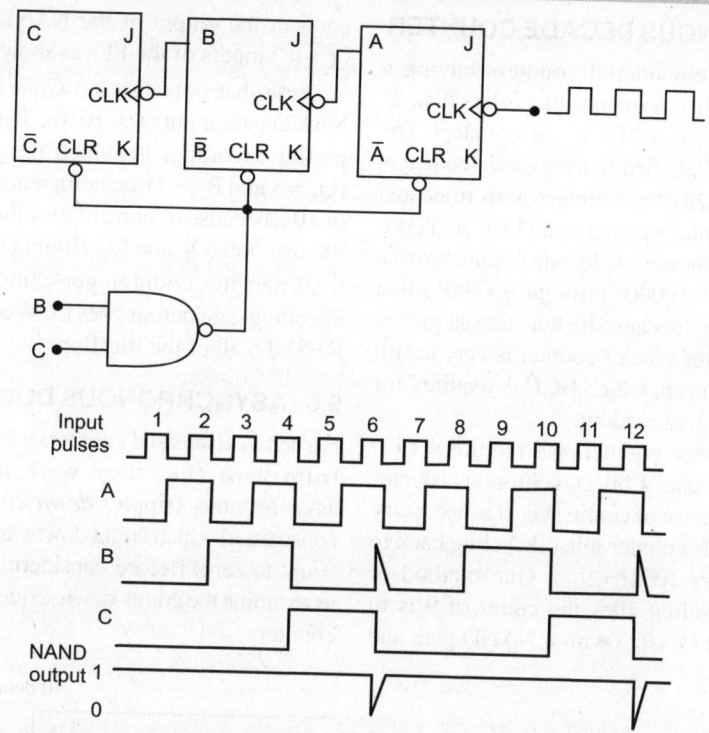

Fig. 9.5: MOD-8 counter converted into MOD-6 counter by clearing when count of six (110) occurs in MOD-8. In above FFs, all J, K inputs are 1

6 on waveforms). The LOW at the NAND output will immediately clear the counter to the 000 state. Once the FFs have been cleared, the NAND output goes back HIGH, since the B = C = 1 condition no longer exists.

3. The counting sequence is therefore

Notice that the waveform at the B output contains a spike caused by the momentary occurrence of the 110 state before clearing. This spike is very narrow and so, would not produce any visible indication on numerical displays. The state transition diagram for the above counter is shown in Fig. 9.5.

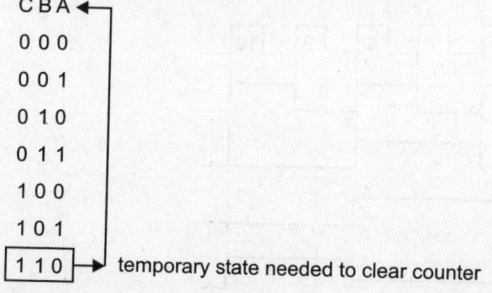

Although the counter does go to the 110 state, it remains there for only a few nanoseconds before it recycles to 000. The counter thus skips 110 and 111 go, it goes through only six different states and it is ∴ a MOD-6 counter.

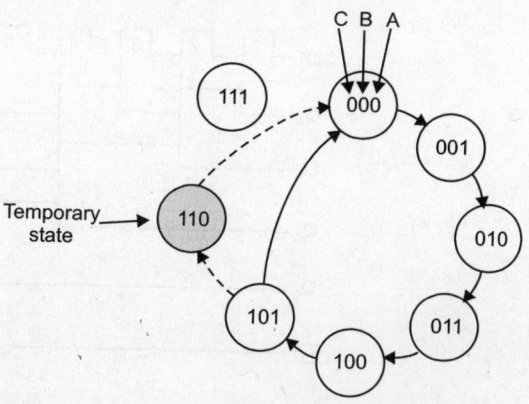

Fig. 9.6: State transition diagram for the MOD-6 counter of Fig. 9.5

9.5 ASYNCHRONOUS DECADE COUNTER

In section 9.4, we considered counters having a number of states in their sequence that is less than 2^N (where N is the number of FF in the counter). The resulting sequence is called a *truncated* sequence. One common modulus for counters with truncated sequence is ten. Counter with ten states in its sequence is called a *decade counter*. A decade counter with a count sequence of 0 (0000) through 9 (1001) is a BCD decade counter, because its ten-state sequence is the BCD code. This type of counter is very useful in display applications in which BCD is required for conversion to a decimal readout.

To obtain a decade counter, we require 4 FFs. ($\because 2^3 = 8$). We will take 4-bit asynchronous counter and modify its sequence of counts, i.e. it is necessary that the BCD decade counter must recycling back to the 0000 state after $1001 (= 9_{10})$. One method of achieving this recycling after the count of 9 is to *decode* count 1010 ($= 10_{10}$) with a NAND gate and connect the output of the NAND gate to the clear ($\overline{\text{CLR}}$) inputs of the FFs as shown in Fig. 9.7a.

Note that only Q_B and Q_D are connected to the NAND gate inputs ($\because 1010$). This is an example of partial decoding, in which the two unique states ($Q_B = 1$ and $B_D = 1$) are sufficient to decode the count of 10_{10} because of none of the other states (0 through 9) have both Q_B and Q_D HIGH (1) at the same time.

When the counter goes into count 1010, the decoding gate output goes LOW and asynchronously RESETS all of the flip-flops.

9.6 ASYNCHRONOUS DOWN COUNTER

Earlier we discussed counters which counted upward from zero (i.e. they were *up counters*). An asynchronous (ripple) *down counters* can also be constructed which count downward from a maximum count to zero. Before considering down counter, let us examine the count-down sequence for a 3-bit down counter:

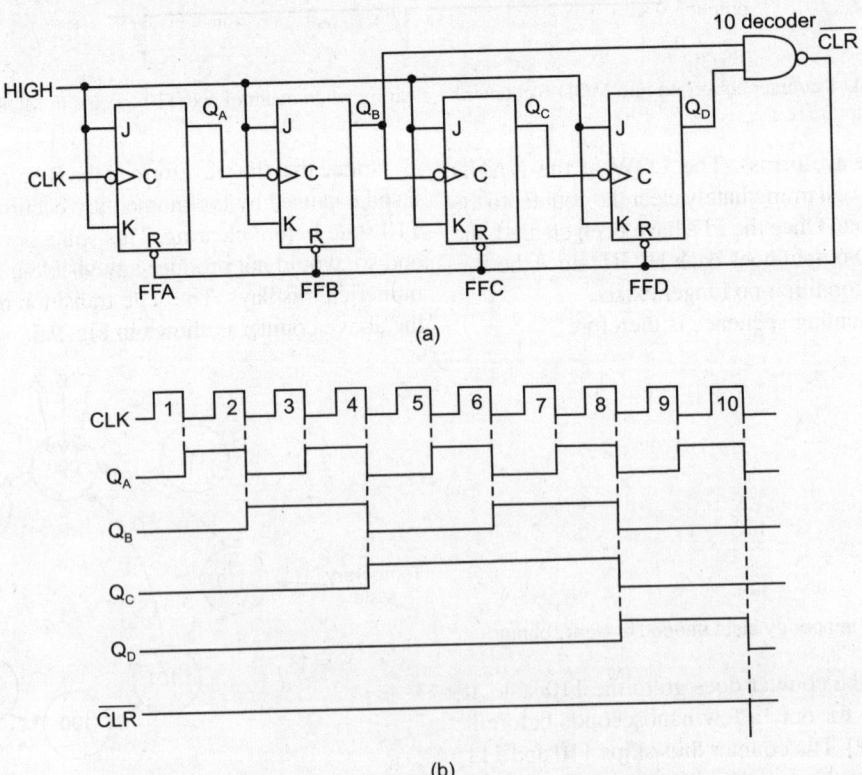

(a)

(b)

Fig. 9.7: Asynchronous clocked decade counter

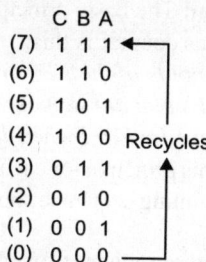

	C B A
(7)	1 1 1
(6)	1 1 0
(5)	1 0 1
(4)	1 0 0 Recycles
(3)	0 1 1
(2)	0 1 0
(1)	0 0 1
(0)	0 0 0

A, B and C represent the FF output states as the counter counts down. It may be noted that A FF (LSB) changes state (toggles) at each step just as in case of the up counter. The B FF changes states each time A goes from LOW to HIGH; C changes states each time

B goes from LOW to HIGH. Thus, in a down counter, each FF except the first, must toggle when the preceding FF goes from LOW to HIGH. If the FFs have CLK inputs that respond to negative transitions (HIGH to LOW), then an inverter can be placed in front of each CLK input; however, the same effect can be accomplished by driving each FF clock input from the *inverted* output of the preceding FF. This is depicted in Fig. 9.8 for a MOD-8 down counter.

The input pulses are applied to the A FF; the $\overline{A}$ output serves as the CLK input for the B FF; the $\overline{B}$ output serves as the CLK input for the C FF. The waveforms at A, B and C show that B toggles whenever A goes LOW to HIGH (so that $\overline{A}$ goes HIGH to LOW) and

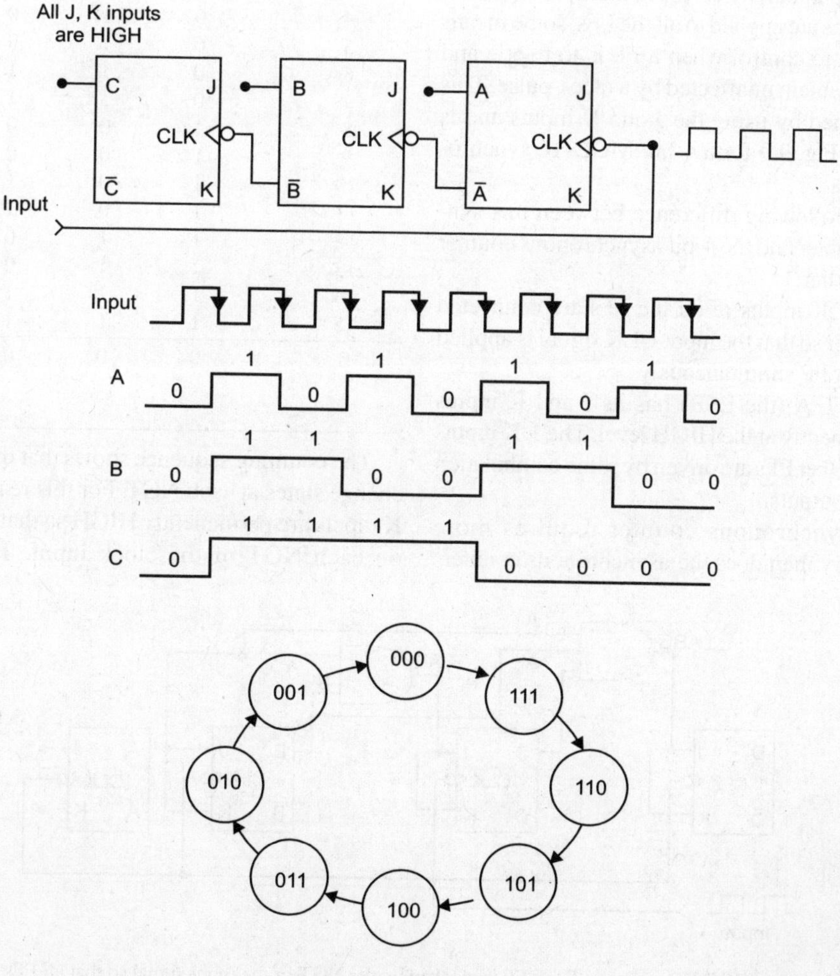

Fig. 9.8: MOD-8 down counter

C toggles whenever B goes LOW to HIGH. This results in the desired down-counting sequence at the C, B and A outputs. The state transition diagram shows the sequence.

9.7 SYNCHRONOUS (PARALLEL) COUNTER (MOD 16 COUNTER)

The problem encountered with ripple (i.e. asynchronous) counters are caused by the accumulated FF propagation delays: in other words, there, the FFs do not all change states simultaneously, in synchronism with the input pulses. This limitation can be overcome with the use of *synchronous* or *parallel* counters where all the FFs are triggered simultaneously (in parallel) by the input CLK pulses. Since the input pulses are applied to all the FFs, some means must be used to control when a FF is to toggle and when it is to remain unaffected by a clock pulse. This is accomplished by using the J and K inputs and is illustrated in Fig. 9.9 for a 4-bit, MOD-16 synchronous counter.

Note the following difference between this synchronous counter and the 4-bit asynchronous counter discussed earlier:

1. The CLK inputs of all the FFs are connected together so that the input CLK signal is applied to each FF simultaneously.
2. Only FF A (the LSB) has its J and K inputs permanently at the HIGH level. The J, K inputs of the other FFs are driven by some combination of FF outputs.
3. The synchronous counter requires more circuitry than does the asynchronous counter.

Circuit operation: The basic principle of operation of the synchronous counter is this:

The J and K inputs of the FFs are connected so that only those FFs that are supposed to toggle on a given NGT will have J = K = 1 when that NGT occurs.

We examine this principle for each of the FFs with the help of the counting sequence shown in Table 9.3

Table 9.3: Counting sequence for MOD-16 counter of Fig. 9.9

Count	D	C	B	A
0	0	0	0	0
1	0	0	0	1
2	0	0	1	0
3	0	0	1	1
4	0	1	0	0
5	0	1	0	1
6	0	1	1	0
7	0	1	1	1
8	1	0	0	0
9	1	0	0	1
10	1	0	1	0
11	1	0	1	1
12	1	1	0	0
13	1	1	0	1
14	1	1	1	0
15	1	1	1	1
0	0	0	0	0
.	.	.	.	.
.	.	.	.	.

The counting sequence shows that the FFA has to change states at each NGT. For this reason, its J and K inputs are permanently HIGH so that it will toggle on each NGT of the clock input. The counting

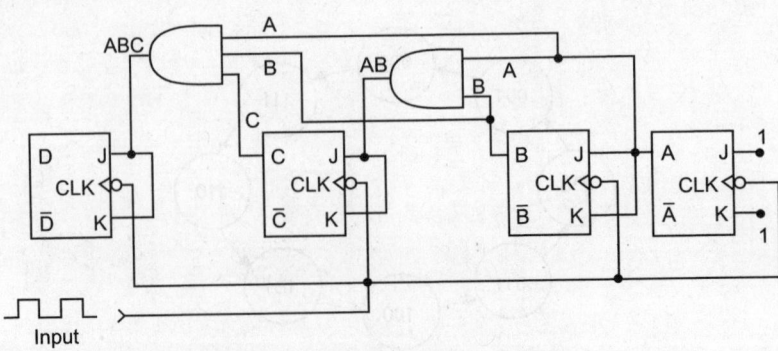

Fig. 9.9: Synchronous MOD-16 counter. (Each FF is clocked by the NGT of the input signal so that all FFs will be toggling at the same time)

sequence shows that FF B has to change states on each NGT that occurs while A = 1. For example, when the count is 0001, the next NGT has to toggle B to the 1 state; when the count is 0011, the next NGT has to toggle B to the 0 state and so on. This operation is accomplished by connecting output A to the J and K inputs of FFB so that J = K = 1 only when A = 1.

The counting sequence shows that FFC has to change states on each NGT that occurs while A = B = 1. For example, when the count is 0011, the next NGT has to toggle C to the 1 state; when the count is 0111, the next NGT has to toggle C to the 0 state and so on. This operation is accomplished by connecting the signal AB to the J and K inputs of FFC.

Similarly, we can see that FFD has to toggle on each NGT that occurs while A = B = C = 1. When the count is 0111, the next NGT has to toggle D to the 1 state; when the count is 1111, the next NGT has to toggle D to the 0 state. This is accomplished by connecting ABC to the J and K inputs of FFD.

9.8 A 4-BIT SYNCHRONOUS DECADE COUNTER (BCD Decade Counter)

As we know, the BCD decade counter exhibits a truncated sequence and goes through a straight binary sequence from 0000 to the 1001 state. Rather than going to the 1010 state, it recycles to the 0000 state. (Fig. 9.10).

The counter operation can be understood by considering the sequence of states in Table 9.4. First note that FFA toggle on each clock pulse, so the logic equation for its J and K inputs is $J_A = K_A = 1$

This is implemented by connecting these inputs to a constant HIGH level. Next notice that FFB changes on the next clock pulse each time $Q_A = 1$ and $Q_D = 0$.

Table 9.4: States of a BCD decade counter

Clock Pulse	Q_D	Q_C	Q_B	Q_A
0	0	0	0	0
1	0	0	0	1
2	0	0	1	0
3	0	0	1	1
4	0	1	0	0
5	0	1	0	1
6	0	1	1	0
7	0	1	1	1
8	1	0	0	0
9	1	0	0	1

So the logic equation for its J and K inputs is

$$J_B = K_B = Q_A \overline{Q_D}$$

This is implemented by ANDing Q_A and $\overline{Q_D}$ and connecting the gate output to the J and K inputs of FFB. FFC changes on the next clock pulse each time both $Q_A = 1$ and $Q_B = 1$. This requires an input logic equation as follows:

$$J_C = K_C = Q_A Q_B$$

This is implemented by ANDing Q_A and Q_B and connecting the gate output to the J and K inputs of FFC. Finally, FFD changes to the opposite state on the next clock pulse each time $Q_A = 1$, $Q_B = 1$ and $Q_C = 1$ (count 7), or when $Q_A = 1$ and $Q_D = 1$ (count 9). The equation for this is

$$J_D = K_D = Q_A Q_B Q_C + Q_A Q_D$$

This function is implemented with AND/OR logic connected to FFD as shown in Fig. 9.10. Note that the only difference between this BCD decade counter and a Mod-16 counter (Fig. 9.9) is the $Q_A Q_D$ AND gate and the OR gate: This essentially detects the occurrence of the 1001 state and causes the counter to recycle on the next (i.e. 10th) clock pulse. The timing diagram is shown in Fig. 9.11.

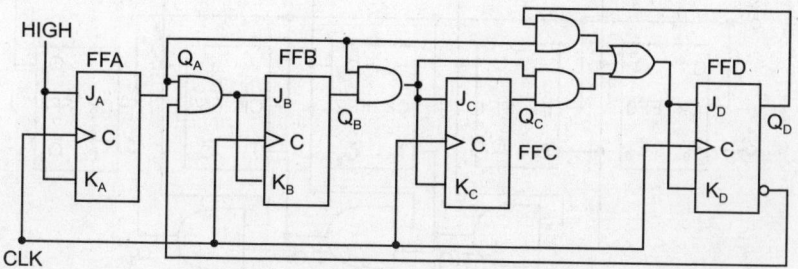

Fig. 9.10: Synchronous BCD decade counter

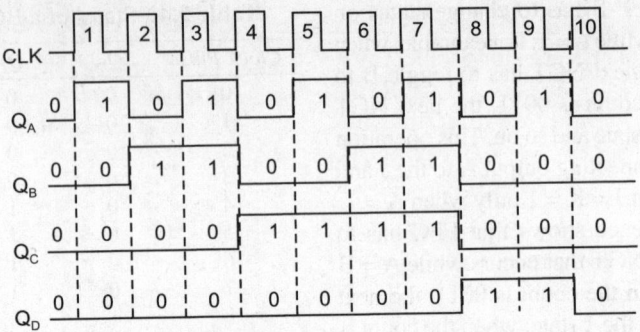

Fig. 9.11: Timing diagram for the BCD decade counter (Q_A is the least significant bit)

9.9 A 5-BIT SYNCHRONOUS COUNTER

A 5-Bit Synchronous Counter is shown in Fig. 9.12. Each flip-flop is a T-type obtained by tying the J terminal to the K terminal of a J-K FF (*see* Fig. 7.16). If T = 0, there is no change of state when the binary is clocked, and if T = 1, the FF output is complemented with each pulse.

The connections to be made to the T inputs are deduced from the waveform chart of Fig. 9.4.

Q_0 toggles with each pulse $\qquad : T_0 = 1$

Q_1 complements only if $Q_0 = 1$ $\quad : T_1 = Q_0$

Q_2 becomes $\overline{Q}_2$ only if $Q_0 = Q_1 = 1 : T_2 = Q_0 Q_1$

Q_3 toggles only if $Q_0 = Q_1 = Q_2 = 1 : T_3 = Q_0 Q_1 Q_2$

Extending this logic to Q_4, we conclude that $T_4 = Q_0 Q_1 Q_2 Q_3$. Therefore, the T logic is given by

$$T_0 = 1, T_1 = Q_0, T_2 = T_1 Q_1, T_3 = T_2 Q_2,$$

$$T_4 = T_3 Q_3. \qquad \qquad ...(1)$$

Clearly, the two-input AND gates of Fig. 9.10 perform this logic.

The minimum time T_{min} between pulses is the interval required for each J and K node to reach its steady-state value and is given by

$$T_{min} = T_F + (n - 2) T_G \qquad ...(2)$$

where T_F is the propagation delay of one flip-flop and T_G is the propagation delay of one AND gate. The maximum pulse frequency for series carry is the reciprocal of T_{min}.

Since the carry passes through all the control gates in series, this is a synchronous counter with *series* or *ripple carry*.

Parallel Carry

In the counter discussed above, the maximum frequency of operation can be improved if we use parallel or *look-ahead, carry*, where the toggle input to each binary comes from a multi-input AND gate excited by the outputs from every preceding FF. From equation (1), it follows that

$$T_1 = Q_0, T_2 = Q_0 Q_1, T_3 = Q_0 Q_1 Q_2,$$

$$T_4 = Q_0 Q_1 Q_2 Q_3 \qquad ...(3)$$

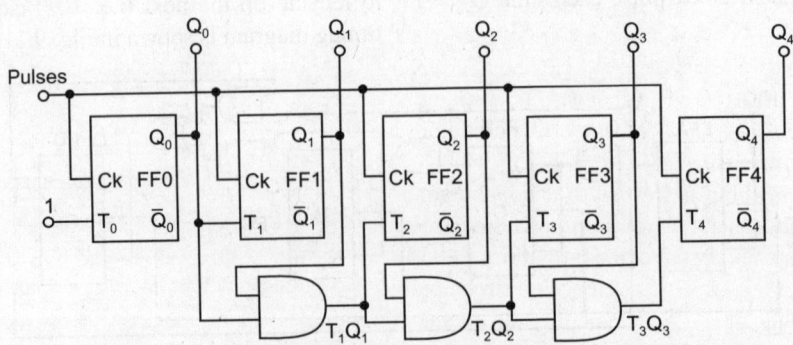

Fig. 9.12: A 5-bit synchronous counter with series carry (J = K = T)

Hence T_4 is obtained from a four input AND gate fed by Q_0, Q_1, Q_2 and Q_3. Clearly, for parallel carry

$$T_{min} = T_F + T_g \qquad \ldots(4)$$

which may be considerably smaller than the corresponding time for series carry given by eqn. (2) particularly if n is large.

9.10 ADVANTAGES OF SYNCHRONOUS COUNTERS OVER ASYNCHRONOUS COUNTER

In a synchronous or parallel counter, all the FFs change states simultaneously, i.e. they are all synchronous to the NGT of the input clock pulses. Thus, unlike asynchronous counters, the propagation delays of the FFs do not add together to produce the overall delay in synchronous counter, i.e.

Total delay = (FF t_{pd}) + (AND gate t_{pd}).

or, in other words, total response time is the time it takes one FF to toggle plus the time for the new logic levels to propagate through a single AND gate to reach the J, K inputs. This total delay is the same no matter

9.11 CASCADING OF COUNTERS

Counters can be connected in cascade to achieve higher modulus of operation. *Cascading* means that the last stage output of one counter drives the input of the next counter. An example of cascading is shown in Fig. 9.13 where output of two bit ripple counter is fed to the input of a three bit ripple counter.

The timing diagram is shown in Fig. 9.14. Notice in the timing diagram that the final output of the mod-8 counter (Q_E) occurs once for every 32 input clock pulses.

The overall modulus of the cascaded counters is 32, i.e. they act as a divide-by-32 counter. Thus, the overall modulus of cascaded counters is equal to the product of individual modulus (here $8 \times 4 = 32$).

When operating synchronous counters in a cascaded configuration, it is necessary to use the *count enable* and the *terminal count* functions to achieve higher modulus operation. (Terminal count is analogous to ripple carry output (RCO) given in ICs).

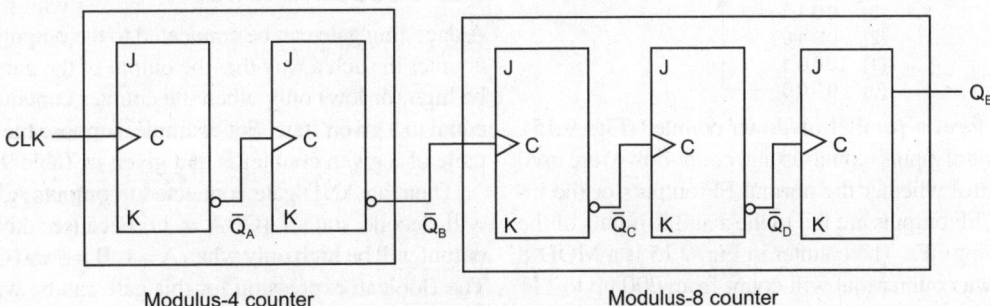

Modulus-4 counter   Modulus-8 counter

Fig. 9.13: Cascaded counters (All J, K inputs are HIGH)

how many FFs are in the counter, and it is generally much lower than an asynchronous counter with the same number of FFs. Thus a synchronous counter can operate at a much higher input frequency. Typically, the maximum frequency of operation of a 4-bit synchronous counter using TTL logic is 32 MHz, which is about twice that of a ripple counter. Another advantage of the synchronous counter is that no decoding spikes appear at the output since all FFs change state at the same time. Hence no strobe pulse is required when decoding a synchronous counter.*

9.12 SYNCHRONOUS DOWN AND UP/DOWN COUNTERS

We have seen in section 9.6 that a ripple counter could be made to count down by using the inverted output of each FF to drive the next FF in the counter. A parallel down counter can be constructed in a similar manner, i.e. by using the inverted FF outputs to drive the following J, K inputs. For example, the parallel up counter of Fig. 9.7 can be converted to a down counter by connecting the $\overline{A}$, $\overline{B}$ and $\overline{C}$ outputs in

---

* In asynchronous counter, the propagation delays of the FFs accumulate, so that the Nth FF cannot change states until a time equal to $(N) \times (t_{pd})$ after the clock transition occurs.

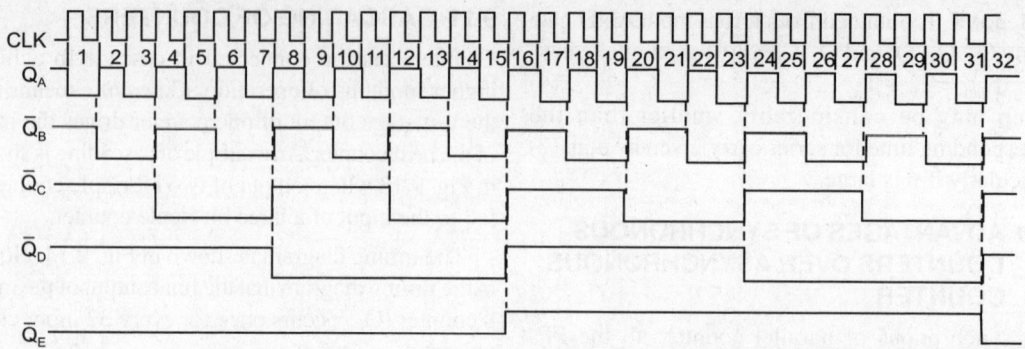

Fig. 9.14: Timing diagram for cascaded configuration of Fig. 9.13

place of A, B and C respectively. The counter will then proceed through the following sequence as input pulses are applied (in place of Fig. 9.8).

```
(15)  1 1 1 1
(14)  1 1 1 0
(13)  1 1 0 1
(12)  1 1 0 0
      - - -
                    Recycles
      - - -
(3)   0 0 1 1
(2)   0 0 1 0
(1)   0 0 0 1
(0)   0 0 0 0
```

To form a parallel up/down counter (Fig. 9.15), the control inputs (count-up and count-down) are used to control whether the normal FF outputs or the inverted FF outputs are fed to the J and K inputs of the following FFs. The counter in Fig. 9.15 is a MOD-8 up/down counter that will count from 000 up to 111 when the count-up control input is 1 and from 111 down to 000 when the count-down control input is 1.

A logical 1 on the count-up line while count-down = 0 enables AND gates 1 and 2 and disables gates 3 and 4. This allows the A and B outputs through to the J and K inputs of the following FFs so that the counter will count up as pulses are applied. The opposite action takes place when count-up = 0 and count-down = 1 ($\overline{A}$ and $\overline{B}$ allowed to pass J and K inputs).

9.13 DECODING GATES

A decoding gate can be connected to the outputs of a counter in such a way that the output of the gate will be high (or low) only when the counter contents are equal to a given state. For example, suppose the truth table of a given counter is that given in Table 9.5.

Then, an AND gate connected to outputs A, B, C will decode state 7 (CBA = 1), because, the gate output will be high only when A = 1, B = 1 and C = 1. The Boolean expression for this gate can be written 7 = CBA. The other seven states of the counter with truth table 9.5 can be decoded in a similar fashion. It

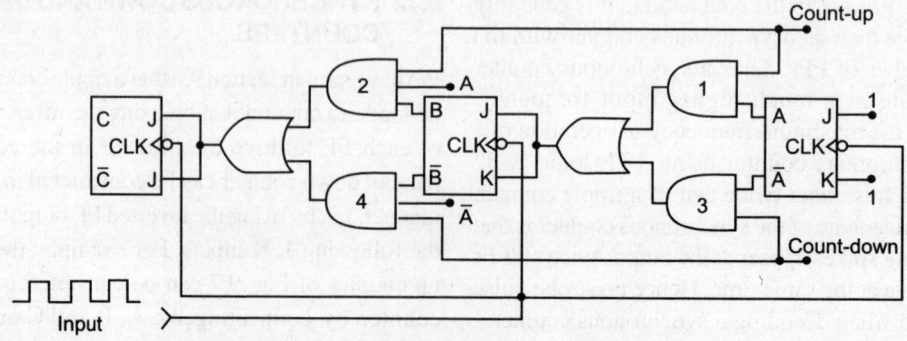

Fig. 9.15: Synchronous up/down MOD-8 counter. (After Ronald J. Tocci, "Digital Systems, Principles and Applications", Prentice Hall of India, (2000))

Table 9.5: Truth table of a down counter

State	C	B	A
7	1	1	1
6	1	1	0
5	1	0	1
4	1	0	0
3	0	1	1
2	0	1	0
1	0	0	1
0	0	0	0
7	1	1	1

requires only to examine the truth table and then writing the proper Boolean expression. For example, to decode state 5, the truth table reveals CBA = 101 and for the gate output to be high during this time, we must use $C, \bar{B}$ and A at the AND gate inputs. The correct Boolean expression is then $5 = C\bar{B}A$. All eight gates necessary to decode the 8 states of the truth table for a given counter are shown in Fig. 9.16.

9.14 SYNCHRONOUS COUNTER DESIGN

Synchronous counters for any given count sequence and modulus can be designed as follows:

1. Find the number of flip-flops required (n)
   $$2^{n-1} \leq N \leq 2^n$$
   where N = counter cycle length
2. Write the count sequence in the tabular form.
3. Determine the FF inputs which must be present for the desired next state from the present state using the excitation table of the flip-flops.
4. Prepare K-map for each FF input in terms of FF outputs as the input variables. Simplify the K-maps and obtain the minimized expressions.
5. Connect the circuit using flip-flops and other gates corresponding to the minimized expressions.

The above design steps will be clear from the following example.

Example 9.2: Design a 3-bit synchronous counter using JK flip-flops

Solution: The number of flip-flops required is 3. Let the flip-flops be FF0, FF1 and FF2 and their inputs and outputs are given below.

Table 9.6 (a)

Flip-flop	Inputs	Outputs
FF0	J_0, K_0	Q_0
Ff1	J_1, K_1	Q_1
FF2	J_2, K_2	Q_2

The count sequence and the required inputs FFs are given in Table 9.6b.

Table 9.6 (b)

Counter state			FF inputs					
Q_2	Q_1	Q_0	FF0		FF1		FF2	
			J_0	K_0	J_1	K_1	J_2	K_2
0	0	0	1	×	0	×	0	×
0	0	1	×	1	1	×	0	×
0	1	0	1	×	×	0	0	×
0	1	1	×	1	×	1	1	×
1	0	0	1	×	0	×	×	0
1	0	1	×	1	1	×	×	0
1	1	0	1	×	×	0	×	0
1	1	1	×	1	×	1	×	1
0	0	0						

The inputs to the flip-flops are determined in the following manner:

Consider one column of the counter state at a time and start from the first row, for example, consider Q_0. Before the first pulse is applied, $Q_0 = 0$ and it is required to be 1 at the end of the first clock pulse. Therefore, to achieve this condition, the values of J_0 and K_0 are 1 and × respectively (from the excitation Table 9.6b). These are entered in the table in the row corresponding to 0 pulse. When the second clock

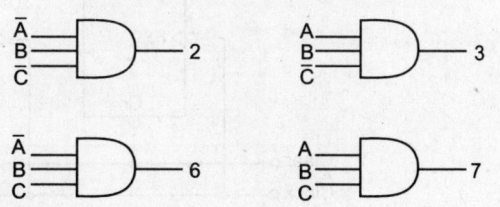

Fig. 9.16: Decoding gates

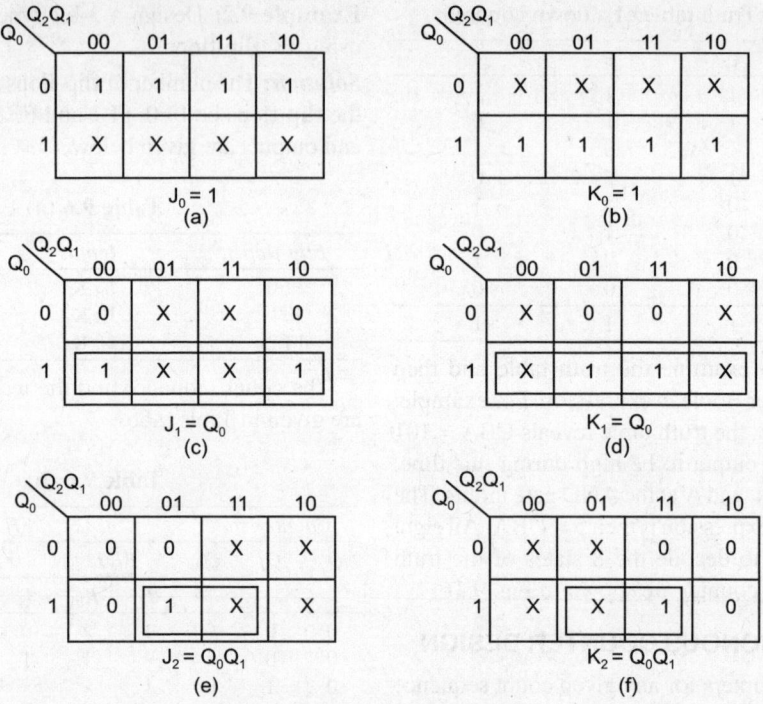

Fig. 9.17: K-maps of example 9.2

pulse is applied, Q_0 is to change from 1 to 0, therefore the required inputs are

$$J_0 = \times, K_0 = 1$$

In a similar manner inputs of each FF are determined. Now we prepare the K-maps (Fig. 9.17) with Q_2, Q_1 and Q_0 as input variables and FF inputs as output variables:

We then minimize the K-maps and the resulting minimized expressions are

$$J_0 = 1, \qquad\qquad K_0 = 1$$
$$J_1 = Q_0, \qquad\qquad K_1 = Q_0$$
$$J_2 = Q_0 Q_1, \qquad\qquad K_2 = Q_0 Q_1$$

The resulting counter circuit is shown in Fig. 9.18.

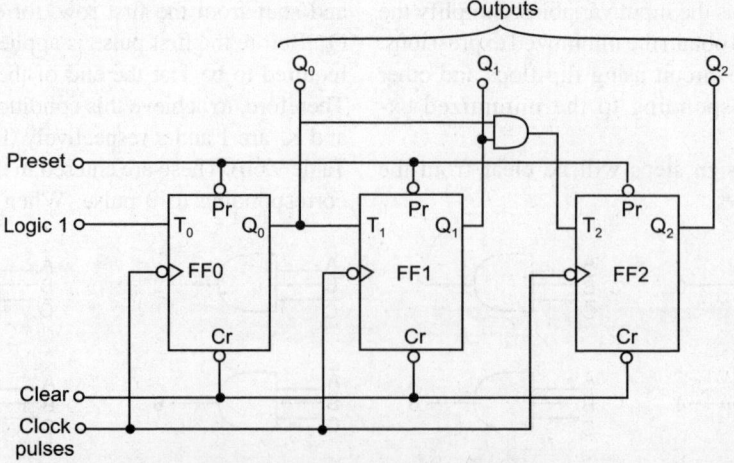

Fig. 9.18: A 3-bit synchronous counter

9.15 USE OF STROBE PULSE IN A RIPPLE COUNTER

In decoding the states of a ripple counter, spikes occur at decode matrix outputs as counter FFs change state (Fig. 9.5). The propagation delay of the FFs creates these false states only for short times. Decode spikes are possible in any counter unless all FFs change state exactly at the same time or only one FF changes state for any clock pulse. To eliminate the spikes, a strobe pulse is used at the decoding gates (Fig. 9.19).

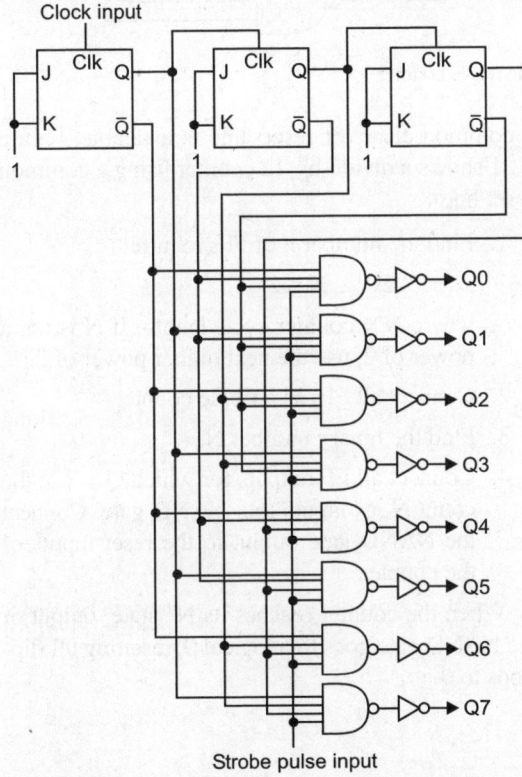

Fig. 9.19: Use of a strobe pulse in a 3-bit ripple counter. Q_0 to Q_7 are decoded outputs

The strobe pulse allows decoding to occur only after all FFs of the counter have become stable.

9.16 FREQUENCY DIVISION IN RIPPLE COUNTERS

Maximum clock frequency f for a counter is given by
$$1/f \leq (N' \times T_p) + T_s$$
where N' = number of flip-flop stages

T_p = propagation delay of one FF.

T_s = Strobe time, width of decoded output pulse.

Consider the example of a 4-stage ripple counter. Assuming each FF has a propagation delay of 50 ns, 200 ns is required for the counter to change from 1111 to 0000. If decoding of any state requires 100 ns then

$$1/f \geq 4 \times 50 + 100 = 300 \text{ ns}$$
$$\Rightarrow \quad f \leq 3.4 \text{ NHz}.$$

First FF in the counter changes state with each clock pulse and therefore divides the input clock frequency by 2. Second FF changes state with every other clock pulse, dividing the frequency by 4. Thus, a 4-stage counter can be used to divide f by 16 or 2^n where n is the number of flip-flops.

Frequency Division by Any Integer

For frequency division by any integer, the following procedure may be adopted:

1. Find the number n of FFs required
   $$2^{n-1} \leq N \leq 2^n$$
   where N = counter cycle length (e.g. 10 for a decade ripple counter).
   If N is not a power of 2, use the next higher power of 2 for 2^n.
2. Connect all flip-flops as a ripple counter (Fig. 9.20).
3. Find the binary number N – 1.
4. Connect all FF outputs that are 1 at the count N – 1 as inputs to a NAND gate. Also feed the clock pulse to the NAND gate.
5. Connect the NAND gate output to the preset inputs of all FFs for which Q = 0 at the count N – 1.

Table 9.7: State-table for counter of Fig. 9.20

State	Q_D	Q_C	Q_B	Q_A
0	0	0	0	0
1	0	0	0	1
2	0	0	1	0
3	0	0	1	1
4	0	1	0	0
5	0	1	0	1
6	0	1	1	0
7	0	1	1	1
8	1	0	0	0
9	1	0	0	1
0	0	1/0	1/0	0

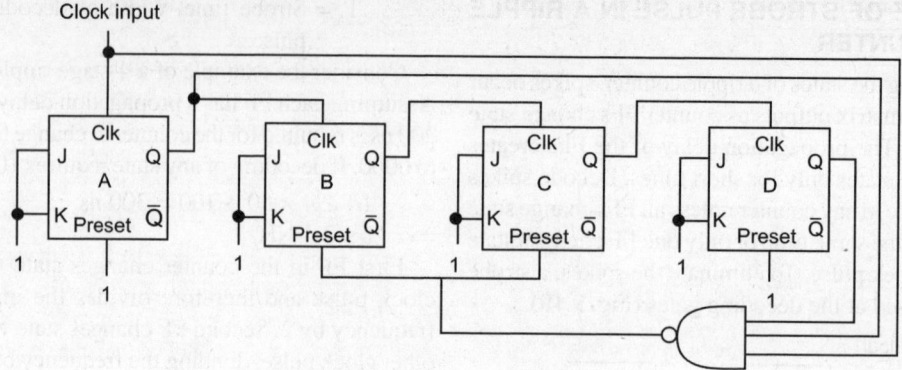

Fig. 9.20: BCD decade ripple counter

The counter resets as follows:

At the positive-going edge of the Nth clock pulse, all FFs are preset to the 1 state. On the NGT of the same clock pulse, all FFs count to the 0 state, i.e. the counter recycles. For N = 10:

1. $2^3 \leq 10 \leq 2^4$ (i.e. 4 FFs are required)
2. N = 10 or 0101 (LSB, is leftmost)
3. N – 1 = 9 or 1001 (LSB leftmost)
4. Connect FFs as shown. Connect outputs of flip-flops A and D as inputs to NAND gate and also clock input to the input of NAND gate.
5. Connect the NAND gate output to the preset inputs of FFs B and C.

This counter will recycle after 10th clock pulse.

9.17 DESIGN OF MOD-12 RIPPLE COUNTER

This counter is designed using following method. (In many IC packages, the preset lines do not exist only a common clear (or reset) line is available. Figure 9.21 shows a divide-by-12 counter using a common reset line).

1. Find the number n of FFs required:
   $$2^{n-1} \leq N \leq 2^n$$
   where N = counter cycle length. If N is not a power of 2, use the next higher power of 2.

2. Connect all FFs as a ripple counter.

3. Find the binary number N.

4. Connect all FF outputs for which Q = 1 at the count N, as inputs to a NAND gate. Connect the NAND gate output to the reset inputs of the counter.

When the counter reaches its Nth state, output of the NAND gate goes to a logical 0, resetting all flip-flops to 0.

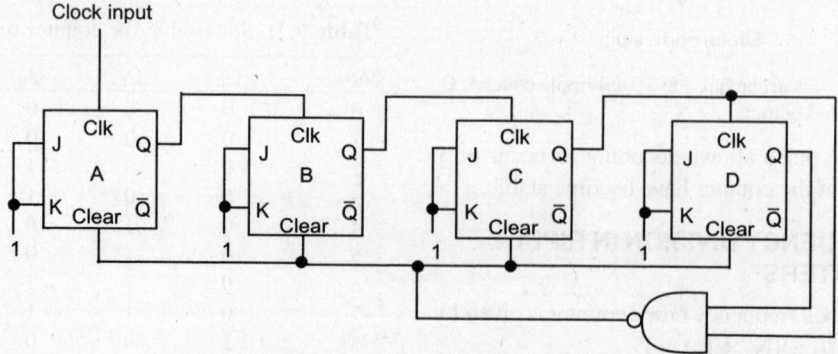

Fig. 9.21: Mod-12 ripple counter

Table 9.8: State-table for counter of Fig. 9.21

State	Q_D	Q_C	Q_B	Q_A
0	0	0	0	0
1	0	0	0	1
2	0	0	1	0
3	0	0	1	1
4	0	1	0	0
5	0	1	0	1
6	0	1	1	0
7	0	1	1	1
8	1	0	0	0
9	1	0	0	1
10	1	0	1	0
11	1	0	1	1
0	1/0	1/0	0	0

Example 9.3: MOD-12 Counter

Since $2^3 < 12 < 2^4$, we require at least four FFs for a modulo 12 counter. With four flip-flops, the counter must be reinitialized when the count reaches 12 or $(1100)_2$, i.e. the FFs must be reset to count again from 0000 to 1011. Figure 9.22 shows the circuit for this.

Since $Q_2 = Q_3 = 1$ is not true for $(0000)_2$ to $(1011)_2$ and is true for $(1100)_2$, the NAND gate will give a 0 at its output when the count reaches 12. This will reset all the four FFs and hence the count will start again.

In modulo-12 counter of Fig. 9.22a, a latch shown in Fig. 9.22b is connected between AB to keep the clear line at 0 till both FFs are cleared. The truth table of the latch is given in Fig. 9.22c.

9.18 JOHNSON COUNTER

In a Johnson counter, the complement of the output of the last FF is connected back to the D input of the first FF. This *feedback* arrangement produces a unique sequence of states as shown in Table 9.9 for a four-bit device. Note that the 4-bit sequence has a total of 8 states.

Table 9.9: Four-bit Johnson counter sequence

Clock pulse	Q_A	Q_B	Q_C	Q_D
0	0	0	0	0
1	1	0	0	0
2	1	1	0	0
3	1	1	1	0
4	1	1	1	0
5	0	1	1	1
6	0	0	1	1
7	0	0	0	1

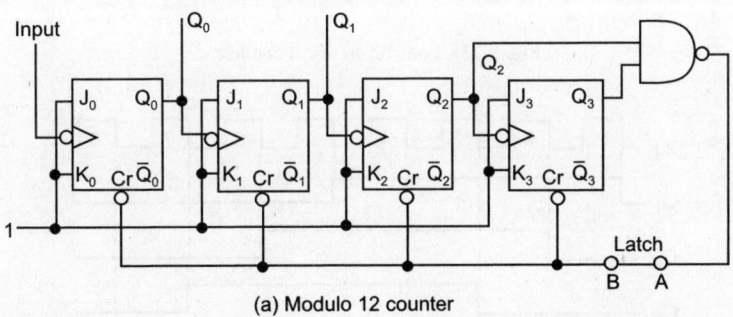

(a) Modulo 12 counter

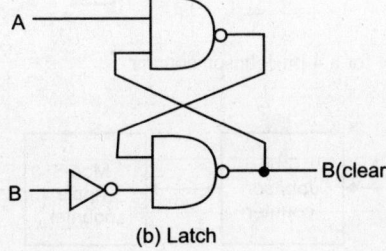

(b) Latch

A	CK	B	
1	1	0	
1	0	1	(after A = 1; CK = 1)
0	0	0	
1	0	0	(after A = 0; CK = 0)
0	1	1	

(c) Truth table of latch

Fig. 9.22

In general, an n-state Johnson counter will produce a modulus of (2n), where n is the number of states in the counter. The implementation of four state In general, an n-state Johnson counter is shown in Fig. 9.23.

The Q output of each state is connected to D input of the next stage (assuming D FF are used) and $\bar{Q}$ output of the last stage is connected to D input of the first stage. The sequence in Table 9.9 shows that the counter "fills up" with 1s from left to right and then "fills up" with 0s from left to right. Diagram of the timing operation are shown in Fig. 9.24.

One advantage of this type of sequence is that it can be readily decoded with two-input AND gates.

Example 9.4: Refer to the multistage counter arrangement of Fig. 9.25. Determine the frequency of output signal.

Solution: The binary decade counter divides the input frequency by 10 so that the output of this counter has a frequency of 100 kHz.

The 10-bit ring counter has a modulus of 10, so this counter divides 100 kHz by 10 to produce an output of 10 kHz.

The 5-bit Johnson counter also has a modulus of 10, thus, its output is 1 kHz.

The Mod-8 ripple counter would divide 1 kHz by 8 to produce a 125 Hz signal at the final output.

Example 9.5: Figure 9.26 shows the circuit represent of a parallel-in/parallel-out 4 bit shift register. It has four parallel data inputs (P_0, P_1, P_2, P_3), an active HIGH parallel enable (PE) input, a serial data input (D), a negative edge triggered clock input ($\overline{CP}$), an active HIGH serial enable input (SE) and four

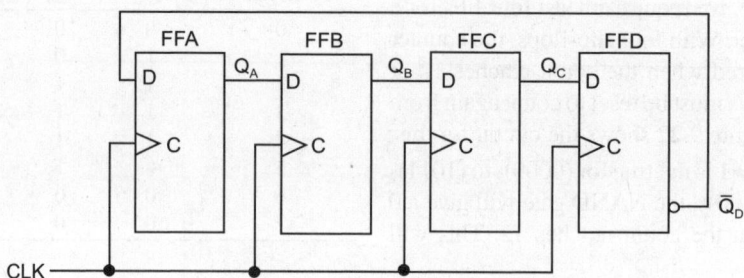

Fig. 9.23: Four-bit Johnson counter

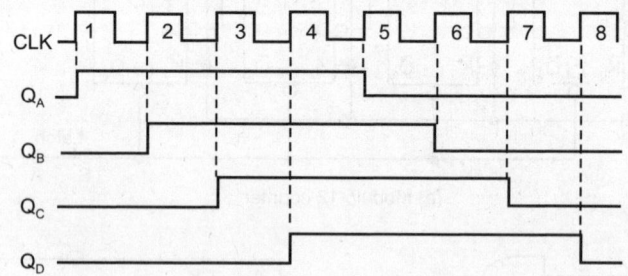

Fig. 9.24: Timing sequence for a 4-bit Johnson counter

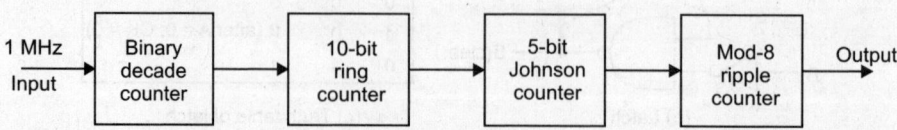

Fig. 9.25

parallel data outputs (Q_0, Q_1, Q_2, Q_3). Show how can we wire this as a shift counter.

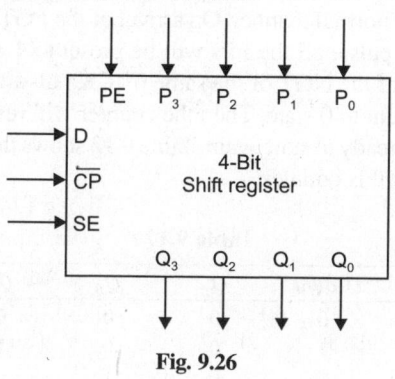

Fig. 9.26

Table 9.10

	Inputs		Function
SE	*PE*	$(\overline{CP})$	
H	×	⅄	Right shift
L	H	⅄	Parallel load
L	L	×	Hold

Solution: The shift counter can be constructed by feeding back $\overline{Q}_0$ (complement of Q_0) to D input. PE input is tied to 1 level, i.e. made active and inputs P_0 to P_3 are tied to 0 level. The counter gets loaded with all 0s with every successive negative going edge of the clock input. Once the SE input is switched over to 1 level, it becomes active and from then onwards, the counter goes to the right shift mode. Figure 9.27 shows the logic circuit. The count sequence would be 0000, 1000, 1100, 1110, 1111, 0111, 0011, 0001 and back to 0000.

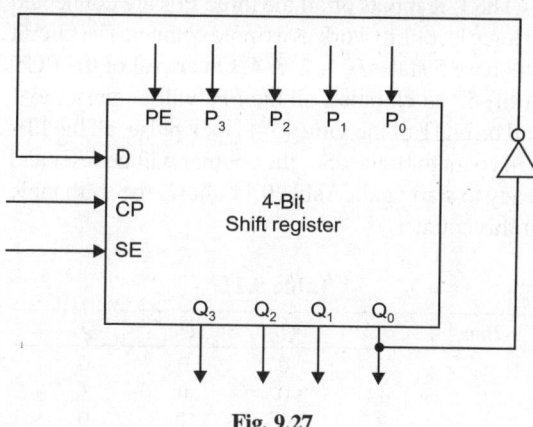

Fig. 9.27

Example 9.6: Design a mod-5 (divide by 5) ripple counter

Solution: For mod-5 ripple counter, value of N = 5

$$\therefore \qquad n = \log_2 N = \log_2 5 = 2.322 \approx 3$$

i.e. 2.322 is rounded off to 3, so we require three J-K negative edge triggered FFs which are to be connected as ripple counter.

The binary equivalent of $N - 1 = (5 - 1) = 4$ is 100. Thus

$$Q = Q_C\ Q_B\ Q_A = 100$$

i.e. only Q_C is high (logic level 1). Hence output Q_C which is high at the count N − 1 is connected as input to the NAND gate as shown in Fig. 9.28. The clock pulse input is supplied as input to this NAND gate. The output of the NAND gate is connected to the preset inputs PR of the J-K FFs A and B for which the output $Q_A = 0$, $Q_B = 0$ were equal to zero at the count N-1 (=100$_2$).

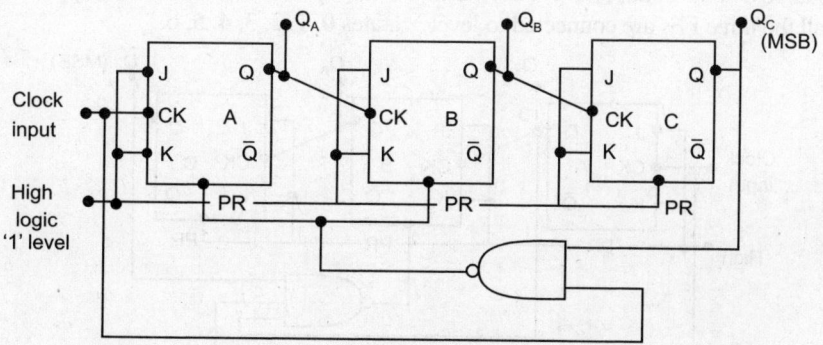

Fig. 9.28: Mod-5 ripple counter

The J, K inputs of all the three FFs are connected to logic level 1 to work as a ripple counter. The circuit will have 5 states 0, 1, 2, 3, 4. On arrival of the PGT of the 5th clock pulse, all the FFs will be preset to 1 and on NGT of the same (5th) clock pulse, all the FFs will count to 0 state. So, the counter will be reset and ready to start again. Table 9.11 shows the truth table of this counter.

Table 9.11

Time	Count	Q_C	Q_B	Q_A
t_0	0	0	0	0
t_1	1	0	0	1
t_2	2	0	1	0
t_3	3	0	1	1
t_4	4	1	0	0
t_5	0	0	0	0

Example 9.7: Design a divide by 6 ripple counter.
Solution: For mod-6, N = 6

∴ $n = \log_2 N = \log_2 6 = 2.585 \approx 3$

So, we require three J-K negative edge triggered FFs which are connected as shown in Fig. 9.29. The binary equivalent of the number

N – 1 = 5 is 101.

Thus output

$$Q = Q_C \, Q_B \, Q_A = 101$$

The outputs Q_C and Q_A are high at logic level 1. Hence outputs Q_C and Q_A which are 1 at the count N–1 are connected as inputs to the NAND gate as shown. The clock pulse input is also to be supplied as input to this NAND gate.

The output of the NAND gate is connected to the preset input PR of J-K FFB for which the output $Q_B = 0$ was equal to 0 at the count N-1 (= 101_2). The J-K inputs of all the three FFs are connected to level

1 to work as a ripple counter, The counter will have six states 0, 1, 2, 3, 4, 5. The circuit counts from 0 to 5 in the normal manner. On arrival of the PGT of the 6th CK pulse, all the FFs will be preset to 1 and on arrival of the NGT of the same 6th CK pulse, all FFs will count to 0 state. Then the counter will reset and will be ready to start again. Table 9.12 shows the truth table of this counter.

Table 9.12

Time	Count	Q_C	Q_B	Q_A
t_0	0	0	0	0
t_1	1	0	0	1
t_2	2	0	1	0
t_3	3	0	1	1
t_4	4	1	0	0
t_5	5	1	0	1
t_6	0	0	0	0

Example 9.8: Design a mod-7 ripple counter
Solution: For mod 7, N = 7

∴ $n = \log_2 N = \log_2 7 = 2.807 \approx 3$

So, we require three J-K FF of negative edge triggered type. The binary equivalent of N-1 = 6 is 110, ∴ output $Q = Q_C \, Q_B \, Q_A = 110$.

The outputs Q_C and Q_B of FFs C and B respectively are high at logic level 1. So these outputs Q_C and Q_B which are 1 at the count N-1 (=110_2) are connected as inputs to the NAND gate. The output of the NAND gate is connected to preset input PR of J-K FFA for which the output $Q_A = 0$ was equal to 0 at the count N-1 (= 110_2) as shown in Fig. 9.30. The J-K inputs of all three FFs A, B and C are connected to logic level 1 (high) so that it works as a ripple counter. It has 7 states 0, 1, 2, 3, 4, 5, 6.

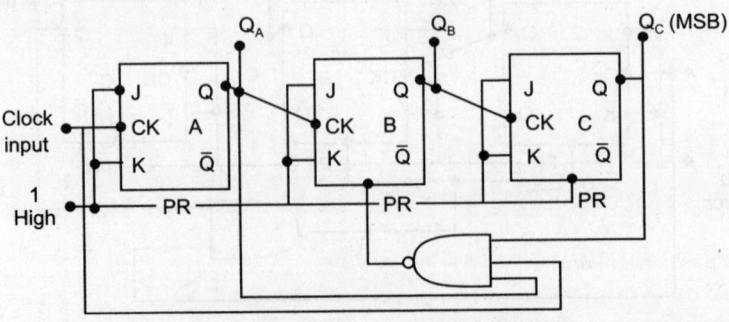

Fig. 9.29: Mod-6 ripple counter

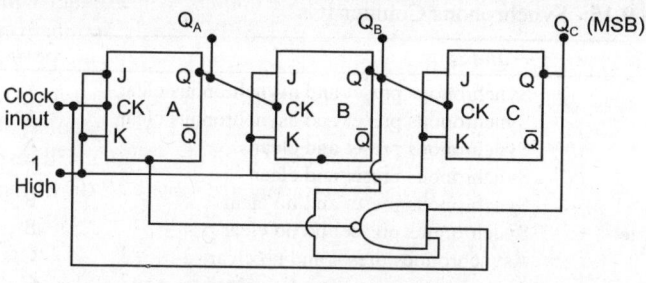

Fig. 9.30: Mod-7 ripple counter

Table 9.13

Time	Count	Q_C	Q_B	Q_A
t_0	0	0	0	0
t_1	1	0	0	1
t_2	2	0	1	0
t_3	3	0	1	1
t_4	4	1	0	0
t_5	5	1	0	1
t_6	6	1	1	0
t_7	0	0	0	0

It counts from 0 to 6 in the usual manner. On arrival of the PGT of the 7th CK pulse, all the FFs will be preset to 1. On arrival 'of the NGT of the same clock pulse, all FFs will count to '0' state. Thus the counter will 'reset' and start again. Table 9.13 shows the truth table of the mod-7 ripple counter.

9.19 APPLICATION OF COUNTERS

We have described sins counter in section 8.15 in the previous chapter. Though Ring counters cannot compete with ripple and synchronous counters when it comes to ordinary counting, yet they are invaluable when it is necessary to control a sequence of operations. Because each ring word has only 1 high bit, one can activate one of several devices. For example, suppose the six small boxes (A to F) of Fig. 9.31 are digital circuits that can be turned on by a high Q-bit. When $\overline{CLR}$ goes low, Q_0 goes high and activates device A. After $\overline{CLR}$ returns to high, successive clock pulse turn on each device for a short time. In other words, as the stored 1-bit shifts left, it turns on B to F in sequence, and then the cycle starts over.

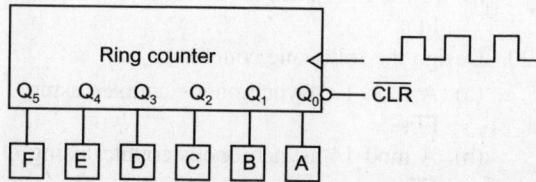

Fig. 9.31: Controlling a sequence of operations using a ring counter

Many digital circuits participate during a computer run. To fetch and execute instructions, a computer has to activate these circuits at precisely the right time and in the right sequence. This is where ring counters are of important use, they produce the ring words for timing different operations during a computer run.

9.20 ASYNCHRONOUS COUNTER ICs

Some asynchronous counters are available in MSI and are given in Table 9.14 along with their features. Depending on these features, these ICs are divided into three groups A, B and C. The group to which a particular IC belongs is also given in the table. All these ICs consist of four MS flip-flops. The load, set, and reset (clear) operations are asynchronous, i.e. independent of the clock pulse.

Table 9.14: Asynchronous counter ICs

IC No.	Description	Features	Group
7490, 74290	BCD counter	Set, reset	A
7492	MOD-12 counter	Reset	B
7493, 74293	4-bit binary counter	Reset	B
74176, 74196	Presettable BCD counter	Reset	C
74177, 74197	Presettable 4-bit binary counter	Reset	C
74390	Dual decade counter	Reset	B
74393	Dual 4-bit binary counter	Reset	B
74490	Dual BCD counters	Set, reset	A

9.21 SYNCHRONOUS COUNTER ICs

Some synchronous counters are available in MSI and are given in Table 9.15 along with some of their features. All these ICs are positive edge triggered, i.e. the change of state, synchronous loading and clearing take place on the positive going transition (PGT) of the input clock pulse. These ICs are grouped as A, B, C and D.

Table 9.15: Synchronous Counter ICs

IC No.	Description	Features	Group
74160	Decade UP counter	Synchronous preset and asynchronous clear	A
74161	4-bit binary UP counter	Synchronous preset and asynchronous clear	A
74162	Decade UP counter	Synchronous preset and clear	A
74163	4-bit binary UP counter	Synchronous preset and clear	A
74168	Decade UP/DOWN counter	Synchronous preset and no clear	B
74169	4-bit binary UP/DOWN counter	Synchronous preset and no clear	B
74100	Decade UP/DOWN counter	Asynchronous preset and no clear	C
74191	4-bit binary UP/DOWN counter	Asynchronous preset and no clear	C
74192	Decade UP/DOWN counter	Asynchronous preset and clear	D
74193	4-bit binary UP/DOWN counter	Asynchronous preset and clear	D

EXERCISE

1. Sketch the block diagram of a twisted (i.e. Johnson) ring counter. Draw the output wave form from each FF of a three stage unit. By what number N does this system divide?

2. Write the count-sequence of a 3-bit binary DOWN counter. Design a ripple counter using Flip-flops for this sequence.

3. Design a 4-bit binary UP/DOWN ripple counter with a control for UP/DOWN counting.

4. Design the following ripple counters using FFs:
   (a) divide-by-5   (b) divide-by-7

5. Design a decade counter to count in the excess-3 code sequence. Use minimum number of J-K FFs.

6. What FF outputs should be connected to the clearing NAND gate to form a MOD-13 counter?

7. What is the difference between the counting sequence of an up counter and a down counter?

8. Describe how an asynchronous down-counter circuit differs from an up counter circuit.

9. What is the advantage of a synchronous counter over an asynchronous counter? What is the disadvantage?

10. Describe various applications of counters.

11. How can a ring counter be converted to a Johnson counter?

12. What is strobing? Explain what is the use of a strobe pulse in a ripple counter.

13. Explain the following terms:
    (a) Modulus of a counter
    (b) Maximum modulus
    (c) Truncated sequence
    (d) Presetting a counter (e) Resetting.

14. Draw the logic diagram of and the timing diagram of a 3-bit binary ripple-up counter and down counter using (a) positive edge triggered FFs (b) negative edge triggered FFs.

15. Write the sequence of stages for a 5-bit Johnson counter.

16. What is the difference between a ring counter and a Johnson counter.

17. How many states does a mod-n counter have?

18. How many states each can an n-bit ring counter and an n-bit Johnson counter have?

19. Design the following counters:
    (a) A mod-7 asynchronous counter using J-K FFs.
    (b) A mod-12 asynchronous counter using J-K FFs.
    (c) A mod-9 synchronous counter using J-K FFs.
    (d) A mod-7 synchronous counter using S-R FFs.

20. Design the following counters:
    (a) A mod-12 asynchronous counter using T FFs.
    (b) A mod-14 asynchronous counter using D FFs.

21. Determine the number of FFs that would be required to build the following counters
    (a) Mod-6, (b) Mod-11, (c) Mod-15, (d) Mod-19, (e) Mod 31.

10

Data-Processing Circuits

10.1 INTRODUCTION

In this chapter, we are going to describe logic circuits that process binary data. These include multiplexers (that select 'one' of many inputs), demultiplexers (circuits with one input and many outputs), decoders (e.g. binary to decimal decoder), encoders (which convert an active input signal into a coded output signal), code converters. Read Only Memories (ROM), programmable ROMs (PROM), CCD memories, PLAs, seven segment visible display and parity checkers and generators.

At the end of the previous chapter, we have seen how decoding gates can be used to give a high output corresponding to a given state. In any digital system, instructions (or numbers) are conveyed by means of binary levels or pulse trains. If say, 4-bits of a character are to convey instructions, then 16 different instructions are possible. This information is *coded* in binary form. Frequently, there arises a need such that for each of the 16 codes, one and only one line is to be excited. This process of identifying a particular code is called *decoding*.

In a short while, we are about to discuss BCD-to-decimal decoder. Though, we have been familiar with BCD system in Chapter 2, we will again consider the BCD system here for sake of convenience.

10.2 BCD-TO-DECIMAL DECODER

10.2.1 Binary-Coded-Decimal System

The code translates decimal numbers by replacing each decimal digit with a combination of 4 binary digits. Though, we can have a wide choice of BCD codes, the most preferred is the "natural binary-coded-decimal" or the 8421 code illustrated in Table 10.1.

The first 4-bit set on the right represent units, the second represent tens, the third hundreds, thus decimal number 264 requires three 4-bit set as shown. Note that this BCD code can represent any number between 0 and 999.

10.2.2 The Decoder

In order to decode a BCD instruction representing a decimal digit say 5, we need a four input AND gate excited by 4 BCD bits, A, B, C, D. The output of the AND gate will be 1 if the BCD inputs are A = 1 (LSB), B = 0, C = 1 and D = 0.

Since this code represents the decimal number 5, we label this output as "line 5". Figure 10.1 represents a BCD-to-decimal decoder. This MSI has four inputs A, B, C, D and 10 output lines (0 to 9). In addition, there must be a ground and a power supply connection, so a 16-pin package is required. The

Table 10.1: BCD representation for the decimal number 264

Weighting factor...	800	400	200	100	80	40	20	10	8	4	2	1
BCD code.....	0	0	1	0	0	1	1	0	0	1	0	0
Decimal digits.....		2				6				4		

complementary inputs $\overline{A}$, $\overline{B}$, $\overline{C}$, $\overline{D}$ are obtained from inverters on the chip. Since NAND gates are used, an output is 0 for the correct BCD code (*). The system in Fig. 10.1 is referred to as "4-to-10 line decoder" since a 4-bit input code selects 1 of 10 output lines.

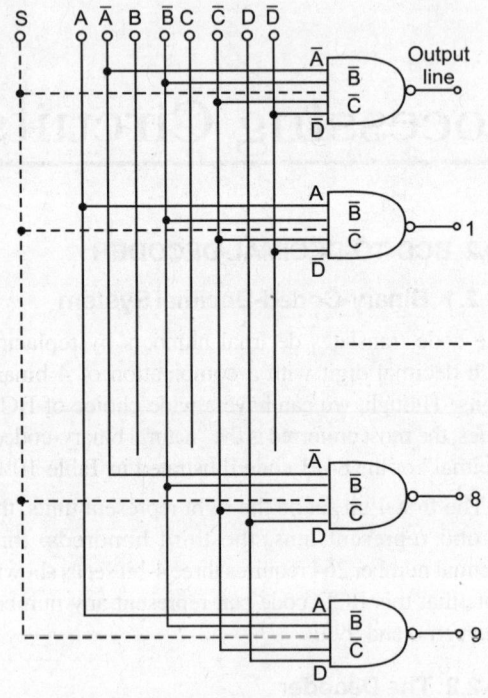

(a)

D	C	B	A	Line	Line in put
0	0	0	0	0	$\overline{D}\,\overline{C}\,\overline{B}\,\overline{A}$
0	0	0	1	1	$\overline{D}\,\overline{C}\,\overline{B}\,A$
0	0	1	0	2	$\overline{D}\,\overline{C}\,B\,\overline{A}$
0	0	1	1	3	$\overline{D}\,\overline{C}\,B\,A$
0	1	0	0	4	$\overline{D}\,C\,\overline{B}\,\overline{A}$
0	1	0	1	5	$\overline{D}\,C\,\overline{B}\,A$
0	1	1	0	6	$\overline{D}\,C\,B\,\overline{A}$
0	1	1	1	7	$\overline{D}\,C\,B\,A$
1	0	0	0	8	$D\,\overline{C}\,\overline{B}\,\overline{A}$
1	0	0	1	9	$D\,\overline{C}\,\overline{B}\,A$

Fig. 10.1: (a) BCD-to-decimal decoder. Assume S = 1. Lines 2 to 7 are not indicated. It is a decoder if dashed lines are not there. The dashed lines convert the system into a demultiplexer if S represents the input signal. (b) Truth table

---

* (and 1 for any other (invalid) code)

When it is described to decode only during certain intervals of time, an additional input, called a *strobe* is added to each NAND gate and all strobe inputs are tied together to a binary signal input S. For S = 1, a gate is enabled and decoding takes place. If S = 0, decoding is inhibited.

10.3 DEMULTIPLEXER

A demultiplexer is a system for transmitting a binary signal (serial data) on one of N lines, the particular line being selected by means of an address, i.e. the control input. A demultiplexer is therefore a logic circuit with one input and many outputs. By applying control signals, we can steer the input signal to one of the many output lines. Figure 10.2 shows a 1-to-16 demultiplexer. The data input bit is labeled D. This data bit is transmitted to one of the output lines. But which line, depends on the value of ABCD, the control input, e.g. when

$$ABCD = 0000$$

The upper AND gate is enabled while all other AND gates are disabled and data bit D is transmitted only to the Y_0 output, giving

$$Y_0 = D$$

If D is low, Y_0 is low. If D is high, Y_0 is high i.g. the value of Y_0 depends on the value of D. All other outputs are in the low state.

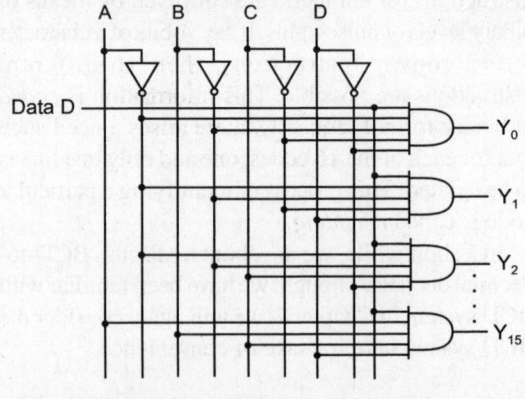

Fig. 10.2: 1-to-16 demultiplexer

If ABCD = 1111, all gates are disabled except the bottom AND gate. Then D is transmitted only to the Y_{15} output and $Y_{15} = D$.

10.4 THE IC74154

This is a 1-to-16 demultiplexer with pin-diagram as shown in Fig. 10.3.

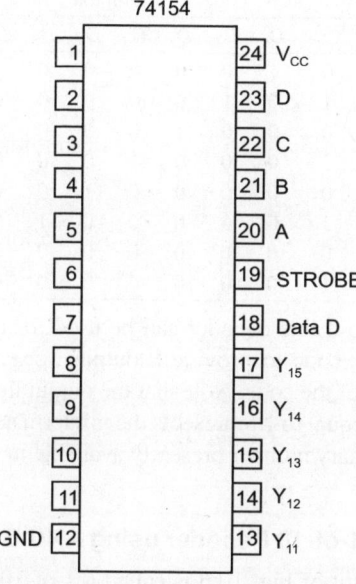

Fig. 10.3: Pin out diagram of 74154

Pin 18 is for the input data D. Pins 20 to 23 are for the control bits ABCD. Pins 1 to 11 and 13 to 17 are for the output bits Y_0 to Y_{15}. Pin 19 is for the strobe (an active low input). Pin 24 is for V_{CC} and pin 12 for ground.

Table 10.2 shows the truth table of a 74154. First note that STROBE input has to be low to activate the IC 74154. When it is low, the control input ABCD determines which of the output lines is low when the data input is low. When the data input is high, all output lines are high. When the strobe is high, all output lines are also high.

Figure 10.4 shows a schematic diagram of a 74154. There is one input data bit (pin 18) under the control of nibble ABCD. The data bit is automatically steered to the output line whose subscript is the decimal equivalent of ABCD. The bubble on the STROBE pin indicates an active low input.

10.5 DECODERS

A decoder is a combinational circuit that converts binary information from n input lines to a maximum of 2^n unique output lines. If the n-bit decoded information has some don't care combinations, the decoded output will have less than 2^n outputs. Thus we have n-to-m line decoders where $m \leq 2^n$. The purpose of a decoder is to generate 2^n (or less) minterms of n input variables.

Table 10.2: 74154 Truth table

Strobe Data		A	B	C	D	Y_0	Y_1	Y_2	Y_3	Y_4	Y_5	Y_6	Y_7	Y_8	Y_9	Y_{10}	Y_{11}	Y_{12}	Y_{13}	Y_{14}	Y_{15}
L	L	0	0	0	0	L	H	H	H	H	H	H	H	H	H	H	H	H	H	H	H
L	L	0	0	0	1	H	L	H	H	H	H	H	H	H	H	H	H	H	H	H	H
L	L	0	0	1	0	H	H	L	H	H	H	H	H	H	H	H	H	H	H	H	H
L	L	0	0	1	1	H	H	H	L	H	H	H	H	H	H	H	H	H	H	H	H
L	L	0	1	0	0	H	H	H	H	L	H	H	H	H	H	H	H	H	H	H	H
L	L	0	1	0	1	H	H	H	H	H	L	H	H	H	H	H	H	H	H	H	H
L	L	1	1	1	0	H	H	H	H	H	H	L	H	H	H	H	H	H	H	H	H
L	L	0	1	1	1	H	H	H	H	H	H	H	L	H	H	H	H	H	H	H	H
L	L	1	0	0	0	H	H	H	H	H	H	H	H	L	H	H	H	H	H	H	H
L	L	1	0	0	1	H	H	H	H	H	H	H	H	H	L	H	H	H	H	H	H
L	L	1	0	1	0	H	H	H	H	H	H	H	H	H	H	L	H	H	H	H	H
L	L	1	0	1	1	H	H	H	H	H	H	H	H	H	H	H	L	H	H	H	H
L	L	1	1	0	0	H	H	H	H	H	H	H	H	H	H	H	H	L	H	H	H
L	L	1	1	0	1	H	H	H	H	H	H	H	H	H	H	H	H	H	L	H	H
L	L	1	1	1	0	H	H	H	H	H	H	H	H	H	H	H	H	H	H	L	H
L	L	1	1	1	1	H	H	H	H	H	H	H	H	H	H	H	H	H	H	H	L
L	L	×	×	×	×	H	H	H	H	H	H	H	H	H	H	H	H	H	H	H	H
H	L	×	×	×	×	H	H	H	H	H	H	H	H	H	H	H	H	H	H	H	H
H	H	×	×	×	×	H	H	H	H	H	H	H	H	H	H	H	H	H	H	H	H

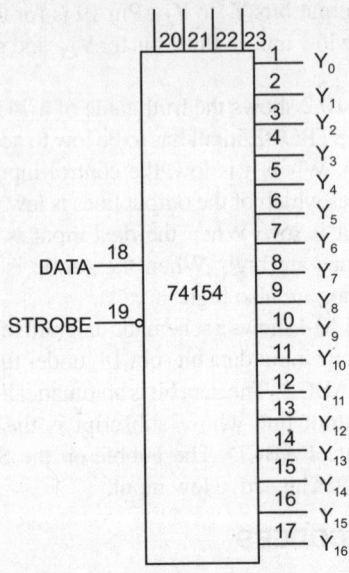

Fig. 10.4: Schematic diagram of IC 74154

10.5.1 3-to-8 Line Decoder

A 3-to-8 line decoder is shown in Fig. 10.5. The three inputs are decoded into eight outputs, each output representing one of the minterms of the three input variables.

The truth table is given in Table 10.3

Table 10.3: Truth table of a 3-to-8 line decoder

Inputs			Outputs							
x	y	z	D_0	D_1	D_2	D_3	D_4	D_5	D_6	D_7
0	0	0	1	0	0	0	0	0	0	0
0	0	1	0	1	0	0	0	0	0	0
0	1	0	0	0	1	0	0	0	0	0
0	1	1	0	0	0	1	0	0	0	0
1	0	0	0	0	0	0	1	0	0	0
1	0	1	0	0	0	0	0	1	0	0
1	1	0	0	0	0	0	0	0	1	0
1	1	1	0	0	0	0	0	0	0	1

A 3-to-8 line decoder can be used for decoding any 3-bit code to provide 8 outputs, one for each element of the code. Note that the output line whose value is equal to 1 represents the minterm equivalent of the binary number presently available in the input lines.

10.5.2 1-of-10 Decoder using AND Gates

The circuit of Fig. 10.6 is called a 1-of-10 decoder because only 1 of the 10 output lines is high. For instance, when ABCD is 0011, only the Y_3 AND gate has all high inputs, therefore only, then Y_3 output is

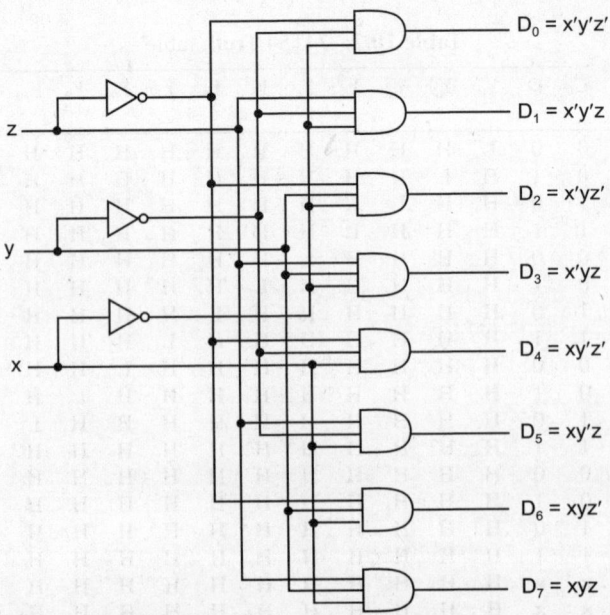

Fig. 10..5: A 3-to-8 line decoder

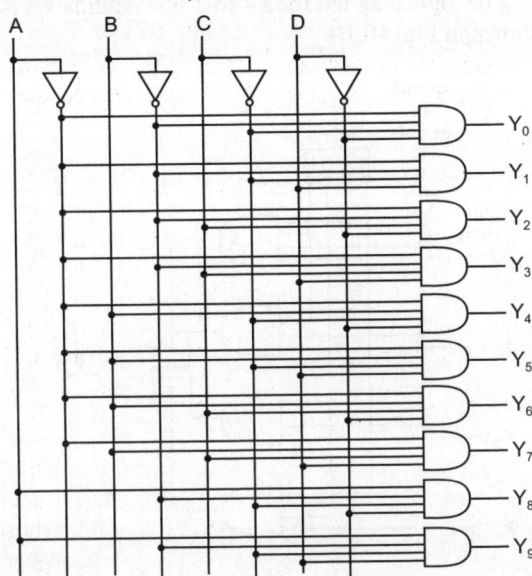

Fig. 10.6: 1-of-10 decoder

high. If ABCD is 1000, only the Y_8 AND gate has all high inputs and only the Y_8 output goes high. Thus, it can be seen that the subscript of the high output always equals the decimal equivalent of the input BCD digit. For this reason, this circuit is also called a BCD-to-decimal decoder.

10.5.3 1-of-16 Decoder

We have seen above that a decoder is similar to a demultiplexer with one exception that there is no data input. The only inputs are the control bits ABCD. A 1-of-16 decoder is shown in Fig. 10.7.

This logic circuit is so called because only 1 of the 16 output lines is high. For example, when ABCD is 0001, only the Y_1 AND gate has all inputs high and so, only Y_1 output is high. If ABCD changes to 0100, only the Y_4 AND gate has all inputs high and so, only Y_4 output goes high.

10.6 CONVERSION OF A 2-TO-4 LINE DECODER INTO A 1 × 4 DEMULTIPLEXER

A decoder with an enable input can function as a demultiplexer. Figure 10.8a shows a 2-to-4 line decoder with an enable input E and Fig. 10.8b shows the truth table.

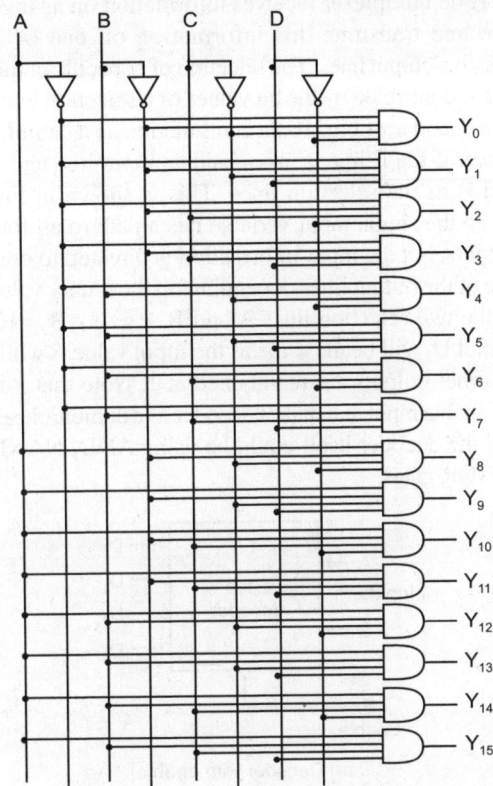

Fig. 10.7: 1-of-16 decoder

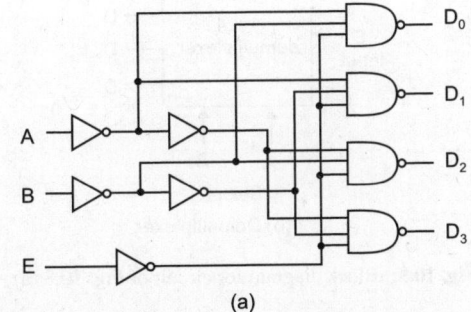

(a)

E	A	B	D_0	D_1	D_2	D_3
1	×	×	1	1	1	1
0	0	0	0	1	1	1
0	0	1	1	0	1	1
0	1	0	1	1	0	1
0	1	1	1	1	1	0

(b)

Fig. 10.8: A 2-to-4 line decoder with enable (E) input (a) Logic diagram, (b) Truth table

A demultiplexer receives information on a single line and transmits this information on one of 2^n possible output lines. The selection of a specific output line is controlled by the bit values of n selection lines. The decoder of Fig. 10.8a can function as a demultiplexer if the E line is taken as data input line and A and B as the selection lines. This is shown in Fig. 10.8b the single input variable has a path to all four outputs, but the input information is directed to only one of the output lines depending on the binary value of the two selection lines A and B, e.g. if AB = 10, output D_2 will be the same as the input value E while all other outputs are maintained at 1. Note that it is the enable input that makes decoder as a demultiplexer and the decoder itself could be using AND, NAND or NOR gates.

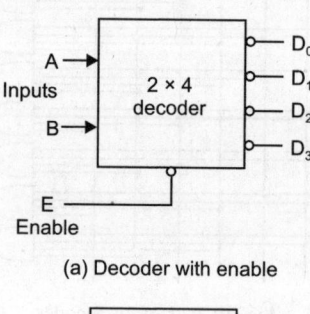

(a) Decoder with enable

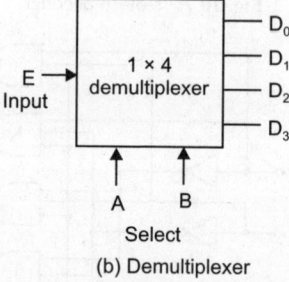

(b) Demultiplexer

Fig. 10.9: Block diagram for circuit of Fig. 10.8 (a)

10.7 MULTIPLEXER

Multiplexer means *many into one*. A multiplexer is a circuit with many inputs but only one output. By applying control signals, we can steer any input to the output. Thus, the function performed by a multiplexer is to select 1 out of N input data sources and to transmit the selected data to a single information channel. (**Note** that a multiplexer performs the inverse process of a demultiplexer).

The logic diagram for a 4-to-1 line multiplexer is shown in Fig. 10.10.

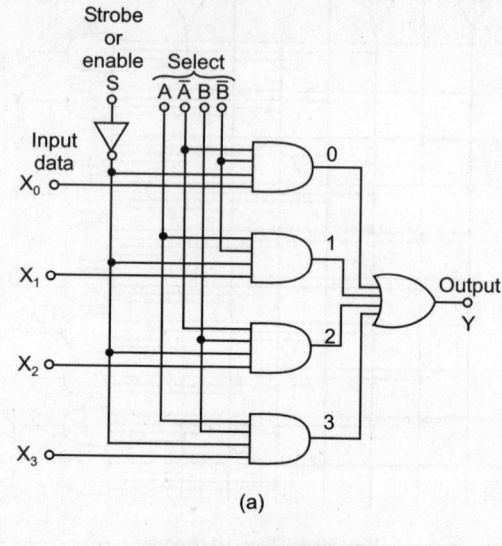

(a)

Input line	Input to the AND gates
0	$\overline{A}\,\overline{B}\overline{S}\ X_0$
1	$A\overline{B}\overline{S}\ X_1$
2	$\overline{A}B\overline{S}\ X_2$
3	$AB\overline{S}\ X_3$

(b)

Fig. 10.10: A 4-to-1 line multiplexer

If the select code is A = 1, B = 0 and strobe is enabled (S = 0), then X_1 appears at the output Y. If the address is A = 1, B = 1, then $Y = x_3$ provided S = 0, etc.

Multiplexers are also available for selecting 1-of-8 or 1-of-16 data sources. TI 74150 is a 24-pin IC with 16 data inputs, a 4-bit select code, a strobe input, one output, a power supply lead and a ground terminal.

10.7.1 A 4-to-1 line Multiplexer

A digital multiplexer is a combinational circuit that selects binary information from one of many input lines and directs it to a single output line. The selection of a particular input line is controlled by a set of selection lines. Normally there are 2^n input lines and n selection lines whose bit combinations determine which input is selected, A 4-line to 1 line multiplexer

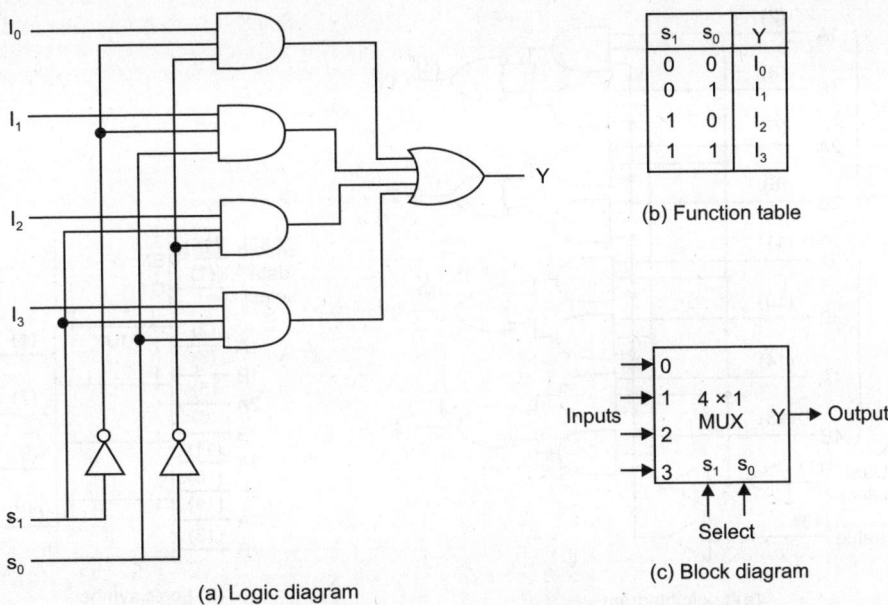

(b) Function table

s_1	s_0	Y
0	0	I_0
0	1	I_1
1	0	I_2
1	1	I_3

(a) Logic diagram

(c) Block diagram

Fig. 10.11: A 4-to-1 line multiplexer

is shown in Fig. 10.11. Each of the four input lines I_0 to I_3 1s applied to one input of an AND gate. Selection lines s_1 and s_0 are decoded to select a particular AND gate. The function table in the figure lists the input-to-output path for each possible bit combination of the selection lines. When this MSI function is used in the design of a digital system, it is represented in block diagram form as shown in Fig. 10.11c.

The size of a multiplexer is specified by the number 2^n of its input lines and the single output line. It is then implied that it also contains n selection lines. A multiplexer is often abbreviated MUX.

10.8 THE IC 74157 MULTIPLEXER

The IC 74157 is a quadrupole, two-input data selector/multiplexer. Each of the four multiplexers shares a common *data-select* line and a common *enable*. This is shown in Fig. 10.12. Because there are only two inputs to be selected in each multiplexer, a single data-select input is sufficient ($2^1 = 2$). It can be seen in the logic diagram that the data select input is ANDed with the B input of each two input multiplexer and the complement of data select is ANDed with each A input.

A LOW on the $\overline{\text{enable}}$ input allows the selected input data to pass through to the output. A HIGH on the $\overline{\text{enable}}$ input prevents data from going through

to the output, i.e. it *disables* the multiplexers. In the logic symbol, note that the four multiplexers are indicated by the partition-lines and the inputs (*Enable* and Data select) which are common to all four multiplexers are indicated as inputs to the block at the top which is called the *common control block*, i.e. all labels within the upper block apply to the other blocks below it.

Note the 1 and $\overline{1}$ labels in the MUX blocks and the G1 level in the common control block. G1 indicates an AND relationship between the data-select input and the data inputs with 1 and $\overline{1}$ labels (Here, the $\overline{1}$ means that AND relationship applies to the complement of the G1 input). When the data select input is HIGH, the B inputs are selected and when the data-select input is LOW, the A inputs are selected, thus G is used to denote AND-dependency.

10.9 A 16-TO-1 MULTIPLEXER

Figure 10.13 shows a 16-to-1 multiplexer and output bit depends on the input data-bit selected. The input bits are labelled $D_0, D_1.....D_{15}$.

Only one of these is transmitted to the output with the help of A, B, C, D the control inputs.

Which one depends on the value of ABCD, the control input, e.g. when

$$ABCD = 0000,$$

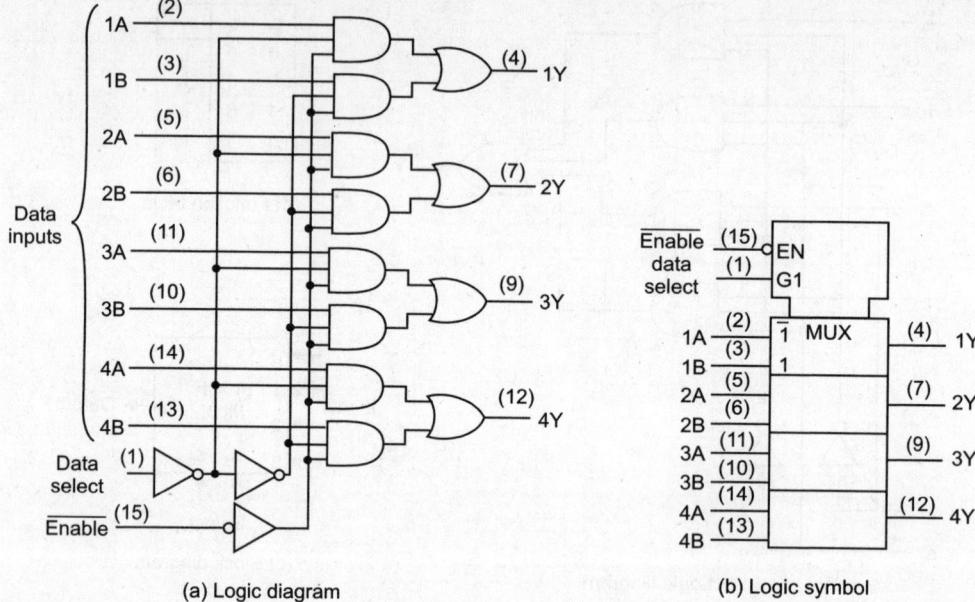

(a) Logic diagram   (b) Logic symbol

Fig. 10.12: The 74157 multiplexer

the upper AND gate is enabled while all other AND gates are disabled. Therefore data D_0 is transmitted to the output giving

$$Y = D_0.$$

If D_0 is low, Y is low, if D_0 is high, Y is high. It is to be noted here that the control nibble determines which of the input data bit is transmitted to the output.

10.9.1 The IC 74150

This is a 16-to-1 multiplexer, with OR gate of Fig. 10.13 replaced by a NOR gate. The pin diagram is shown in Fig. 10.14.

Pins 1 to 8 and 16 to 23 are for the input data bits D_0 to D_{15}. Pins 11, 13, 14 and 15, are for control bits A, B, C, D. Pin 10 is the output. Pin 9 is for *strobe*, i.e. an input signal that disables or enables the multiplexer. A low strobe enables the multiplexer, so that output Y equals the complement of the input data bit

$$Y = \bar{D}_n$$

Where n is the decimal equivalent of ABCD. On the other hand, a high strobe disables the multiplexer and forces the output into the high state. With a high strobe, the control bits are ineffective. The truth table is given in Table 10.4.

Table 10.4: 74150 Truth table

Strobe	A	$\bar{B}$	C	D	Y
L	0	0	0	0	$\bar{D}_0$
L	0	0	0	1	$\bar{D}_1$
L	0	0	1	0	$\bar{D}_2$
L	0	0	1	1	$\bar{D}_3$
L	0	1	0	0	$\bar{D}_4$
L	0	1	0	1	$\bar{D}_5$
L	0	1	1	0	$\bar{D}_6$
L	0	1	1	1	$\bar{D}_7$
L	1	0	0	0	$\bar{D}_8$
L	1	0	0	1	$\bar{D}_9$
L	1	0	1	0	$\bar{D}_{10}$
L	1	0	1	1	$\bar{D}_{11}$
L	1	1	0	0	$\bar{D}_{12}$
L	1	1	0	1	$\bar{D}_{13}$
L	1	1	1	0	$\bar{D}_{14}$
L	1	1	1	1	$\bar{D}_{15}$
H	×	×	×	×	H

Tables 10.5 and 10.6 show some of the available multiplexer ICs and demultiplexer ICs respectively.

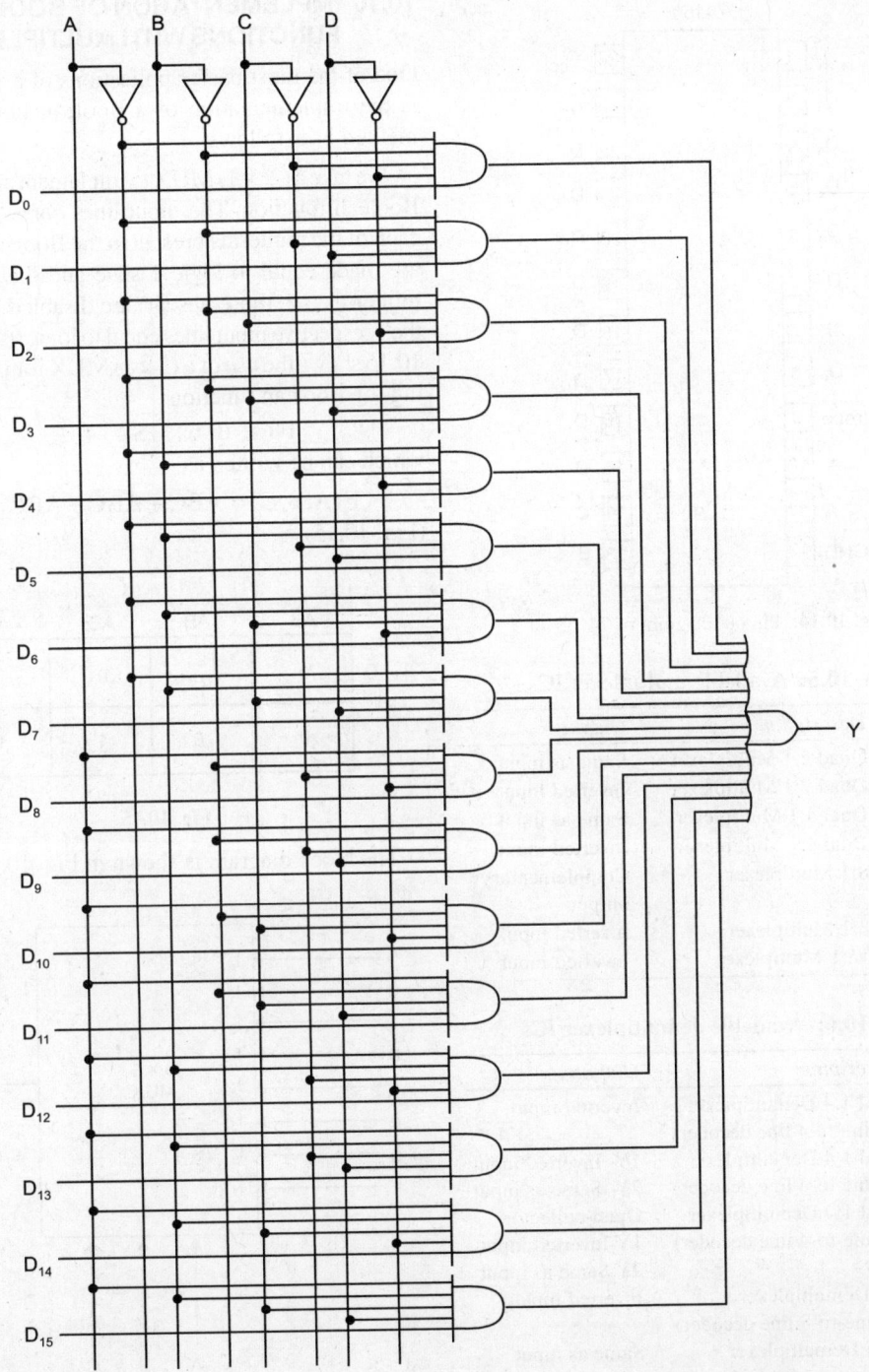

Fig. 10.13: 16-to-1 multiplexer (After A.P. Malvino and D.P. Leach "Digital Principles and Applications" McGraw-Hill (1986))

74150

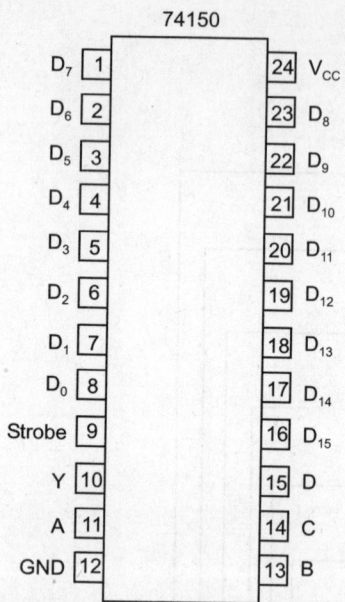

Fig. 10.14: Pin out diagram of 74150

Table 10.5: Available multiplexer ICs

IC No.	Description	Output
74157	Quad 2:1 Multiplexer	Same as input
74158	Quad 2:1 Multiplexer	Inverted input
74153	Dual 4:1 Multiplexer	Same as input
74352	Dual 4:1 Multiplexer	Inverted input
74151A	8:1 Multiplexer	Complementary output
74152	8:1 Multiplexer	Inverted input
74150	16:1 Multiplexer	Inverted input

Table 10.6: Available demultiplexer ICs

IC No.	Description	Output
74139	Dual 1:4 Demultiplexer (2-line-to-4 line decoder)	Inverted input
74155	Dual 1:4 Demultiplexer (2-line-to-4 line decoder)	1Y- Inverted input 2Y- Same as input
74156	Dual 1:4 Demultiplexer (2-line-to-4 line decoder)	Open-collector 1Y-Inverted input 2Y-Same as input
74138	1:8 Demultiplexer (3-line-to-8 line decoder)	Inverted input
74154	1:16 Demultiplexer (4-line-to-16 line decoder)	Same as input
74159	1:16 Demultiplexer (4-line-to-16 line decoder)	Same as input Open-collector

10.10 IMPLEMENTATION OF BOOLEAN FUNCTIONS WITH MULTIPLEXER

One of the most basic applications of a multiplexer is the implementation of a Boolean function. The method is as follows:

We take a $(2^n \times 1)$ MUX to implement an n variable Boolean function. The input lines corresponding to each of the minterms present in the Boolean function are made equal to logic 1 state and the rest of the minterms (i.e. those absent) are disabled by making their respective input lines equal to logic 0. e.g. Figure 10.16 shows the use of a (3×1) MUX for implementing the Boolean function

$$F (A, B, C) = S\ 2, 4, 7.$$

which can be written as

$$F(A, B, C) = \bar{A}B\bar{C} + A\bar{B}\bar{C} + ABC \qquad (i)$$

(Fig. 10.15).

	$\bar{A}\bar{B}$	$\bar{A}B$	AB	$A\bar{B}$
$\bar{C}$	₁0	₂1	₃0	₄1
C	₅0	₆0	₇1	₈0

Fig. 10.15

The block diagram is shown in Fig. 10.16

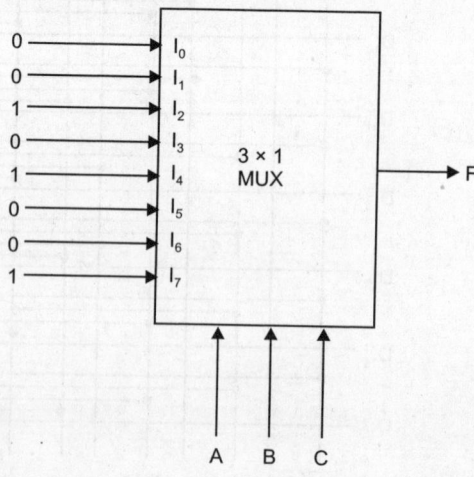

Fig. 10.16: A 3 ×1 multiplexer used to implement Boolean function of eqn. (1)

10.11 MULTIPLEXERS FOR PARALLEL TO-SERIAL DATA CONVERSION

As the data is processed parallely in various digital systems, the processing performed is faster. When the data is to be sent to large distances, it is done so serially, because otherwise, it would require a large number of transmission-lines. A multiplexer can be used for parallel-to-serial conversion. Figure 10.17 shows one such arrangement. An 8×1 MUX has been used to convert 8-bit parallel binary data to serial form. A 3-bit Mod-8 counter is used to control the selection inputs.

When the counter goes through 000 to 111, the multiplexer output goes through X_0 to X_7. This process takes a total of 8 clock cycles.

10.12 CONTROLLED SHIFT REGISTER USING MULTIPLEXERS

Controlled shift registers can be made using MUXs and JK flip-flop. Such a shift register can be an integrated circuit in a single chip. We will illustrate the logical design of such a shift register. A block diagram of such a shift register is given as Fig. 10.18. Table 10.7 gives the functions to be performed by this shift register.

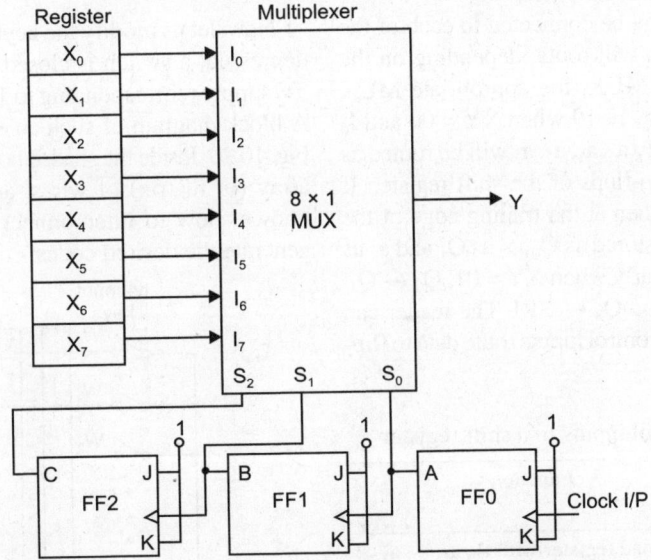

Fig. 10.17: A multiplexer used for parallel to serial conversion

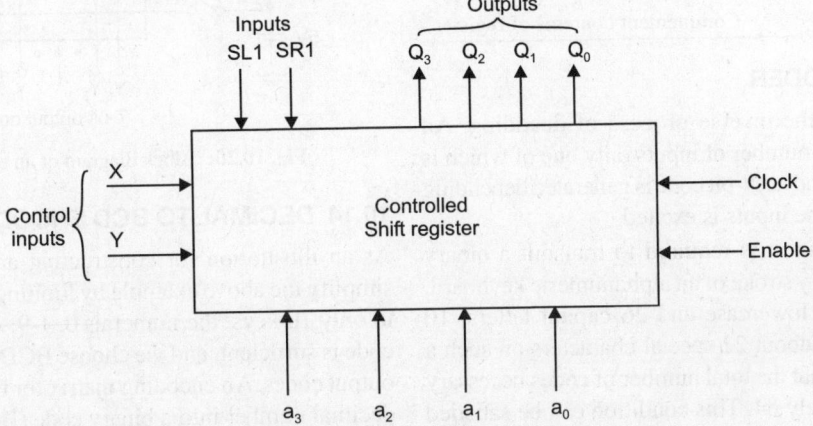

Fig. 10.18: A controlled shift register

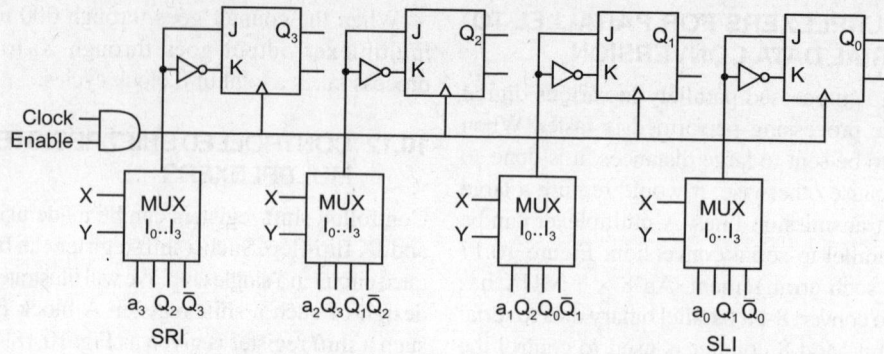

Fig. 10.19: Realization of controlled shift register with MUXs

A MUX output may be connected to each of the flip-flop inputs which will route, depending on the control inputs to the MUX, the appropriate MUX input. Referring to Fig. 10.19 when $XY = 00$ and I_0 input to MUXs, namely a_0, a_1, a_2, a_3 will be routed as the inputs to the 4 flip-flops of the shift register. If the enable input is 1 then at the trailing edge of the clock pulse a_0 will be stored as Q_0, a_1 as Q_1 and a_2 as Q_2 and a_3 as Q_3. Similarly when $XY = 01$, $Q_0 \leftarrow Q_1$, $Q_1 \leftarrow Q_2$, $Q_2 \leftarrow Q_3$ and $Q_3 \leftarrow$ SRI. The reader can deduce how the other control inputs route data to flip-flops.

Table 10.7: Control inputs to a shift register

Control inputs		Function
X	Y	
0	0	Load register with a_1, a_2, a_3
0	1	Shift right register
1	0	Shift left register
1	1	Complement contents of register

10.13 ENCODER

Encoding is the inverse process of decoding. An encoder has a number of inputs only one of which is in the 1 state and an N-bit code is generated depending on which of the inputs is excited.

Suppose that it is required to transmit a binary code with every stroke of an alphanumeric keyboard. There are 26 lowercase and 26 capital letters, 10 numerals and about 22 special characters on such a keyboard so that the total number of codes necessary is approximately 84. This condition can be satisfied with a minimum of 7 bits ($2^7 = 128$) ($\because 2^6 = 64$ only).

Now, let us modify the keyboard so that if a key is depressed, a switch is closed, thereby connecting a 5V supply (corresponding to 1 state) to an input line. A block diagram of such an encoder is indicated in Fig. 10.20. Inside the shaded box, there is a rectangular array (or matrix) of wires, and this requires to be known, how to interconnect these wires so as to generate the desired codes.

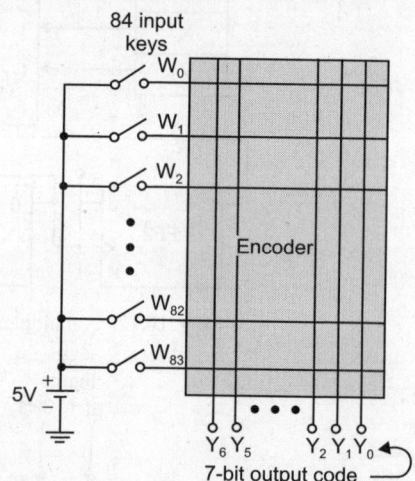

Fig. 10.20: Block diagram of an encoder

10.14 DECIMAL TO BCD ENCODER

As an illustration for constructing an encoder, we simplify the above example by limiting the keyboard to only 10 keys, the numerals 0, 1-9. A 4-bit output code is sufficient, and we choose BCD words for the output codes. An encoding matrix for transforming a decimal number into a binary code (BCD) is shown in Fig. 10.21.

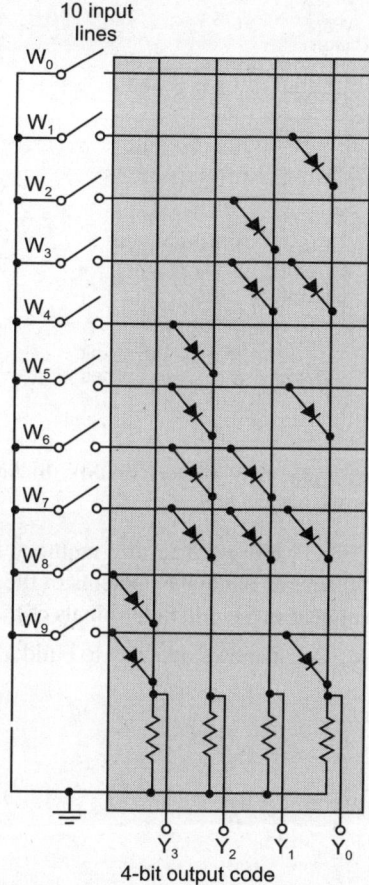

10 input lines

W_0
W_1
W_2
W_3
W_4
W_5
W_6
W_7
W_8
W_9

Y_3 Y_2 Y_1 Y_0

4-bit output code

Fig. 10.21: An encoder to transform a decimal number into a binary code

The truth table corresponding to this encoding matrix is given in Table 10.8.

In Fig. 10.21, input W_n (n = 0, 1, 2.....9) represents the nth key (It is assumed that no more than one key is depressed at a time).

Table 10.8: Truth table for encoding the decimal numbers 0 to 9

Inputs										Outputs			
W_9	W_8	W_7	W_6	W_5	W_4	W_3	W_2	W_1	W_0	Y_3	Y_2	Y_1	Y_0
0	0	0	0	0	0	0	0	0	1	0	0	0	0
0	0	0	0	0	0	0	0	1	0	0	0	0	1
0	0	0	0	0	0	0	1	0	0	0	0	1	0
0	0	0	0	0	0	1	0	0	0	0	0	1	1

(Contd.)

Table 10.8 contd.

Inputs										Outputs			
W_9	W_8	W_7	W_6	W_5	W_4	W_3	W_2	W_1	W_0	Y_3	Y_2	Y_1	Y_0
0	0	0	0	0	1	0	0	0	0	0	1	0	0
0	0	0	0	1	0	0	0	0	0	0	1	0	1
0	0	0	1	0	0	0	0	0	0	0	1	1	0
0	0	1	0	0	0	0	0	0	0	0	1	1	1
0	1	0	0	0	0	0	0	0	0	1	0	0	0
1	0	0	0	0	0	0	0	0	0	1	0	0	1

From the truth table, we conclude that $Y_0 = 1$ if $W_1 = 1$ or if $W_3 = 1$ or if $W_5 = 1$ or if $W_7 = 1$ or if $W_9 = 1$. Hence, in Boolean notation

$$Y_0 = W_1 + W_3 + W_5 + W_7 + W_9 \qquad ...(1)$$

Similarly

$$Y_1 = W_2 + W_3 + W_6 + W_7$$
$$Y_2 = W_4 + W_5 + W_6 + W_7 \qquad ...(2)$$
$$Y_3 = W_8 + W_9$$

The OR gates in above eqn. (1) and (2) are implemented with diodes in Fig. 10.21. An encoder array such as that in Fig. 10.21 is called a rectangular diode matrix.

10.15 READ ONLY MEMORY (ROM)

Consider the problem of converting one binary code into another. Such a code-conversion system (designated ROM and given in Fig. 10.22a has M-inputs (X_0, X_1,....X_{M-1}) and N outputs (Y_0, Y_1.....Y_{N-1}) where N may be greater than, equal to, or less than M. A definite M-bit code is to result in a specific N-bit output code. This code translation is achieved, by first decoding the M inputs onto $2^M \equiv \mu$ word lines (W_0, W_1....W_{m-1}) and then encoding each line into the desired output word. If the inputs assume all possible combinations of 1s and 0s, then mN-bit words are "read" at the output (not all these 2^M words need be unique, since it may be desirable to have the same output code for several different input words).

The functional relationship between output and input words is built into hardware in the encoder block of Fig. 10.22b since the information is stored permanently, the system is said to have a *memory*.

(The memory, elements are, e.g. diodes in Fig. 10.21). Further, since the stored relationship between output and input codes cannot be modified without adding or subtracting memory elements (hardware), this system is called a *read only memory* (ROM).

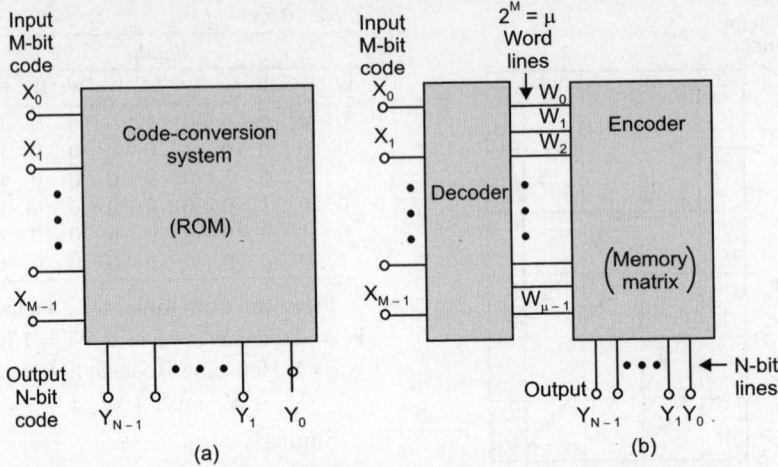

Fig. 10.22: (a) A block diagram for converting one code into another, a read only memory (ROM). (b) An ROM is considered to be a decoder for the input code followed by an encoder for the output code.

10.16 DIODE ROM

A read only memory (ROM) is the simplest kind of memory. It is equivalent to a group of registers, each permanently storing a word. By applying control signals, we can "read" the word in any memory location. ["Read" means to make the contents of the memory location appear at the output terminals of the ROM].

Figure 10.23a shows one way to build a ROM.

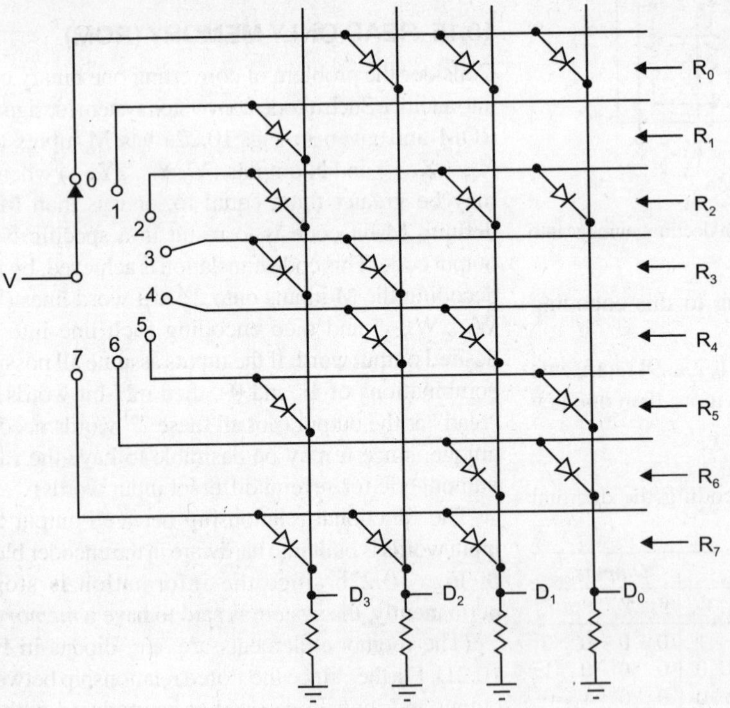

Register	Address	Word
R_0	0	0111
R_1	1	1000
R_2	2	1011
R_3	3	1100
R_4	4	0110
R_5	5	1001
R_6	6	0011
R_7	7	1110

(b)

(a)

Fig. 10.23: (a) Simple diode ROM (b) 'contents' of registers

Each horizontal row is a register or memory location. The R_0 register has three diodes, R_1 has one diode, R_2 has three diodes, R_3 has 2 diodes, R_4 has 2 diodes R_5 and R_6 have 2 diodes each and R_7 has three diodes. The output of the ROM is the word

$$D = D_3 \, D_2 \, D_1 \, D_0$$

In switch position 0, a high voltage turns on the diodes in the R_0 register; all other diodes are off. This means that a high output appears at D_2, D_1 and D_0. Therefore the word stored at memory location 0 is

$$D = 0111$$

When the switch is moved to position 1, the diode in the R_1 register conducts, forcing D_3 to go high. Because all other diodes are off, the output becomes

$$D = 1000$$

So, the contents of memory location 1 are 1000. As we move the switch to other position, we will read the contents of other memory locations. These contents are shown in Fig. 10.23b.

With discrete circuits, we can change the contents of a memory location by adding or removing diodes.

With ICs, the manufactures stores the words at the time of fabrication. In either case, the words are permanently stored once the diodes are wired in place.

Addresses in ROMs

The *address* and *contents* of a memory are two different things. As shown in table of Fig. 10.23 (b). The address of a memory location is the same as the subscript of the register storing the word. Consequently, register 0 has an address of 0 and contents of 0111, register 1 has an address of 1 and contents of 1000 and so on.

The idea of addresses applies to ROMs of any size. For example, a ROM with 256 memory locations has decimal addresses running from 0 to 255. A ROM with 1024 memory locations has decimal addresses from o to 1023.

10.17 ON CHIP DECODING

Rather than switch-select the memory location, IC manufacturers use *on-chip decoding*. Figure 10.24

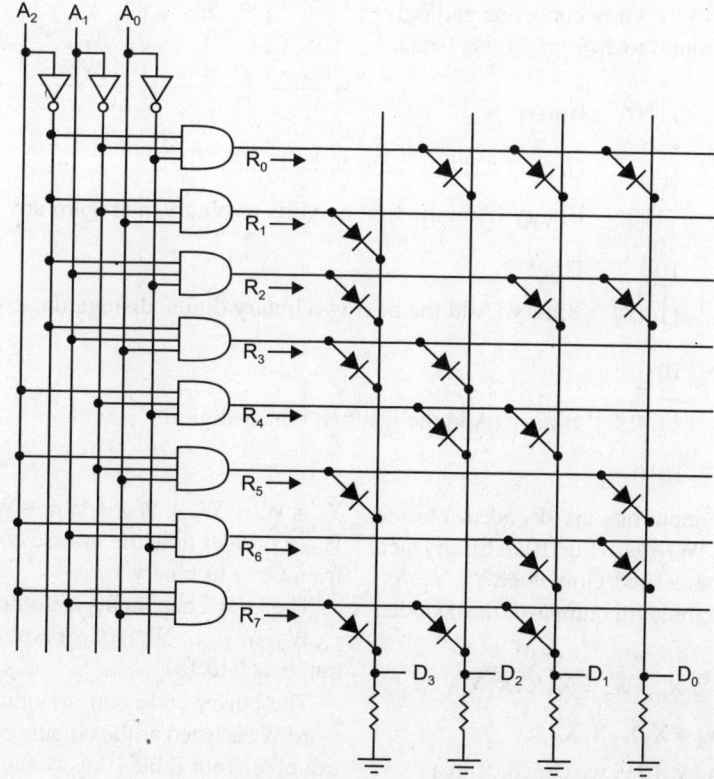

Fig. 10.24: ROM with on-chip decoding

gives this idea. The three input pins (A_2, A_1 and A_0) supply the binary address of the stored word. Then a 1-of-8 decoder produces a high output to one of the registers. For example, if

$$ADDRESS = A_2 A_1 A_0 = 100$$

the 1-of-8 decoder applies a high voltage to the R_4 register and the ROM output is

$$D = 0110$$

If we change the address word to

$$ADDRESS = 110$$

We will read the contents of memory location 6 which is

$$D = 0011$$

The circuit of Fig. 10.24 is a 32-bit ROM organized as 8 words of 4-bit each. It has 3 address (input) lines and four data (output) lines.

10.18 CODE CONVERTERS

We have discussed Gray code in chapter 3 (article 3.5.2). The truth table for translating from a Binary to a Gray code is given in Table 10.9. In going from one line to the next of the Gray code, one and only one bit is changed from 0 to 1 or vice versa (see as follows)

Step 1:	1100	Binary
	↓	
	1	Gray (1st digit, same as 1st binary digit)
	⊓	
Step 2:	1100	Binary (Add the first two bits of binary, disregard any carry)
	↓↓	
	10	Gray
	⊓	
Step 3:	1100	Binary (Add the next two binary digits, disregard carry)
	↓	
	101	
	⊓	
Step 4:	1100	Binary (Add the last two binary digits).
	↓	
	1010	

This equation is implemented by connecting eight diodes with their cathodes all tied to Y_0 and anodes connected to the decoder lines W_1, W_2, W_5, W_6, W_9, W_{10}, W_{13} and W_{14}. Similarly from Table 10.9, we may write for the other output bits, e.g.

Table 10.9: Conversion from a binary to a Gray code

Binary code inputs				Decoded Word	Gray code outputs			
X_3	X_2	X_1	X_0	W_n	Y_3	Y_2	Y_1	Y_0
0	0	0	0	W_0	0	0	0	0
0	0	0	1	W_1	0	0	0	1
0	0	1	0	W_2	0	0	1	1
0	0	1	1	W_3	0	0	1	0
0	1	0	0	W_4	0	1	1	0
0	1	0	1	W_5	0	1	1	1
0	1	1	0	W_6	0	1	0	1
0	1	1	1	W_7	0	1	0	0
1	0	0	0	W_8	1	1	0	0
1	0	0	1	W_9	1	1	0	1
1	0	1	0	W_{10}	1	1	1	1
1	0	1	1	W_{11}	1	1	1	0
1	1	0	0	W_{12}	1	0	1	0
1	1	0	1	W_{13}	1	0	1	1
1	1	1	0	W_{14}	1	0	0	1
1	1	1	1	W_{15}	1	0	0	0

In an ROM, the input bits are decoded into the word lines W_0, W_1....W_{15} (as in Fig. 10.22b), and then are encoded into the desired Gray code Y_3, Y_2, Y_1, Y_0. The W's are the minterm outputs of the decoder. For example:

$$W_0 = \bar{X}_3\bar{X}_2\bar{X}_1\bar{X}_0, \quad W_5 = \bar{X}_3 X_2 \bar{X}_1 \bar{X}_0,$$
$$W_9 = X_3 \bar{X}_2 \bar{X}_1 X_0$$

From the truth table 10.9, we conclude that

$$Y_0 = W_1 + W_2 + W_5 + W_6 + W_9 + W_{10} + W_{13} + W_{14}.$$

$Y_3 = W_8 + W_9 + W_{10} + W_{11} + W_{12} + W_{13} + W_{14} + W_{15}$ Consider now the inverse code translation, viz. from Gray to binary.

The Gray code inputs are arranged in the order W_0, W_1.......W_{15} (Corresponding to decimal numbers 0 to 15).

The binary code corresponding to a given input word W_n is listed as the output code for that line. For example, from Table 10.9, we find that the Gray code 1001 corresponds to the binary code 1110 and this relationship is maintained in Table 10.10 on line W_9. From this latter table, we obtain the relationship between output and input bits, For example.

$$Y_0 = W_1 + W_2 + W_4 + W_7 + W_8 + W_{11} + W_{13} + W_{14}.$$

Table 10.10: Conversion from Gray to a binary code

Gray code inputs				Decoded Word	Binary code outputs			
X_3	X_2	X_1	X_0	W_n	Y_3	Y_2	Y_1	Y_0
0	0	0	0	W_0	0	0	0	0
0	0	0	1	W_1	0	0	0	1
0	0	1	0	W_2	0	0	1	1
0	0	1	1	W_3	0	0	1	0
0	1	0	0	W_4	0	1	1	1
0	1	0	1	W_5	0	1	1	0
0	1	1	0	W_6	0	1	0	0
0	1	1	1	W_7	0	1	0	1
1	0	0	0	W_8	1	1	1	1
1	0	0	1	W_9	1	1	1	0
1	0	1	0	W_{10}	1	1	0	0
1	0	1	1	W_{11}	1	1	0	1
1	1	0	0	W_{12}	1	0	0	0
1	1	0	1	W_{13}	1	0	0	1
1	1	1	0	W_{14}	1	0	1	1
1	1	1	1	W_{15}	1	0	1	0

The such equations define how the memory elements are to be arranged in the encoder. Note that the ROM for Table 10.10 uses the same decoding arrangement as that for Table 10.9 but the encoders are completely different.

10.19 DECIMAL-TO-BCD PRIORITY ENCODER

One of the most commonly used input device for a digital system is a set of ten switches, one for each numeral between 0 and 9. These switches generate 1 or 0 logic levels in response to turning them OFF or ON. When a particular number is to be fed to the digital circuit in BCD code, the switch corresponding to that number is to be pressed. There is an IC available for performing this function, viz. IC 74147 which is a priority encoder. The block diagram of this IC is given in Fig. 10.25 and Table 10.11 gives its truth table. It has active low inputs and outputs. The meaning of the word priority becomes clear from the truth table. For example, if inputs 2 and 5 are LOW, the output will be corresponding to 5 which has a higher priority than 2, i.e. the highest numbered input has priority over lower numbered inputs.

Table 10.11: Truth table of IC 74147

Active-low decimal inputs									Active-low BCD outputs			
1	2	3	4	5	6	7	8	9	D	C	B	A
1	1	1	1	1	1	1	1	1	1	1	1	1
0	1	1	1	1	1	1	1	1	1	1	1	0
×	0	1	1	1	1	1	1	1	1	1	0	1
×	×	0	1	1	1	1	1	1	1	1	0	0
×	×	×	0	1	1	1	1	1	1	0	1	1
×	×	×	×	0	1	1	1	1	1	0	1	0
×	×	×	×	×	0	1	1	1	1	0	0	1
×	×	×	×	×	×	0	1	1	1	0	0	0
×	×	×	×	×	×	×	0	1	0	1	1	1
×	×	×	×	×	×	×	×	0	0	1	1	0

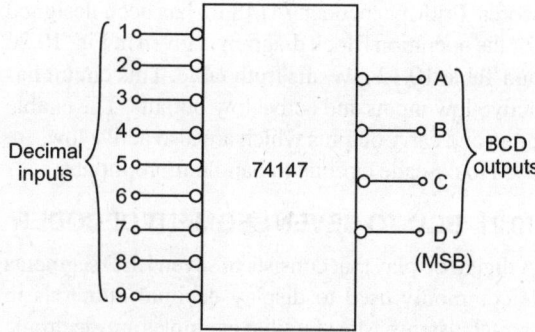

Fig. 10.25: Block diagram of 74147 decimal to-BCD priority encoder

10.20 OCTAL-TO-BINARY PRIORITY ENCODER

The octal code is often used at the inputs of digital circuits that require manual entering of long binary

Table 10.12: Truth table of IC 74148

EI	Inputs								Outputs				
	0	1	2	3	4	5	6	7	C	B	A	GS	E0
1	×	×	×	×	×	×	×	×	1	1	1	1	1
0	0	1	1	1	1	1	1	1	1	1	1	0	1
0	×	0	1	1	1	1	1	1	1	1	0	0	1
0	×	×	0	1	1	1	1	1	1	0	1	0	1
0	×	×	×	0	1	1	1	1	1	0	0	0	1
0	×	×	×	×	0	1	1	1	0	1	1	0	1
0	×	×	×	×	×	0	1	1	0	1	0	0	1
0	×	×	×	×	×	×	0	1	0	0	1	0	1
0	×	×	×	×	×	×	×	0	0	0	0	0	1
0	1	1	1	1	1	1	1	1	1	1	1	1	0

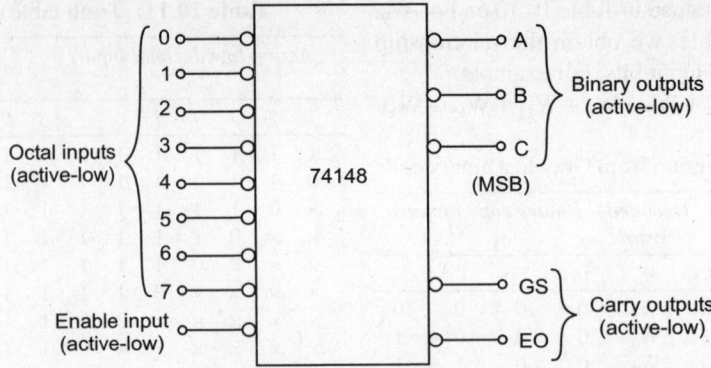

Fig. 10.26: Block diagram of 74148 octal-to-binary priority encoder

words. Priority encoder 74148 IC has been designed for the operation Block diagram is given in Fig. 10.26 and Table 10.12 gives its truth table. This circuit has active-low inputs and active low outputs. The enable input and carry outputs which are also active low, are used to cascade circuits to handle more inputs.

10.21 BCD-TO-SEVEN SEGMENT DECODER

A digital display that consists of seven LED segments is commonly used to display decimal numerals in digital systems. Most familiar examples are electronic calculators and digital watches where one 7-segment display device is used for displaying one numeral 0 through 9. For using this display device, the data has to be converted from some binary code to the code required for the display. Usually, the binary code used is natural BCD. Figure 10.27a shows the display device. Figure 10.27b shows the segments which must be illuminated for each of the numerals and Fig. 10.27c gives the display system.

Table 10.13 gives the truth table of BCD-to-7 segment decoder. Here ABCD is the natural BCD code for numerals 0 through 9. The K-maps for each of the outputs *a* through *g* are given in Fig. 10.28. The entries in the K-map corresponding to six binary combinations not used in the truth table are X (don't care). The K-maps are simplified and the minimum expressions are as eqns (1) through (7).

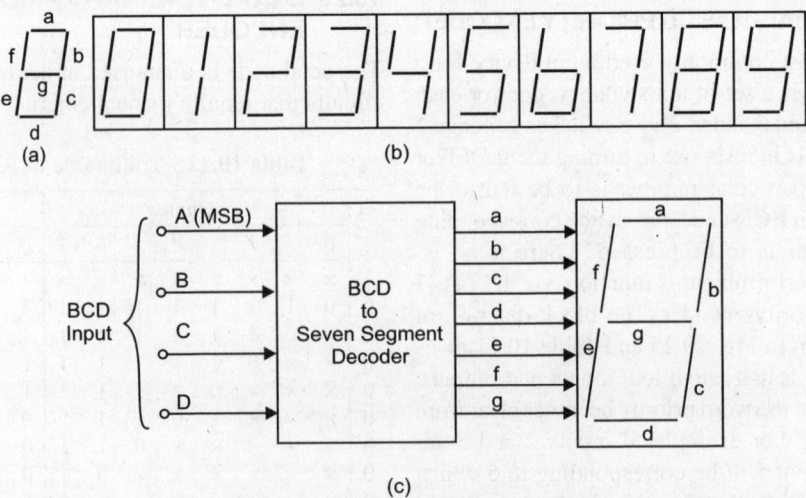

Fig. 10.27: (a) Seven segment display (b) display of numerals (c) display system

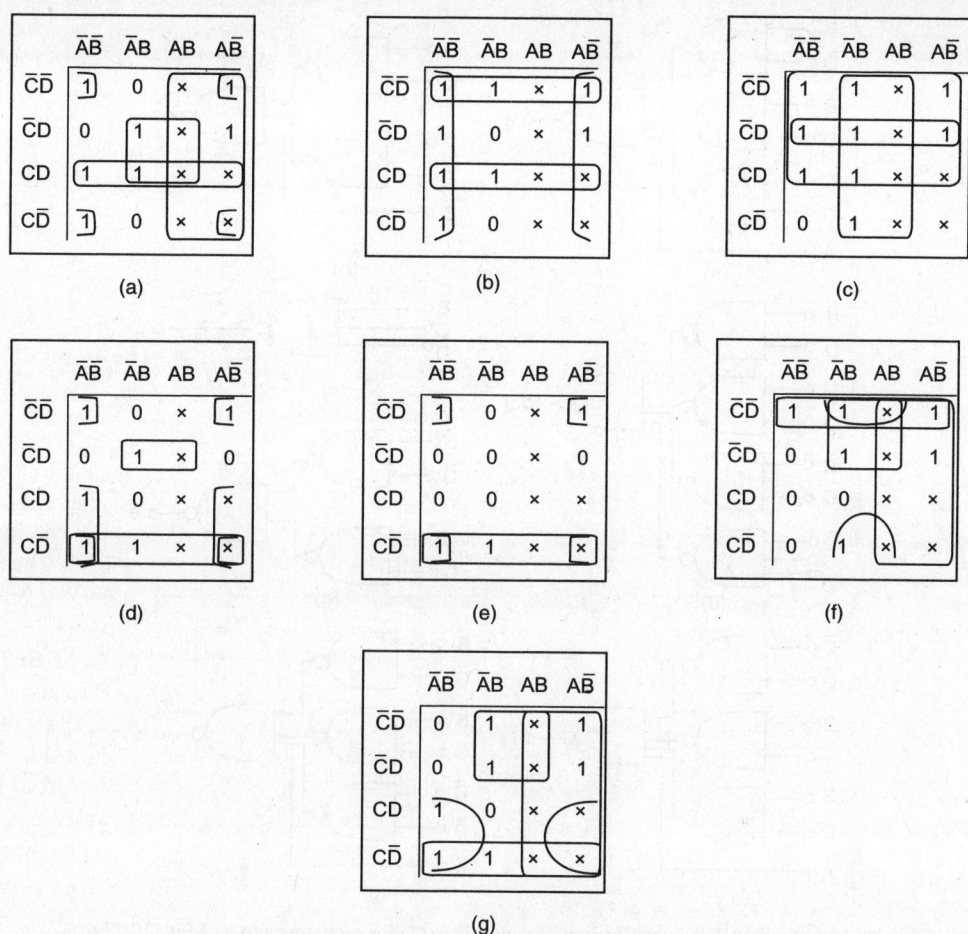

Fig. 10.28: Karnaugh-map of Table 10.13

Table 10.13: Truth table of BCD-to 7 segment decoder

Decimal digit displayed	Inputs				Outputs						
	A	B	C	D	a	b	c	d	e	f	g
0	0	0	0	0	1	1	1	1	1	1	0
1	0	0	0	1	0	1	1	0	0	0	0
2	0	0	1	0	1	1	0	1	1	0	1
3	0	0	1	1	1	1	1	1	0	0	1
4	0	1	0	0	0	1	1	0	0	1	1
5	0	1	0	1	1	0	1	1	0	1	1
6	0	1	1	0	0	0	1	1	1	1	1
7	0	1	1	1	1	1	1	0	0	0	0
8	1	0	0	0	1	1	1	1	1	1	1
9	1	0	0	1	1	1	1	0	0	1	1

$$a = \overline{B}\overline{D} + BD + CD + A \qquad \ldots(1)$$

$$b = \overline{B} + \overline{C}\overline{D} + CD \qquad \ldots(2)$$

$$c = B + \overline{C} + D = \overline{\overline{B}C\overline{D}} \qquad \ldots(3)$$

$$d = \overline{B}\overline{D} + C\overline{D} + \overline{B}C + BC D \qquad \ldots(4)$$

$$e = \overline{B}\overline{D} + C\overline{D} \qquad \ldots(5)$$

$$f = A + \overline{C} + B\overline{C} + B\overline{D} \qquad \ldots(6)$$

$$g = A + B\overline{C} + \overline{B}C + C\overline{D} \qquad \ldots(7)$$

The NAND gate realizations of eqns. (1) through (7) are shown in Fig. 10.29.

10.22 SEVEN SEGMENT DISPLAY SYSTEM

The decimal outputs of digital voltmeters and frequency counters are often displayed using seven

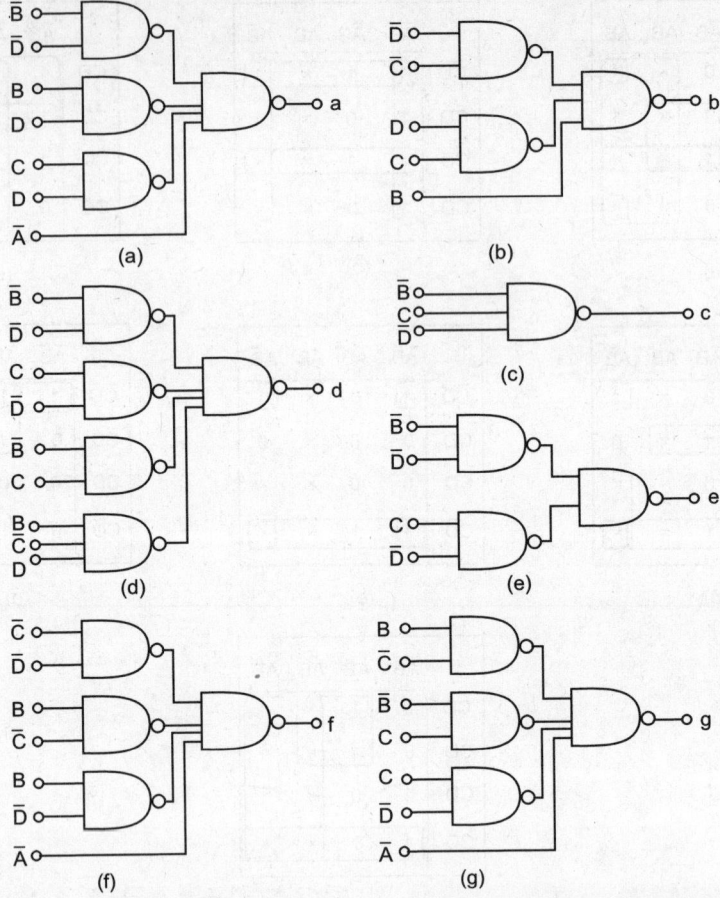

Fig. 10.29: NAND gate realization of eqns. (1) to (7) for segments a to g of Fig. 10.27 (c)

segment indicators. For indicators, light emitting diodes (LEDs) are convenient because they are directly compatible with TTL circuits, do not require higher voltages as by fluorescents and are generally brighter than liquid crystals.

The circuit in Fig. 10.30 is a common-anode, LED-type, 7-segment indicator used to display a single decimal digit. The 7447 BCD-to-7-segment decoder is used to drive the indicator, and the four inputs to the 7447 are the four FF outputs of the 7490 decade counter.

Since the 7447 has active low outputs so the equivalent circuit of an illuminated segment appears as in Fig. 10.30b. A 1 Hz square wave applied at the clock input of the 7490 will cause the counter to count upward, advancing one count each second, and the equivalent decimal number will appear on the display.

10.23 IC DECODERS/DRIVERS FOR DISPLAY DEVICES

BCD-to-Decimal Decoder/Driver: In may digital systems, it is preferable to see the output in a decimal format. The outputs can either be displayed using display, e.g. LEDs, Nixie tubes, etc. or can be used

Table 10.14: Available BCD-to-Decimal decoder/driver ICs

IC No.	Output circuit	Application
7441	Open-collector	Nixie tubes driver
7442	Totem-pole	LED driver
7445	Open-collector	Indicator/relay driver
74141	Open-collector	Nixie tubes driver
74145	Open-collector	Indicator/relay driver
74445	Open-collector	Indicator/relay driver

Table 10.15: Available BCD-to-7-segment decoder/driver ICs

IC No.	Output	Rating (Max. voltage, Sink current)	Facilities available			
			Lamp Test LT	Ripple RBI blanking Input	Output RBO Ripple blanking	Blanking Input BI
7446, 74246	Active-low: Open-collector	30V, 40mA	Yes	Yes	Yes	Yes
7446, 74247	Active-low: Open-collector	15V, 40mA	Yes	Yes	Yes	Yes
7448, 74248	Active-high: Pull-upresistor = 2kΩ	5.5V, 6.4mA	Yes	Yes	Yes	Yes
7449, 74249	Active-high: Open-collector	5.5V, 8mA	No	No	No	Yes

to actuate some indicators or relays. Table 10.14 gives the available BCD-to-decimal decoder/driver ICs. All these ICs have active-high inputs and active-low outputs.

BCD-to-7 segment Decoder/Driver: Seven segment display is the most popular display device used in digital systems. For displaying data using this device, the data have to be converted from BCD to-7-segment code. A number of MSI ICs are available for performing this function. These are given in Table 10.15.

The decoder/driver circuit has 4-input lines for BCD data and 7 output lines to drive a 7-segment display. Output terminals *a* through *g* of the decoder are to be connected to *a* through *g* terminals of the display respectively. If the outputs are active-low, then the 7 segment LED must be of the common anode type, whereas if the outputs are active-high then the 7 segment LED must be of the common cathode type.

Table 10.16: Summary of BCD-to-7 segment decoder functions

LT	RBI	BI/RBO	BCD inputs	Display mode
0	×	1 output	×	Lamp test
×	×	0 input	×	Display blank
1	1	1 output	Any number	Normal decoding
1	0	Normally at logic 1output. Goes to logic 0 during zero blanking interval	Any number	Normal decoding with zero blanking

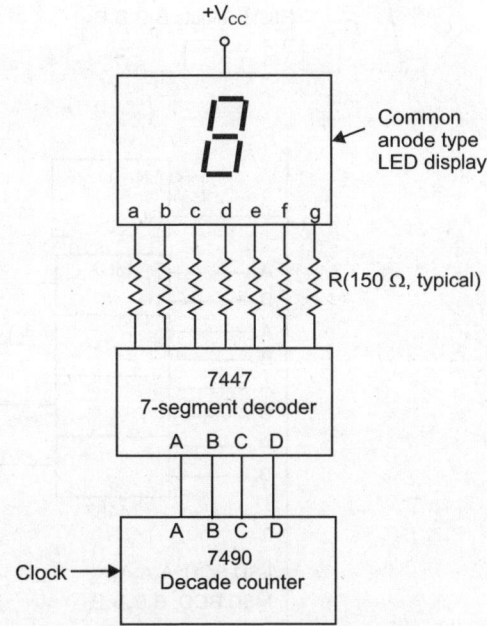

(a) Single decimal-digit display

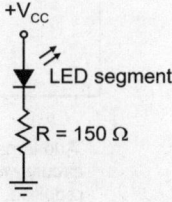

(b) Equivalent circuit for an illuminated segment

Fig. 10.30: Display system for a single digit of a 7-segment indicator

The functions of LT, RBI, RBO and BI are given in Table 10.16.

LT: This is used to check the segments of LED. If it is connected to logic 0 level, all the segments of the display connected to the decoder will be ON. For normal decoding operation, this terminal is to be connected to logic 1 level.

RBI: It is to be connected to logic 1 for normal decoding operation. This is used for blanking out leading zeros in multidigit displays.

BI: If it is connected to logic 0 level, the display is switched off irrespective of BCD inputs. This is used for conserving power in multiplexed displays.

RBO: This output, which is normally at logic 1 goes to logic 0 during zero blanking interval. This is used for cascading purposes and is connected to RBI of the succeeding stage.

10.24 THE SEVEN SEGMENT DISPLAY MULTIPLEXER

Figure 10.31 shows a simplified method of multiplexing BCD numbers to a 7-segment display. Here, two digit numbers are displayed on the 7-segment readout using a single BCD-to-7-segment decoder. This method of display multiplexing can be extended to display any number of digits. The basic operation is as follows:

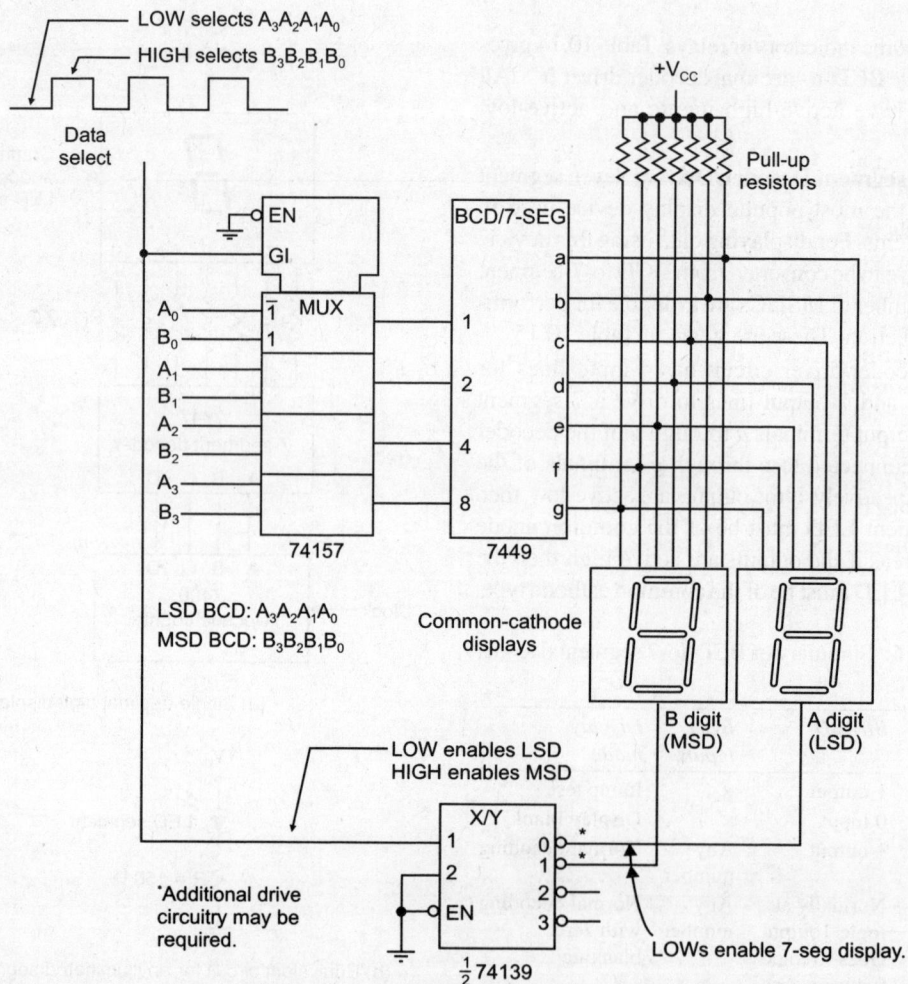

Fig. 10.31: 7-Segment display multiplexing logic (After Thomas L. Floyd, "Digital Fundamentals", Universal Book, New Delhi (2000))

Two BCD digits A and B are applied to the multiplexer inputs. A square wave is applied to the data-select line and when it is LOW, the A bits (A_3 A_2 A_1 A_0) are passed through to the inputs of the 7449 BCD-to-7-segment decoder. The LOW on the data-select also puts a LOW on the 1 input of the 74139 2 line-to-4 line decoder, thus activating its 0 output and enabling the A-digit display by effectively connecting its common terminal to ground. The A digit is now on and the B digit is off.

When the data-select line goes HIGH, the B bits (B_3 B_2 B_1 B_0) are passed through to the inputs of the BCD-to-7-segment decoder. Also, the 74139 decoder's 1 output is activated, thus enabling the B digit display. The B digit is now on and the A digit if off.

The cycle repeats at the frequency of the data-select square wave. This frequency must be high enough (~30 Hz) to prevent visual flicker during display.

10.25 THE ROM FAMILY

Semiconductor ROMs are manufactured with bipolar technology such as TTL or with MOS technology. Figure 10.32 shows how ROMs are categorized. In mask ROM data are permanently stored in the memory during manufacturing process. In PROM (progra-mmable ROM) data are electrically stored by the user. Note that both the mask ROM and the PROM can be of either technology. But, the EPROM (erasable PROM) is strictly a MOS device.

The EPROM is electrically programmable by the user and the stored data can be erased by exposure to u.v. light or by electrical means (electrically erasable), therefore also called EEPROM.

10.26 COMMERCIALLY AVAILABLE ROMS

A binary number is sometimes referred to, as a *word*. In a computer, binary numbers (or words) represent instructions, alphabet letters, decimal numbers, etc. The circuit given in Fig. 10.23 is a 32-bit ROM organized as 8 words with 4-bits at each address (an 8×4 ROM). The ROM of Fig. 10.23 is only for instructional purpose and in actual practice, commercially available ROMs are used. Here are some available TTL ROMs:

7488 : 256 bits (organized as 32×8)

74187 : 1024 bits (organized as 256×4)

74S370 : 2048 bits (organized as 512×8)

One way to change the stored numbers of a ROM is by adding or removing diodes. With discrete circuits, one would have to solder or unsolder diodes to change the stored nibbles. With ICs, however, one can send a list of the data to be stored to the IC manufacturer, who then produces a *mask* (a photographic template of the circuit), which is used for mass production of ROMs.

10.27 PROGRAMMABLE ROMS

A programmable ROM (PROM) allows the user (instead of the manufacturer) to store the data. It is a type of chip that allows the user to program it with an instrument called *PROM programmer*. This PROM

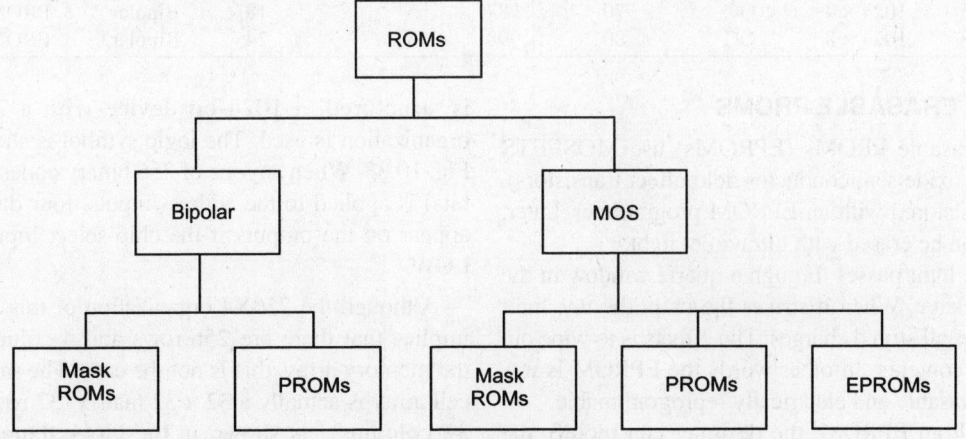

Fig. 10.32: The ROM family

programmer stores the words by "burning in". Here is how a PROM programmer works:

Originally all diodes are connected at the cross points. For example in Fig. 10.23, there would be a total of 32 diodes (8 rows and 4 columns). Each of these diodes has fusible link (a small fuse). The PROM programmer sends destructively high currents through all diodes to be removed. In this way only the desired diodes remain connected afterwards.

Programming like this is permanent because the data can not be erased after it has been burned in. Here are some commercially available PROMS:

74 S188 : 256 bits (organized as 32×8)
74 S287 : 1024 bits (organized as 256×4)
74 S472 : 4096 bits (organized as 512×8)
PROMS are useful for small production sums.

design is finalized, the data can be burned into PROMs (small runs) or sent to an IC manufacturer for large productions.

10.29 ROM ICS

Some of the commonly used ROM ICs with their characteristics are given in table 10.17. The inputs and outputs are TTL compatible. EPROMs 27128 ($16k \times 8$), 27256 ($32k \times 8$), 27512 ($64k \times 8$) and 271024 ($128k \times 8$) are also available and are commonly used in microprocessor based systems.

10.30 INTERNAL ROM STRUCTURE

Most IC ROMs have a somewhat more complex internal structure than that in the basic simplified example just covered. To illustrate how an IC ROM

Table 10.17: ROM ICs with their characteristics

IC No.	Organization	Output	Access time ns	Max. Power dissipation mW	Power supplies V	No. of pins	Technology	Type of ROM
1702A	256×8	TS	1000	885	5, –9	24	MOS	EPROM
2308	1024×8	TS	450	840	5, 12, –5	24	MOS	ROM
2316E	2048×8	TS	450	630	5	24	MOS	ROM
2704	512×8	TS	450	800	5, 12, –5	24	MOS	EPROM
2708	1024×8	TS	450	800	5, 12, –5	24	MOS	EPROM
2716	2048×8	TS	450	525	5	24	MOS	EPROM
2732A	4096×8	TS	250	790	5	24	MOS	EPROM
2764	8192×8	TS	250	790	5	28	MOS	EPROM
2816	2048×8	TS	250	495	5	24	MOS	E^2PROM
3601	256×4	OC	70	685	5	16	Bipolar	PROM
3602A	512×4	OC	70	735	5	16	Bipolar	PROM
3604A	512×8	OC	70	998	5	24	Bipolar	PROM
3605	1024×4	OC	70	787	5	18	Bipolar	PROM
3608	1024×8	OC	80	998	5	24	Bipolar	PROM

10.28 ERASABLE PROMS

The erasable PROMs (EPROMs) use MOSFETS (metal oxide semiconductor field effect transistors). Data is stored with an EPROM programmer. Later, data can be erased with ultraviolet light.

The light passes through a quartz window in the IC package. When it strikes the chip, the u.v. light releases all stored charges. The effects is to wipe out stored contents. In other words the EPROM is u.v. light erasable and electrically reprogrammable.

With an EPROM, the designer can modify the contents until the stored data is perfect. When the

is structured, a 1024-bit device with a 256X4 organization is used. The logic symbol is shown in Fig. 10.33. When anyone of 256 binary codes (eight bits) is applied to the address inputs, four data bits appear on the outputs if the chip-select inputs are LOW.

Although the 256X4 organization of this device implies that there are 256 rows and 4 columns in the memory array, this is not the case. The memory cell array is actually a 32×32 matrix (32 rows and 32 columns), as shown in the block diagram in Fig. 10.34.

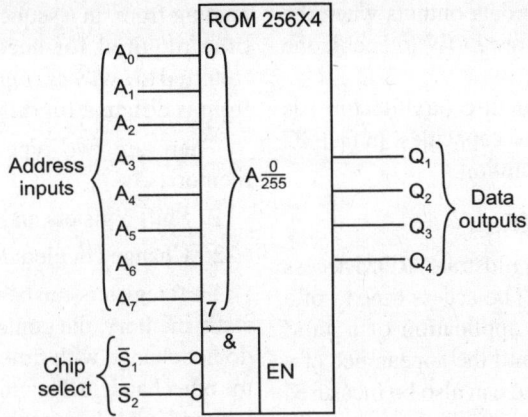

Fig. 10.33: A256X4 ROM logical symbol

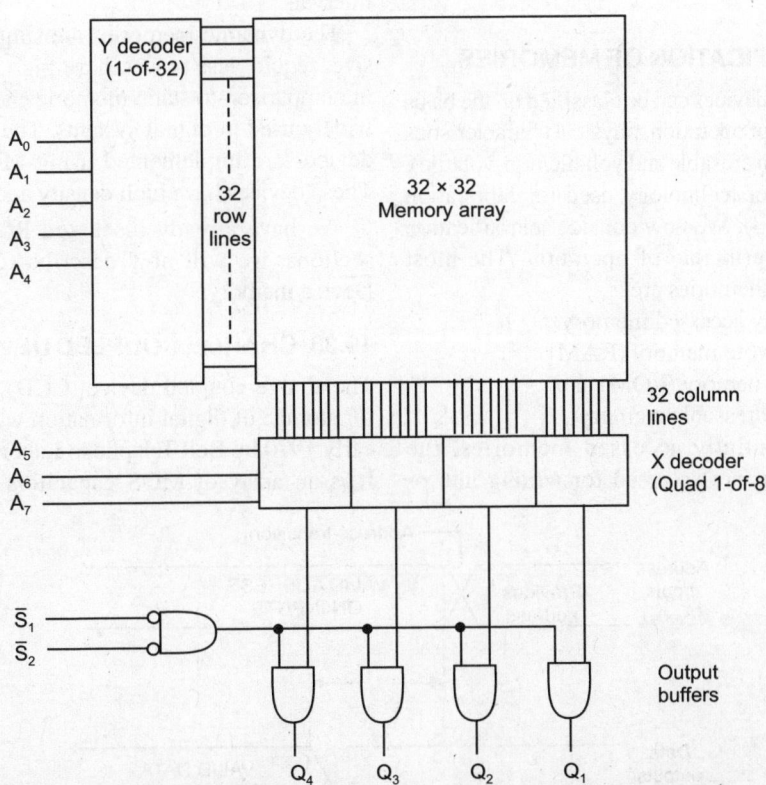

Fig. 10.34: Internal structure of a typical ROM. (This particular example is a 1024-bit ROM with 256 × 4 organization)

The ROM works as follows: Five of the eight address lines (A_0 through A_4) are decoded by the ROW decoder (often called the Y decoder) to select one of the 32 rows. Three of the eight address lines (A_5 through A_7) are decoded by the column decoder (often called the X decoder) to select four of the 32 columns. Actually, the X decoder consists off our l-of-8 decoders (data selectors) as shown.

The result of this structure is that when an eight-bit address code (A_0 through A_7) is applied, a four-bit data word appears on the data outputs when the chip-select lines (S_1 and S_2) are LOW to enable the output buffers.

This type of internal structure (architecture) is typical of IC ROMs of various capacities. In fact, IC 74187 ROM has this configuration.

10.31 ROM ACCESS TIME

A typical timing diagram that illustrates ROM access time is shown in Fig. 10.35. The access time t_a, of a ROM is the time from the application of a valid address code on the inputs until the appearance of a valid output data. Access-time can also be measured from the activation of *chip select* ($\overline{S}$) to the occurrence of valid output data when a valid address is already on the inputs.

10.32 CLASSIFICATION OF MEMORIES

Various memory devices can be classified on the basis of their principle of operation, physical characteristics (e.g.) erasable/non erasable and volatile/non-volatile), mode of access or technology used for fabrication (e.g. bipolar/MOS). We now consider classification on the basis of principle of operation. The most commonly used memories are:

1. Sequentially accessed memory
2. Read and write memory (RAM)
3. Read only memory (ROM) and
4. Content addressable memory

In the sequentially accessed memories, the memory locations are accessed for writing into or reading from, in a sequential fashion. Therefore, the time required for accessing a memory location (referred to as *access time*) for writing into or reading from is different for different locations.

There are two types of sequentially accessed memories.

1. Shift registers and
2. Change coupled devices (CCD)

Shift registers can be either static or dynamic, In a static memory, the contents of the memory location do not change with time as long as power in on. On the other hand, in dynamic memories, the information is stored in MOS capacitors which changes with time and, therefore, it has to be refreshed at regular intervals.

The dynamic memories are simpler, less expensive, require less power, have high packing density in comparison to static memories, and are therefore widely used in digital systems. The charge coupled devices are implemented using MOS technology. These devices have high density and low cost.

We have already discussed ROM in previous sections, we will now describe Charge Coupled Device memory.

10.33 CHARGE COUPLED DEVICE MEMORY

The charge coupled device (CCD), a new concept for storage of digital information was announced in early 1970 by Bell Telephone Laboratories of USA. It is an array of MOS capacitors operating as a

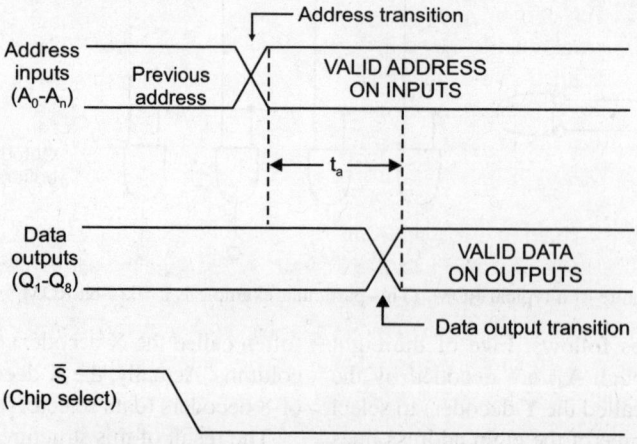

Fig. 10.35: ROM access time t_a from address change to data output when chip select already active

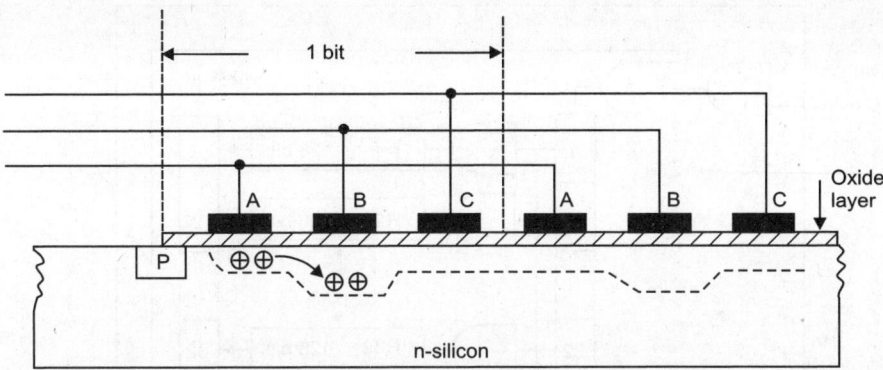

Fig. 10.36: Cross section of a CCD shift register memory

dynamic shift register. These are simple, versatile and low cost devices that can be used wherever a serially accessed memory is required.

These are of the class known as *direct access memories* and unlike RAM, each address stores a fairly large number of bits (~1K) and these bits are retrieved serially. The access times are of the order of 100 microseconds.

In these devices information is stored as electrical charge in potential wells which are formed at the surface of a semiconductor. These wells are created by fabricating a string of capacitors by oxidizing a low resistivity Si-wafer by a 10^{-5} an thick oxide layer and placing metal electrodes over it. A pn-junction is used to introduce a charge underneath the first capacitor (under electrode A in Fig. 10.36).

This capacitor has a voltage-V applied to its electrode and a depletion layer is created below it and the charge from the pn junction is trapped in this layer. If a voltage –2V is applied to the electrode B of the next capacitor, a deeper depletion region is created and the charge from under spills into this region. B is now taken to -V, pushing the charge under the next electrode C which is taken to -2V and so on. Three electrodes forming a three phase system are required for each bit location since the charge can move in either direction and we need two electrodes to trap a charge bundle between them and then move them in a controlled fashion. In short, we see that this device functions as a shift register with a 1 being represented by a charge injection and a 0 by no charge injection.

These devices are easy to fabricate. However, fabricating larger memories using CCD poses many problems. As three capacitances are required per bit, the number of bits per chip is restricted. A more serious problem is bit transfer efficiency. Due to traps in the semiconductor surface and in the interface states, only 99.9 to 99.8 percent of charges move from under one electrode to the next and for a memory of 1000 electrodes (333 bits) the last bit position receives only $0.999^{1000} = 0.3676$ of the injected charge. If a limit of 0.5 is put for charge attenuation, for reliable detection of a 1 at the output of the register, the maximum register length would be 23. Thus, we need a number of registers in an appropriate organization to increase the capacity of the memory. We will subsequently see how 64 K bit chips are organized using shorter CCD shift registers. Fairchild corporation and Texas instruments market 64 K bit CCD chips which contain 16 addressable shift registers each with 4K bits. The access time of the memory is 100ms with a shift rate of 2 megabits per second. The chip is internally organized as depicted in Figs. 10.37 and 10.38.

32 individual shift registers of 126 bits each are connected to two 32 bit parallel out/in shift registers (Fig. 10.37). First, the 32 bits are serially entered starting at t_0. At time (t_0 + 32 clock intervals) these bits are simultaneously shifted into 32 serial shift registers. Then, the outputs of these shift registers are entered in parallel into a 32-bit shift register and serially shifted out. The total serial memory is thus $32 + 32 \times 126 + 32 = 4096$ bits. Sixteen such circulating shift registers are connected into an addressable memory which has a 4-bit address (Fig. 10.38).

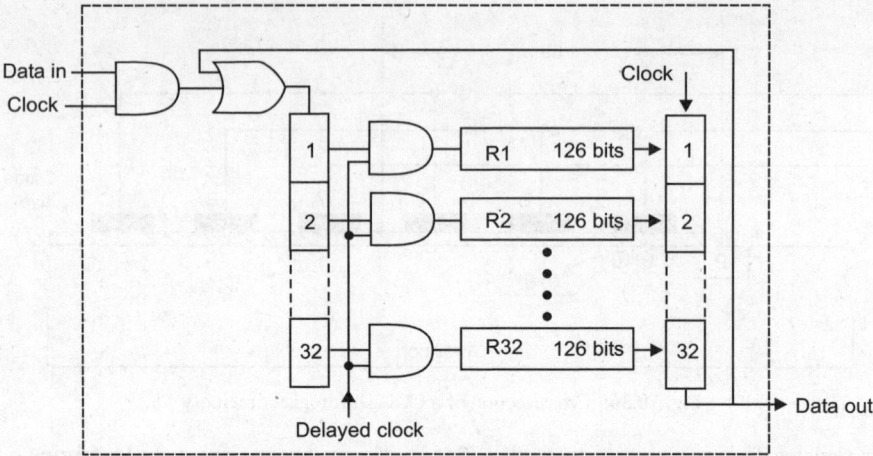

Fig. 10.37: A 4K-bit serial circulating shift register

Each of these 4K registers goes on circulating data within themselves except when new information is written in. The entire circuit shown in Figs. 10.37 and 10.38 are integrated into one chip having externally 4-bit address terminals, data in and out terminals, read, write, clock, voltage, and ground terminals.

10.34 RANDOM ACCESS MEMORY (RAM) USING SEMICONDUCTOR FLIP-FLOPS

For memory systems constructed with IC FFs, the access time is independent of the address of the word. Such a memory is known as a random access memory.

Memory systems in which the stored information is lost when the power is turned off are known as volatile memories. IC FFs are volatile whereas a magnetic core is non-volatile.

With the advent of ICs and the steady decline in their prices, it has become feasible (from cost view point) to fabricate memories with semiconductor flip-flops. The configuration of a binary cell which is capable of storing one-bit of information is shown in Fig. 10.39. The cell consists of an RS FF with associated gates.

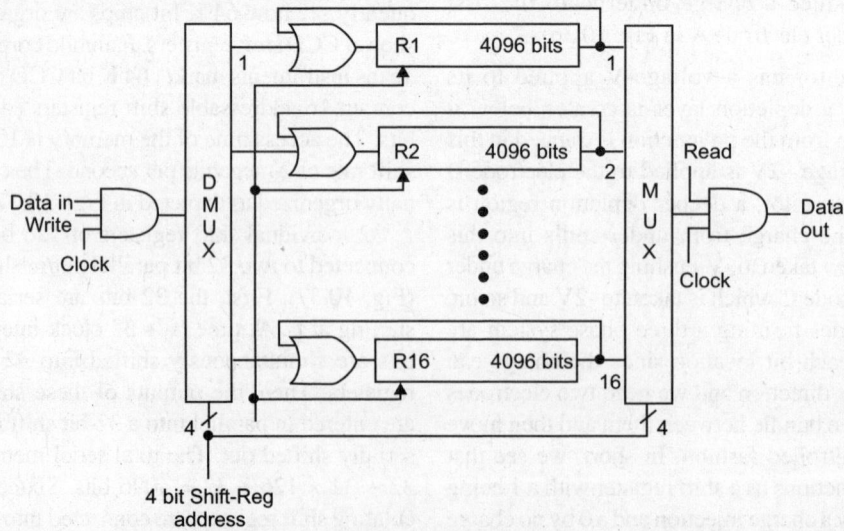

Fig. 10.38: A 64 K bit serial memory using registers of Fig. 10.37

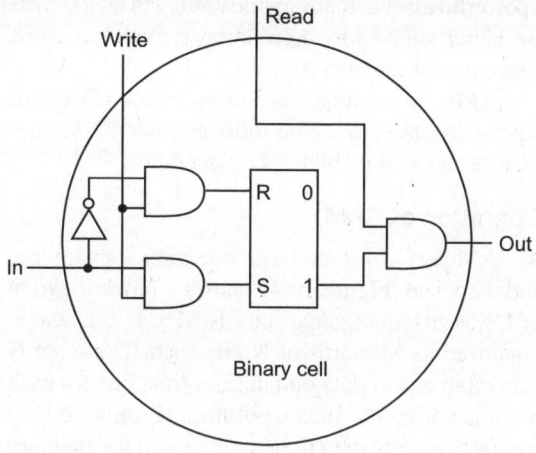

Fig. 10.39: A binary cell

A '1' applied to the 'In' terminal puts the FF in the '1' state if the 'Write' signal is also '1'. A '0' applied at 'In' puts the FF in the '0' state if the 'Write' signal is '1'. Information is read at the 'Out' terminal when a 'Read' signal is applied. The organization of a group of such binary cells to form a 4 word (3 bits per word) memory is shown in Fig. 10.40. The memory works as follows:

For reading a word, its address is entered in MAR.

The decoder connected to the MAR energizes the READ wire connected to all the cells belonging to the selected word. The outputs of these cells are entered in the MBR.

For writing, the information is entered in the MBR. The outputs of the MBR are connected to the inputs of the appropriate binary cells of all the words. The

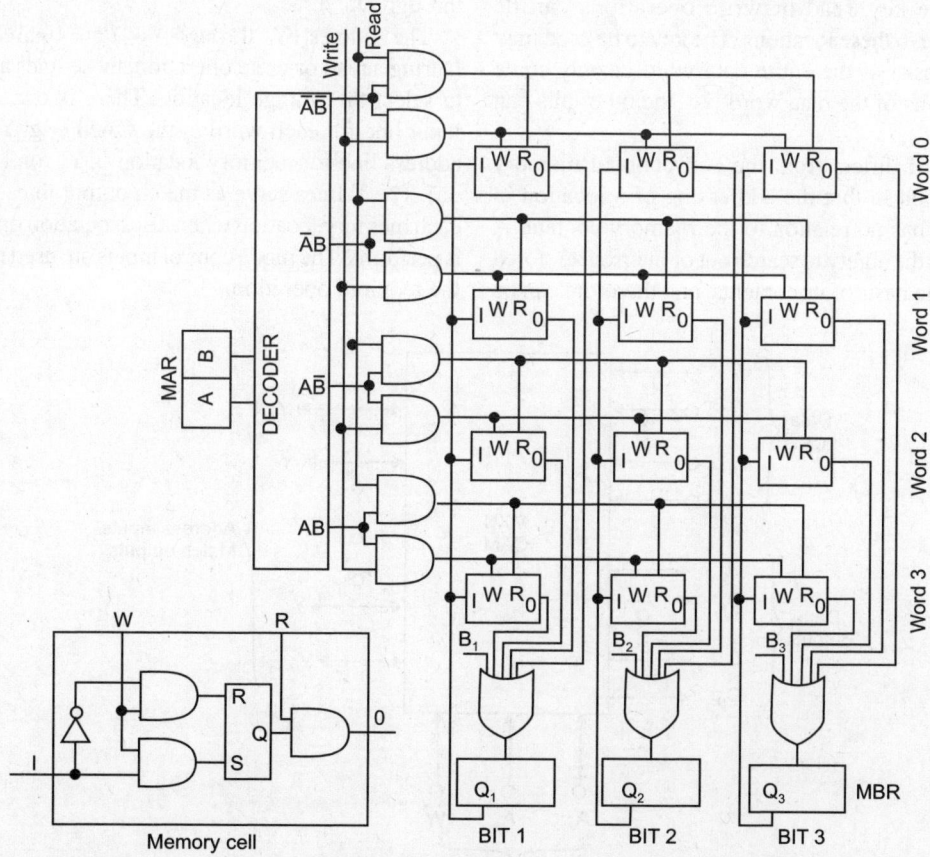

Fig. 10.40: A random access memory (RAM) using flip-flops. (After T.R. Viswanathan, G.K. Mehta and V. Rajaraman, "Electronics for Scientists and Engineers", Prentice Hall of India, N. Delhi (1989))

address where the information is to be written is entered in MAR. The decoder energizes the 'write' input of the word addressed. This enables the MBR contents to be written into the cells of the selected word.

10.35 CONTENT ADDRESSABLE MEMORY

The content addressable memory (CAM) is a special purpose random access memory device that can be accessed by searching for data content. For this purpose, it is addressed by associating the input data, referred to as key, simultaneously with all the stored words and produces output signals to indicate the match conditions between the key and the stored words. This operation is referred to as *association* and this type of memory is also known as associative memory.

After identifying the locations where contents match the key, read or write operations can be performed to these locations. The key to be used may either consist of the entire data word or only some specific bits of the data word (i.e. the other bits can be masked).

A CAM differs from the conventional memory organization in that the addressing of a location in the latter has no relation to the memory content. A CAM has the ability to search out or interrogate stored data on the basis of its contents, and therefore can be a powerful asset in many applications. (In fact, CAMs are better suited for information retrieval than the conventional memories).

CAMs are manufactured using MOS, CMOS or bipolar technologies. The most popular CAMs use ECL because of its high speed operation.

Operation of CAM

A CAM perform three basic operations read, write and associate. Figure 10.41 shows a block diagram of CAM. Its storage capacity is M × N bits and is organized as M words of N bits each. These are N data input and N data output lines (one line for each bit of a word). The data input lines (I_0 through I_{N-1}) are used to input data to be written into the memory and for key word in case of associate operation. Data are read out of the CAM at the data output lines (D_0 through D_{N-1}).

The Y lines (Y_0 through Y_{N-1}) are bidirectional. During a read or write operation, these lines are used to select the storage location. There is one address input line for each word in the CAM (e.g. Y_0 is the address line for memory location 0, Y_1 for 1 and so on). The Y lines serve as match output lines one for each memory location when an association operation is performs. The mode control inputs are used to select the required operation.

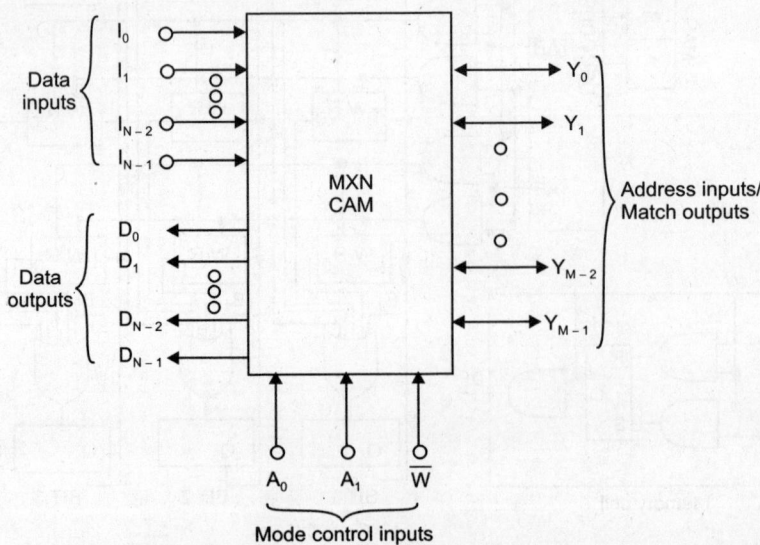

Fig. 10.41: Block diagram of a CAM

10.36 PROGRAMMABLE ARRAY LOGIC (PAL)

PAL is a programmable array of logic gates on a single chip. PALs are another design solution like SOP solution, POS solution and multiplexer logic. Before discussing PAL, we need to discuss some key ideas about the PROM.

Simplified Drawing of PROM

Consider again, Fig. 10.42. Assume this Fig. represents a PROM that has been burned with the diodes at the places shown. The AND gate array produces all the fundamental products $\overline{A}\,\overline{B}\,\overline{C}$ to ABC. Then depending on which diodes are connected, some of these fundamental products are ORed. by the diodes to get outputs Y_3 to Y_0.

It is very difficult to draw larger PROMs because it would involve many diodes. Therefore, drawing is streamlined, as shown in Fig. 10.42.

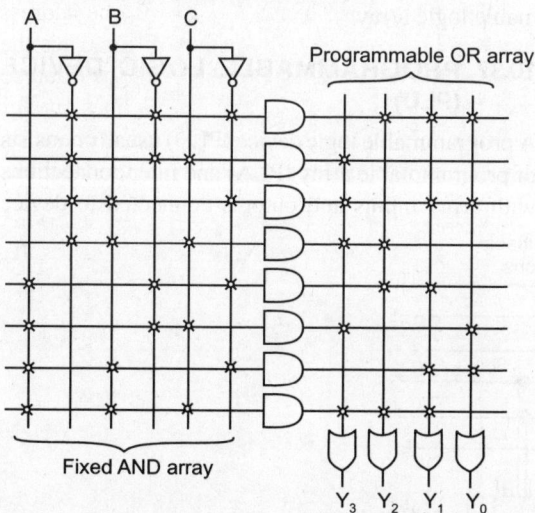

Fig. 10.42: Streamlined drawing of PROM

In this drawing, the crosses (×) indicate connections to the gate inputs. Further, each AND gate has 3 inputs indicated by the ×s on its input line. Similarly each OR gate has several inputs given by ×s on its input line.

It is to be noted that the input side of the Fig.10.42 is fixed AND array (meaning the inputs to the AND gates are not programmable in a PROM). On the other hand, the output side of the circuit is programmable

OR array, (because of fusible links at each diode location). A fixed AND array and programmable OR array is characteristic of all PROMs. With this approach, any desired combination of fundamental products can be obtained.

Programmable Array Logic

A PAL has a programmable AND array and a fixed OR array (i.e. it is different from a PROM which has fixed AND array and programmable OR array).

Figure 10.43 shows a PAL with 4 inputs and 4 outputs. The ×s on the input side are fusible links while ×s on the output side are fixed connections. With a PROM programmer, one can bush in desired fundamental products, which are ORed on the output side by fixed output connections.

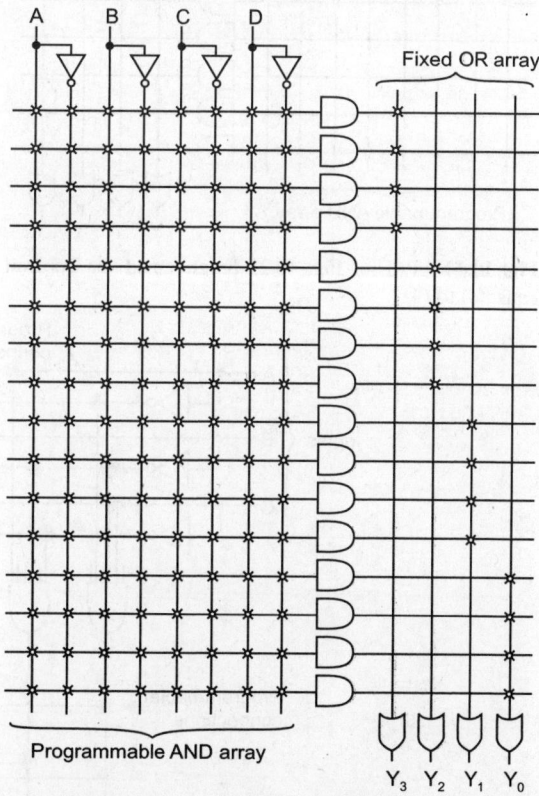

Fig. 10.43: Structure of programmable array logic (PAL)

Now we consider an example, how to program a PAL suppose we have to generate the following Boolean functions:

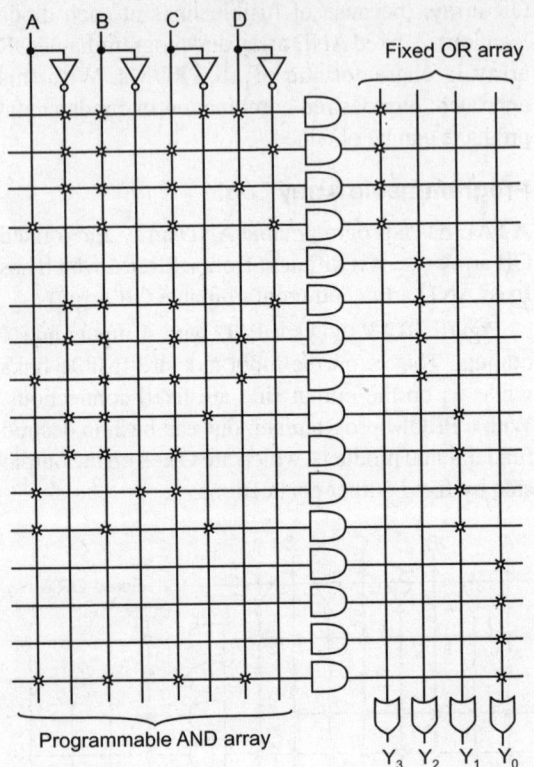

Fig. 10.44: PAL of Fig. 10.23 programmed for Boolean eqns. (1) to (4)

$$Y_3 = \overline{A}B\overline{C}D + \overline{A}BC\overline{D} + \overline{A}BCD + ABC\overline{D} \quad ...(1)$$

$$Y_2 = \overline{A}BC\overline{D} + \overline{A}BCD + ABCD \quad\quad ...(2)$$

$$Y_1 = \overline{A}B\overline{C} + \overline{A}BC + A\overline{B}C + AB\overline{C} \quad ...(3)$$

$$Y_0 = ABCD \quad\quad\quad\quad ...(4)$$

Start with eqn. (1). The 1st desired product is $\overline{A}B\overline{C}D$, on the top input line of Fig. 10.43. For this, we have to remove the first ×, the fourth ×, the fifth × and the eighth ×. As a result, the top AND gate has an input $\overline{A}B\overline{C}D$.

Then by similarly, removing ×s on the next three input lines, we can make the top four AND gates produce the fundamental products of eqn (1). The fixed OR connections on the output side imply that the first OR gate produces an output given by eqn. (1).

Similarly, we can remove ×s as needed to generate Y_2, Y_1 and Y_0. Finally we get Fig. 10.44 for the PAL which will produce Y outputs given by eqn. (1) to (4). Figure 10.45 is a simple example of a programmable logic array.

10.37 PROGRAMMABLE LOGIC DEVICE (PLD)

A programmable logic device (PLD) usually consists of programmable array (PLA) and inter connections with array inputs and outputs connected to device

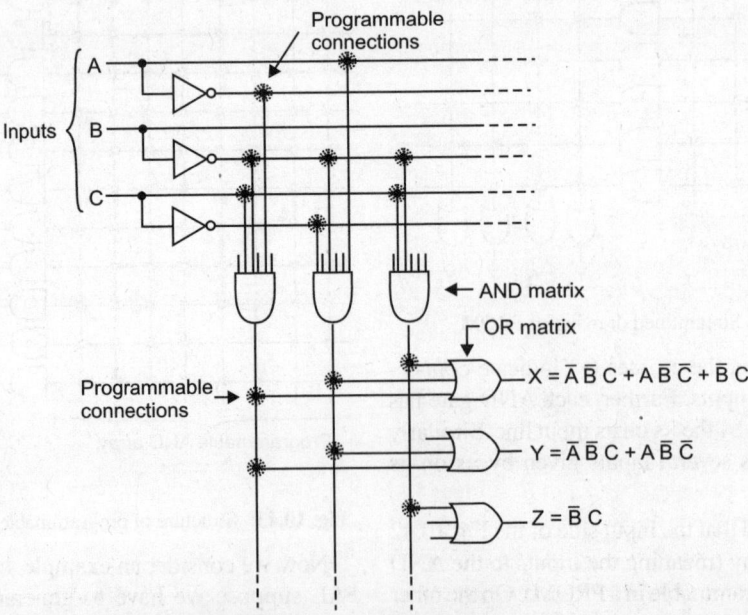

Fig. 10.45: An example of PAL

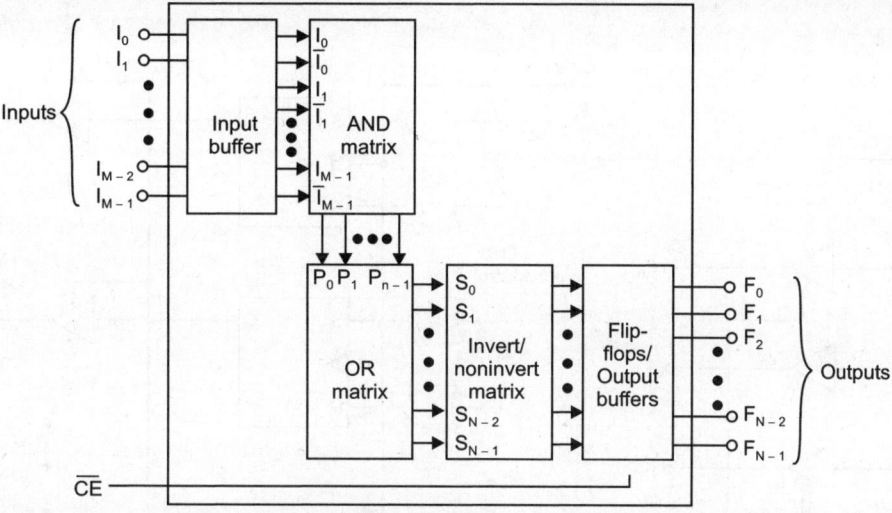

Fig. 10.46: Block diagram of a PLD

pins through fixed logic elements such as buffers and Flip-Flops.

We know that a PAL consists of two-level AND-OR circuit on a single chip. The number of AND and OR gates and their inputs are fixed for a given PAL chip. The AND gates provide the product terms, the OR gates logically sum these product terms and thereby generate a SOP expression. It has M inputs, n product terms and N outputs with $n < 2^M$ and can be used to implement a logic function of M variables with N outputs. Since all of the possible 2^M minterms are not available, therefore, logic minimization is required for a given logic function. Figure 10.46 shows block diagram of a PLA device.

10.37.1 Input Buffers

The buffer circuits at the input are required to limit loading of the sources that drive the inputs. These produce inverted as well as non-inverted inputs at the outputs as shown in Fig. 10.47 for one input.

Similar buffers are there for each of the M inputs. Figure 10.48 illustrates a convenient method of showing the input buffers and the AND matrix with the interconnections marked as ×s. Each AND gate has 2M inputs indicated by a single line in each case. For programming to implement a particular logic, (as previously discussed), the desired interconnections are left with × marks and the remaining unwanted

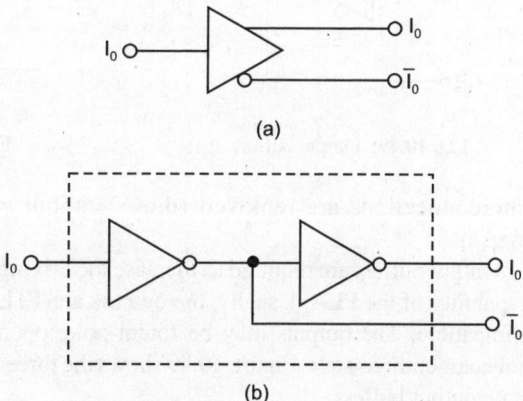

Fig. 10.47: An input buffer

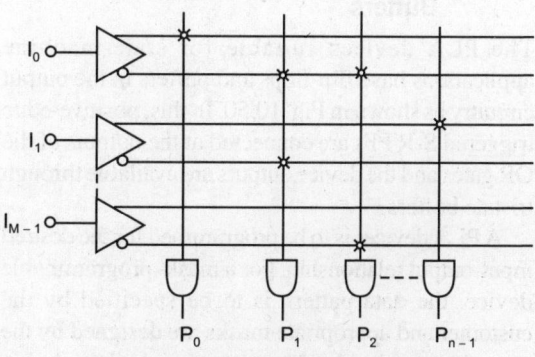

Fig. 10.48: Input buffers and AND matrix

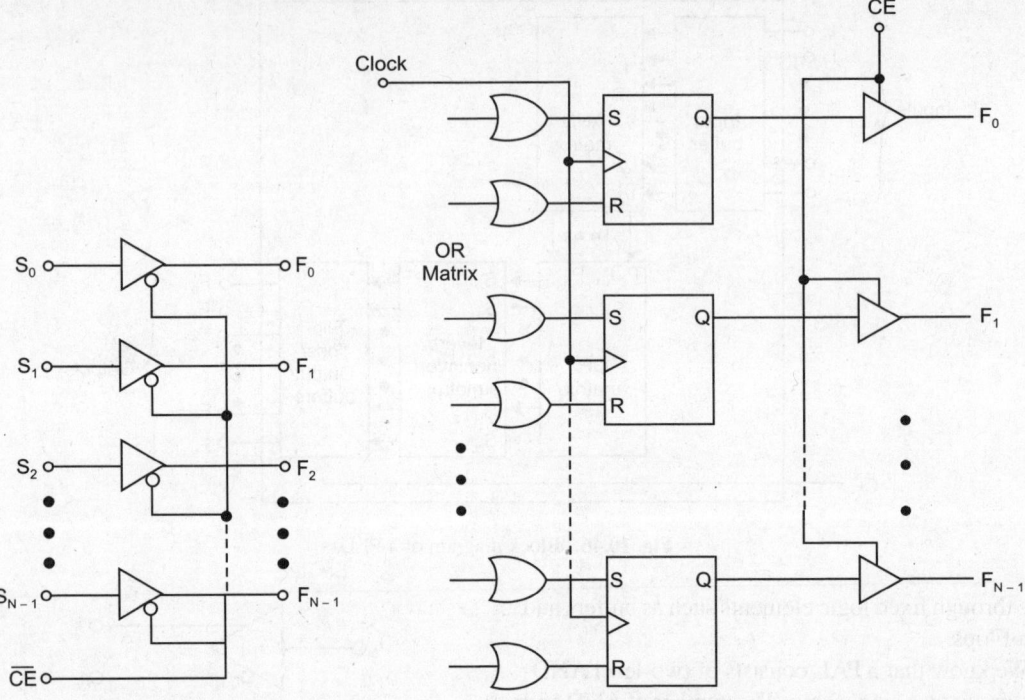

Fig. 10.49: Output buffers   **Fig. 10.50:** A section of PLA with flip-flops in the output

interconnections are removed (those without × marks).

Output buffers are required to increase the driving capability of the PLA. Usually, the outputs are TTL compatible. The outputs may be totem-pole, open collector or three-state. Figure 10.49 shows the three-state output buffers.

10.37.2 Output Through Flip-flops and Buffers

The PLA devices suitable for state-machine applications have flip-flops and buffers in the output circuitry as shown in Fig. 10.50. In this, positive-edge triggered S-R FFs are connected at the outputs of the OR gates and the device outputs are available through tristate buffers.

A PLA devices is to be programmed for the desired input-output relationship. For a mask-programmable device, the data pattern is to be specified by the customer and appropriate masks are designed by the manufacturer, i.e. the data pattern are built in during manufacturing process.

An FPLA has all its nichrome fuse links intact at the time of manufacturing and the unwanted links are electrically open circuited during programming. The links to be opened are accessed by applying voltages at the inputs and outputs of the device. The FPLAs are not reprogrammable.

10.37.3 Expanding PLA Capacity

Some applications require a capacity that exceeds the capacity of a single PLA. The required capacity can be achieved by suitably connecting several identical devices together as shown in Fig. 10.51.

The number of inputs can be increased by making the connections as depicted. The circuit has (M + Q) inputs and N outputs. This circuit also increases the number of product terms. The number of product terms is P× (2^Q–1). The outputs are allowed to be connected together for the devices with only passive pull-up.

10.37.4 Applications of PALs

The PALs can be used to implement combinational and sequential logic circuits. For the design of

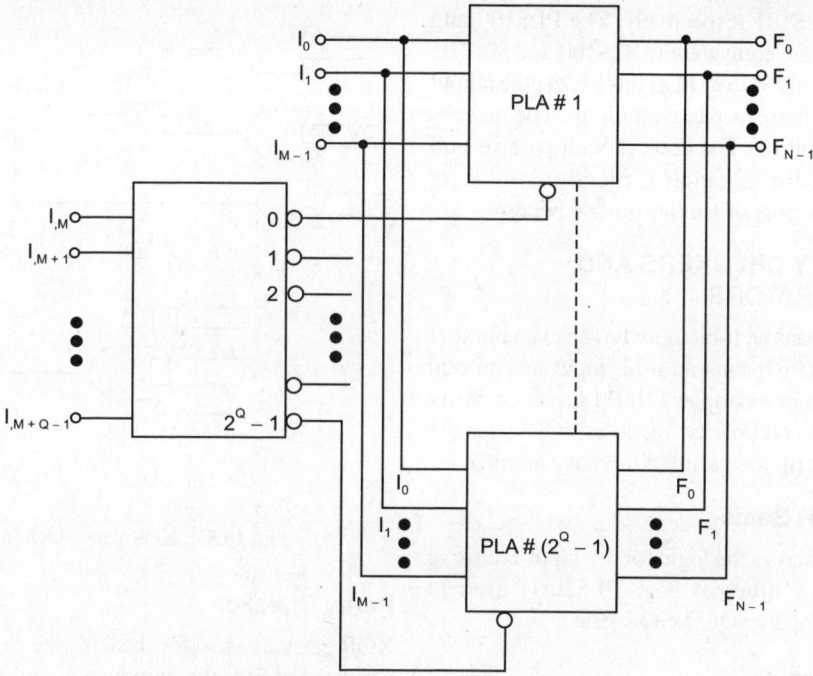

Fig. 10.51: Expanding the input word length of PLA

combinational circuits, PLAs with output circuit as shown in Fig. 10.50 are used. The following steps can be used for implementing combinational logic functions:

1. Prepare the truth table.
2. Write the Boolean equations in SOP form.
3. Simplify the equations to obtain minimum SOP. (The main criterion is to minimize the number of product terms).
4. Determine the input connections of AND matrix to generate the required product terms.
5. Determine the input connections of OR matrix to generate the required sum terms.

6. Determine the connections required for INVERT/NON-INVERT matrix to set the active logic levels of the outputs.
7. Program the PLA.

The devices having flip-flops in the output circuit (as shown in Fig. 10.50) are required for the design of sequential circuits.

10.37.5 Available PLAs

Some of the commercially available PLA ICs are given in Table 10.18. The 82S200 and 82S201 are pin-for-pin mask programmable replacements for the

Table 10.18: Some available PLAs with their features

IC No.	PLA/ FPLA	Number of			Out- put	Access time ns	Supply voltage V	Power dis- sipation mW	Packaging	Input/Output logic levels
		Input	Product terms	Outputs						
82S200	PLA	16	48	8	TS	80	+5	600	28-pin DIP	TTL compatible
82S201	PLA	16	48	8	OC	80	+5	600	28-pin DIP	TTL compatible
82S100	F PLA	16	48	8	TS	80	+5	600	28-pin DIP	TTL compatible
82S101	F PLA	16	48	8	OC	80	+5	600	28-pin DIP	TTL compatible
DM7575	PLA	14	96	8	TS	150	+5	550	24-pin DIP	TTL compatible
DM7576	PLA	14	96	8	OC	150	+5	550	24-pin DIP	TTL compatible

82S100 and 82S101 respectively. The PLS100 and PLS105 PLAs are equivalent to 82S100 and 82S105 respectively. The DM7575 PLA has totem pole output whereas DM 7576 has passive pull up. The devices with passive pull up. The devices with passive pull up are useful for expanding functions by wire ANDing the outputs of similar other devices.

10.38 PARITY CHECKERS AND GENERATORS

Even parity means an n-bit input has an even number of 1s. *Odd parity* means an n-bit input has an odd number of 1s. For example: 110011 has even parity and 110001 has odd parity. Exclusive-OR gates are ideal for checking the parity of a binary number.

Exclusive OR Gates

Figure 10.52a shows the logic for a 4-input exclusive OR gate and symbol of Fig. 10.52b is used to represent a 4-input exclusive OR gate.

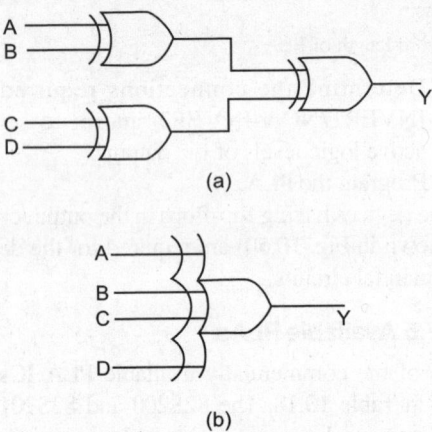

(a)

(b)

Fig. 10.52: Four inputs XOR gate

When your see this symbol, remember that the gate produces an output 1 when the ABCD input has an odd number of 1s.

Using 2-input XOR gates, one can produce XOR gates with any number of inputs. For example, Fig. 10.53a shows a circuit with 6 inputs and 1 output.

If one analyses this circuit, one will find that it produces an output 1 only when the 6-bit input has an odd number of 1s. Figure 10.53b shows an abbreviated symbol for 6-input XOR gate.

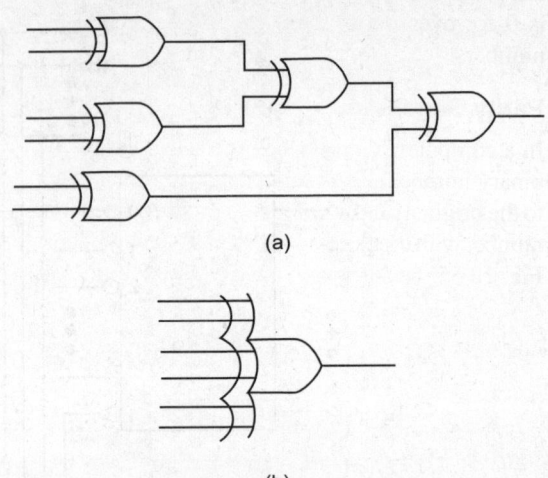

(a)

(b)

Fig. 10.53: XOR gate with 6 inputs

Parity Checker

XOR gates are ideal for checking the parity of a binary number because they produce an output 1 when the input has an odd number of 1s. Therefore, an even parity input to an XOR gate produces a low output whereas an odd parity input produces a high output.

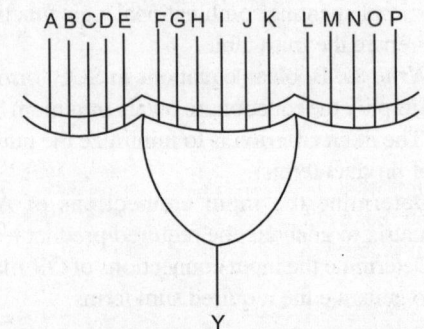

Fig. 10.54: XOR gate with 16 inputs

Figure 10.54 shows a 16-input XOR gate. Consider inputs

(i) 1111 0000 1111 0011 (even parity)

(ii) 1111 0000 1111 0111 (odd parity)

The first binary number has even parity and the second has odd parity. (Longer binary numbers can be much easier to read if they are split into nibbles, i.e. groups of four, as done above). For input (i), the output Y will be low (even parity) and for input (ii) the output will be high (odd parity). Thus, the output is 0 for even-parity numbers and 1 for odd-parity numbers.

Parity Generation

In a computer, a data to be processed may be like a binary number and sometimes, an extra bit is added to the original binary number to produce a new binary number with even or odd parity. For instance, consider Fig. 10.55, which shows an 8-bit binary number

$$X_7\, X_6\, X_5\, X_4\, X_3\, X_2\, X_1\, X_0$$

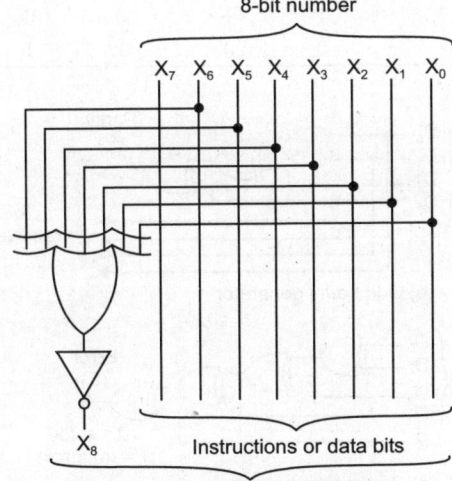

Fig. 10.55: Odd-parity generation

Suppose this number equals 01000001. Then the number has even parity. The X-OR gate produces an output 0, and because of the inverter

$$X_8 = 1.$$

and the final 9-bit output is 1 0100 000 1. Notice that this has odd parity.

Now suppose, we change the 8 bit input to 0110 0001. Now this number has odd parity. In this case, the X-OR gate produces an output 1. But the inverter produces a 0, so that the final 9-bit output is 001100001 (again, odd parity).

Therefore, the circuit given in Fig. 10.55 is called an *odd-parity generator*, (because it always produces a 9-bit output number with odd-parity. (To get an even parity generator, the inverter is to be deleted).

10.39 EVEN/ODD PARITY GENERATORS AND CHECKERS

We have seen above that, to detect bit error, an additional bit called parity bit is added to data bits and the combined word is transmitted. At the receiving end, the number of 1s in the word received are counted and the error, if any, is detected. This parity clock, however detects single bit errors.

A parity bit (a 0 or a 1) is attached to the data bits such that the total number of 1s is even for even parity and odd for odd parity. (The parity bit can be attached either at the beginning or at the end depending on the system design). *A system operates with either even or odd parity but not both.* (So, a word always actually contain either even or odd number of 1s)

In order to check or generate the proper parity bit in a given code word, the basic principle used is *the modulo sum of an even number of 1s is always a 0 and the modulo sum of an odd number of 1s is always a 1, therefore, in order to check for an error, all the bits in the received word are added. If the modulo sum is a 0 for an odd parity system or a 1 for an even parity system, an error is detected. (parity failure).* (Fig. 10.56).

Odd Parity Checking

The parity bit associated with each code-group in an *odd-parity-bit* checking system has such a value that the total number of 1s in each code group plus the parity bit is always odd. Consider (for example) Table 10.19. The example shown uses an 8421 code and has an odd parity bit which makes the sum of the 1s in each code group an odd number.

Table 10.19

Decimal	BCD	Odd Parity Bit
0	0000	1
1	0001	0
2	0010	0
3	0011	1
4	0100	0
5	0101	1
6	0110	1
7	0111	0
8	1000	0
9	1001	1

Now, if a single error occurs in transmitting a code group-for instance, if 0011, 1 is erroneously changed to 0010, 1-the fact that there is an even number of 1s in the code group plus the parity bit will indicate that an error has occurred.

Even Parity Checking

Suppose that an even-parity check is used like Table 10.19 in place of odd parity check. Suppose, for instance the code group to be sent is 0010; the parity bit in this case will be a 1. If the code group is erroneously read as 0011; the number of 1s in the code group plus the parity bit will be odd, and the error will be detected. Note that such a parity checking will detect only a single bit error.

the truth table operation of this IC. This device can be used to check for odd or even parity on a 9-bit code (8 data bits and one parity bit) or can be used to generate a 9-bit odd or even parity code.

Table 10.20: Truth table of IC 74180 (× are don't cares)

| Σ of 1s at | Inputs | | Outputs | |
A through H	Even	Odd	Σ Even	Σ Odd
Even	1	0	1	0
Odd	1	0	0	1
Even	0	1	0	1
Odd	0	1	1	0
×	1	1	0	0
×	0	0	1	1

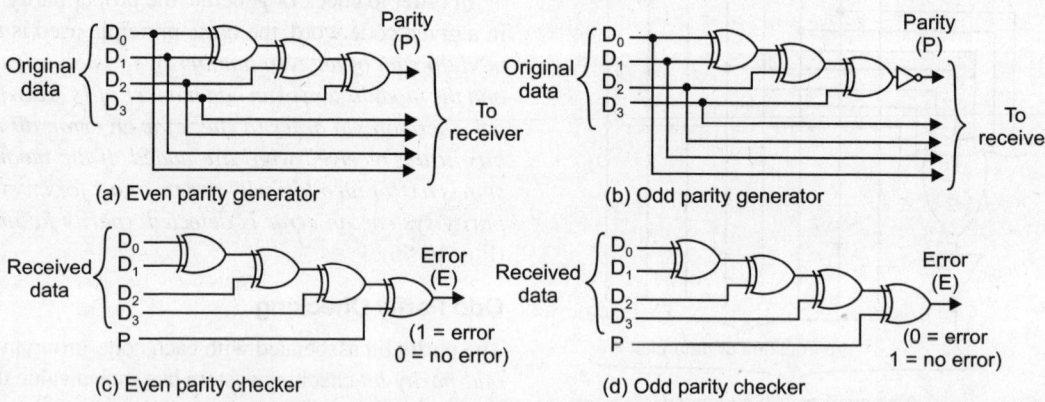

(a) Even parity generator (b) Odd parity generator

(c) Even parity checker (d) Odd parity checker

Fig. 10.56: Even parity generator/checker (a) and (c)) and odd parity generator/checker (b) and (d)) (After A. Anand Kumar, "Fundamentals of Digital Circuits" Prentice Hall of India, N. Delhi (2001)

Figure 10.57 shows the logic symbol of IC 74180, a 9-bit parity generator/checker. Table 10.20 gives

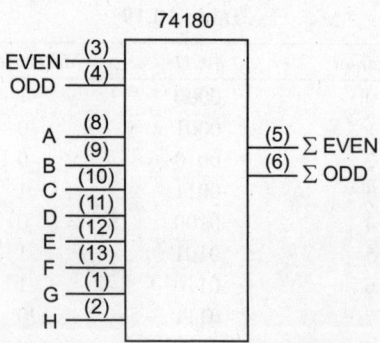

Fig. 10.57: IC 74180 parity generator/checker

10.40 FOUR BIT PARITY CHECKER/ GENERATOR

We now consider the design of a circuit which generates/checks the even-parity of a 4-bit number.

The circuit contains 5 flip-flops (four to store the number and one to store the parity-bit. The control signals necessary to check parity along with the circuit diagram are given in Fig. 10.58.

The signal PG is set to 1 if parity is being generated. It is set to 0, when parity is to be checked.

The parity checker works as follows:

During the first clock pulse LD is "1". The parity bit P is loaded into the parity FF (P). The data is entered in the shift register through the set direct (SD)

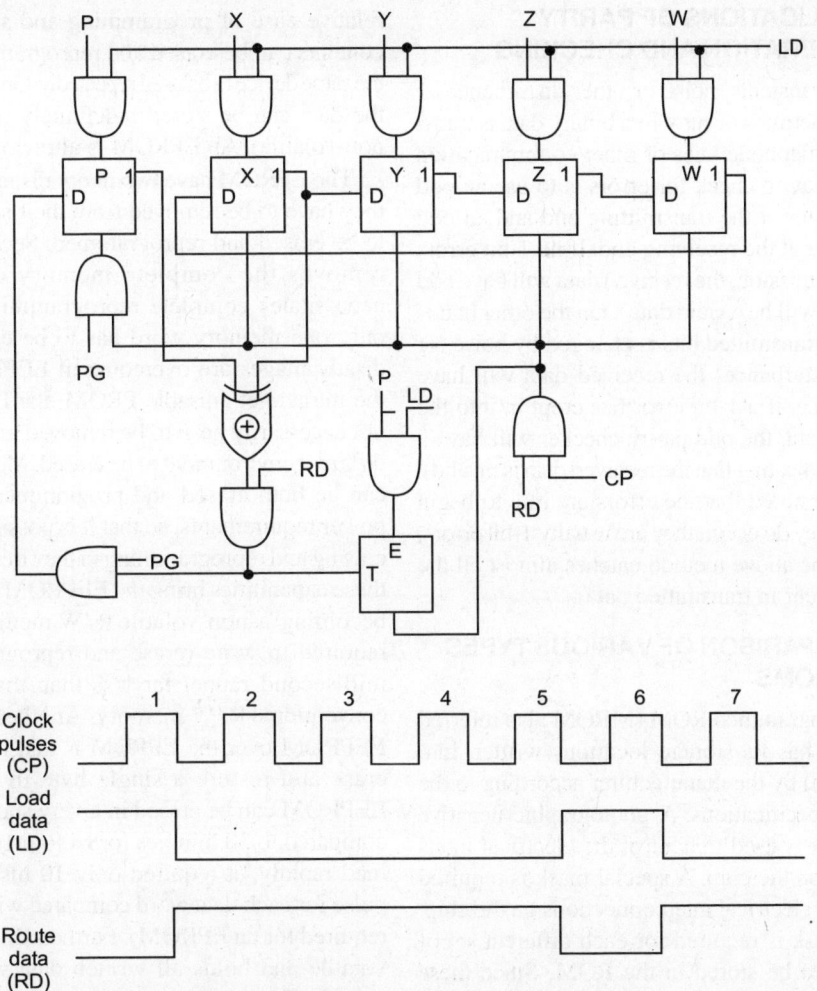

Fig. 10.58: A 4-bit (even) parity checker/generator (After T.R. Viswanathan, G.K. Mehta and V. Rajaraman, "Electronics for Scientists and Engineers", Prentice Hall of India, N. Delhi (1989))

inputs. The parity error indicating FFE is also set equal to the value of P.

The 'route data' signal RD is '1' for four clock periods. During this time the contents of the data registers (X, Y, Z, W) are shifted four times. As PG is OFF when parity is being checked, the FF P is unaffected.

If P = 0, then the FFE is toggled by the output of the EXCLUSIVE-OR gate. The number of times it is complemented is equal to the number of 1s in X, Y, Z, W. Thus E will be set to 1 if the parity is ODD, which will indicate parity failure.

If P = 1, then FFE is toggled. The number of times it is toggle is equal to the number of zeros in X, Y, Z, W. As E is originally '1', at the end of toggling, E will be 1 only if the number of 0s is even indicating a parity failure (odd no. of 1s).

If parity is to be generated, PG is set = 1 and P is set = 0. X, Y, Z, W are entered with LD = 1. The output of the X-OR gate is feedback to the P FF. If an odd number of 1s are in X, Y, Z, W, then P is set to 1 at the end of 5th CP. If an even number of 1s are in X, Y, Z, W, then P is finally in the '0' state. Thus the correct parity bit is generated and entered in the P flip-flop.

10.41 APPLICATIONS OF PARITY GENERATION AND CHECKING

Because of transients, noise or other disturbances 1 bit errors sometimes occur when binary data is transmitted over telephone lines or other communication paths. One way to check for errors is to use an odd parity generator at the transmitting end and an odd parity checker at the receiving end. If no 1-bit errors occur in transmission, the received data will have odd parity (and it will be a valid data). On the other hands if one of the transmitted bits is changed by noise (or any other disturbance) the received data will have even parity (i.e. if a 1-bit error has crept in into the transmitted data, the odd parity checker will have a low output, indicating that the received data is invalid).

(It is to be noted that the errors are rare to begin with, and if they do occur, they are usually 1-bit errors, that is why the above method catches almost all the errors that occur in transmitted data).

10.42 COMPARISON OF VARIOUS TYPES OF ROMS

The mask programmed ROM (MROM also referred to as ROM) has its storage locations written into (programmed) by the manufacturer according to the customer's specifications. A photographic negative called a *mask* is used to control the electrical inter-connections on the chip. A special mask is required to control the electrical interconnections on the chip. A special mask is required for each different set of information to be stored in the ROM. Since these masks are expensive, this type of ROM is economical only if manufactured in large quantities. A major disadvantage is that it cannot be reprogrammed if a change in design is required.

The field programmable ROM or PROM is ideally suited for development work and small production quantities. These are not programmed during manufacturing process but are custom-programmed by the user. Once programmed, a PROM cannot be erased and reprogrammed, i.e. if the program is faulty or has to be changed; it has to be thrown away. These are, however, used during testing and refining new system designs because they can be replaced with relatively little effort and cost.

EPROM (i.e. erasable PROM) is also useful for development and experimental work, because of relative ease of programming and small cost. The contents can be erased and reprogrammed, enabling the same device to be used repeatedly. Once programmed, the data can be stored indefinitely (i.e. EPROM is non-volatile). An EPROM is ultraviolet erasable.

The EPROM have two major disadvantages: First they have to be removed from their sockets in order to be erased and reprogrammed. Second, the eraser removes the complete memory contents; this necessitates complete reprogramming even when only one memory word has to be changed. These disadvantages are overcome in EEPROMs. Unlike the ultraviolet erasable PROM, the EEPROM does not necessarily have to be removed and exposed to a different environment to be erased. Many EEPROMs can be both erased and programmed with modest power requirements, so that it is possible to integrate erasing and reprogramming in the circuitry. Although these capabilities bring the EEPROM a step closer to becoming a non-volatile $R/\overline{W}$ memory. The time required to write (erase and reprogram) is still in millisecond range, far less than that required by conventional R/W memory. Another advantage of EEPROM over the EPROM is that it is possible to erase and restore a single byte in an array. The EEPROM can be erased in a very short time (10 ms compared to 30 minutes for EPROM) and programmed rapidly (it required only 10 ms programming pulse for each data word compared with 50 ms pulse required for an EPROM). Further, EEPROM is non-volatile and holds all written data when power is turned off.

10.43 STATIC AND DYNAMIC RANDOM ACCESS MEMORIES

Large scale random access memory (RAM), also known as read/write memory is used for temporary storage of data and program instruction in microprocessor based systems.

The term random means that the contents of any memory location can be accessed randomly. Many types of memory can be classified as having random access, but when the term RAM is used with semiconductor memories it is taken, to wean read/write memory (RWM).

A major drawback of RAMs is that they are volatile and will lose all stored data if power is switched off.

Figure 10.59 shows the block diagram to illustrate how data are written into and read out of RAM device. This RAM device can store 4 bit words into 16 individual memory locations. To write data into a memory location, a high or 1 is applied to the write enable input and 0 to $\overline{CS}$ (Chip Select). To select a memory location where data is to be written, a binary number equivalent to the desired location is placed at the address lines. The data to be stored into the selected location are applied to the data input lines. To read data from memory requires that a 1 be applied to read enable input 0 to chip select $\overline{CS}$. The specific memory location from where data are read out, is determined by binary number applied to the address line. The data appears at the data output lines and as the information is read, the flip-flop retain their data.

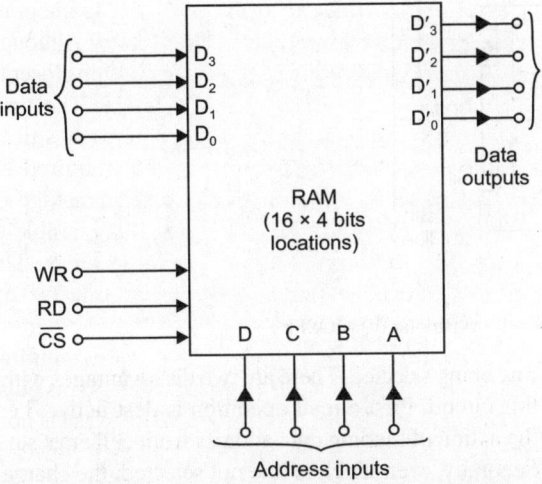

Fig. 10.59: Block diagram of 4 × 16 RAM

The RAM devices include internal address decoders and output buffers. Also output amplifiers are used to sense low voltage levels of data stored at each memory location so that information can be increased to a certain useable level at the output of the IC. There are two general categories of R/W (RAM) memories:

(i) Static RAM.   (ii) Dynamic RAM

10.43.1 Static RAM

SRAM is classified as volatile memory because it depends upon the application of continuous power to maintain the stored data. If power is interrupted, the memory contents are destroyed. A SRAM is a matrix of static, volatile memory cells and address decoding functions integrated on chip to allow access to each cell for read/write operation.

BJT static RAMs are extremely fast but they are less compact and less energy efficient. They are TTL compatible and are used as small "scratch pad" memories for data being processed. The RAM cell shown in Fig. 10.60a employs two cross coupled triple emitter transistors in a flip-flop. One emitter of each transistor is connected to bit line to provide for data transfer and storage. The other emitters are connected to address lines for coincident selection. The address lines are normally low and the currents from all conducting transistors in the matrix flow out along these lines.

Assuming Q_2 is on, voltage at P_2 is low, Q_1 is off and voltage at P_1 is high. To address cell a – b, the corresponding address lines X_a and Y_a are taken high. As these lines go high, the connected emitters rise in potential and current in Q_2 is transferred to the 1 bit line. This current activates the sense amplifier and a logic 1 appears at the output. To write data into, particular address lines are made high, write enable goes low and 1 (say) is input to the write amplifier. The amplifier output takes the 1 bit line low and Q_2 is turned on. After write enable goes high, the bit line returns to normal.

The high packing density and low power consumption of MOS devices have led to their wide use. The typical MOS memory cell contains $6n$ channel enhancement mode MOSFETs. The transistor Q_1 is the driver and Q_3 the load for an inverter that is cross coupled to second inverter Q_2 - Q_4 to make the flip-flop. When Q_2 is on and Q_1 is off, output Q is 1. With row select line low, transistors Q_5 and Q_6 are off and the cell is isolated from the bit line.

In read operation, when row select goes high Q_5 and Q_6 provide coupling and Q appears on bit line D. A sense amplifier connected to bit line provides buffering and the proper logic level appears at the output. In write operation, the .selected row of cells is connected to the bit lines and Q is set or reset by a 1 placed on bit line D or $\overline{D}$, by write amplifier.

10.43.2 Dynamic RAM

Each static RAM cell of type in Fig.10.60 has six MOSFETs, which clearly limits the number of cells which can be packaged into one IC.

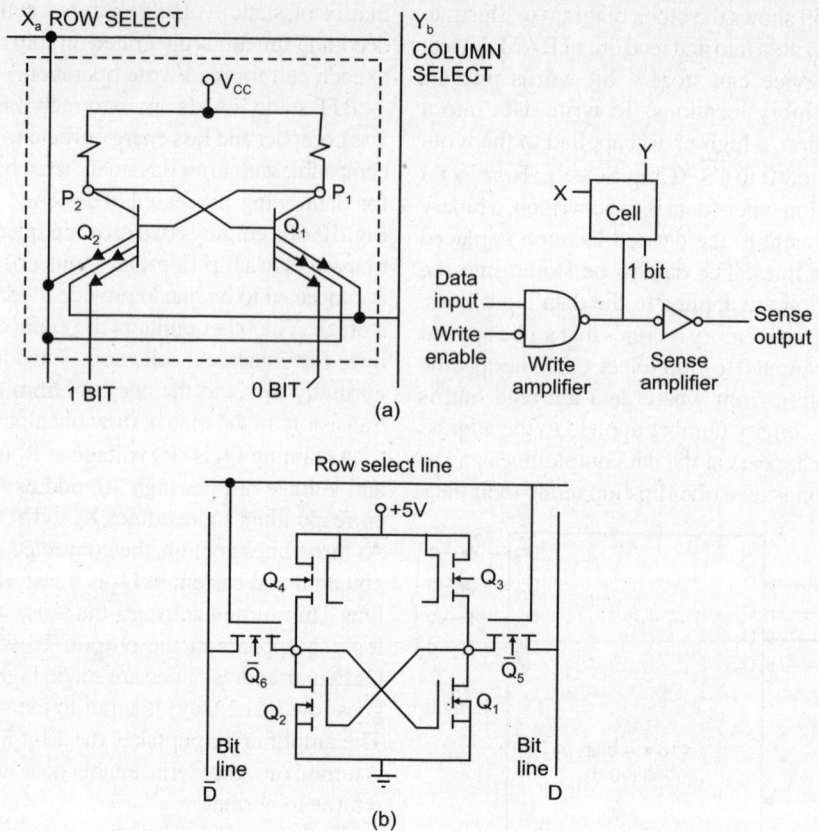

Fig. 10.60: Static RAM memory cell (a) Bipolar (b) MOSFET

Dynamic RAMs use charge storage on a capacitor to represent binary data values. These are called dynamic because the stored charge leaks away even with power continuously applied. Therefore these cells must be read and refreshed continuously at periodic intervals. Despite this complex operating mode, the advantages of cost per bit and high density have made DRAMs the most widely used semiconductor memories in commercial applications.

Dynamic RAM cell uses a capacitive element for storing the data bit. Binary information is stored as charge. If charge is present at a capacitive element indicates a logic 1, whereas the absence of the charge represents indicates logic 0. Figure 10.61 has only one MOSFET which is switched on when ROW line is selected. Each column has a sensing amplifier and in a read operation, the charge in C is sensed when the row is selected. For a write operation data on Col/data line charges C when the MOSFET is opened by

row being selected. There are two disadvantages with this circuit. First a read operation is destructive, i.e. the action of reading data address from cell erases it. Secondly, even if cell is line not selected, the charge on C is lost through leakage.

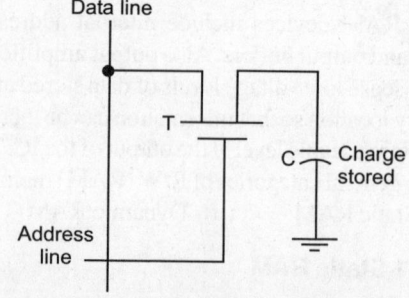

Fig. 10.61: Dynamic RAM cell

The leakage problem is solved by refreshing operation which consists of reading every cell once every few milliseconds and rewriting back into the cell the data read out. Similarly, destructive read out is followed by a rewrite back into cell.

DRAM disadvantages are overweighed by its cell simplicity. DRAMs have at least four times the packing density as compared to SRAM. DRAMs consume less power as compared to SRAMs.

10.44 RAM IC

A common type of Read/write RAM is the 7489 IC It has 64 memory cells and can store 16 four bit words. A block diagram showing the input and output lines is shown in Fig. 10.62.

Suppose it is desired to write the 4 bit word 0001 into memory location 8. The following procedure is required:

(i) Place the binary equivalent of 8, i.e. 1000 at the address lines by applying $D = 1$, $C = 0$, $B = 0$, $A = 0$.

(ii) Apply the desired 4 bit word 0001 into the memory by making $D_{IN} = 0$, $C_{IN} = 0$, $B_{IN} = 0$ and, $A_{IN} = 1$ to the input data lines.

(iii) Apply 0 to write enable line ($\overline{WE}$).

(iv) Apply 0 to memory enable line ($\overline{ME}$).

To read data out of memory, a similar procedure is necessary to read out the data out of memory location. It consists of placing the address of memory location on address lines, enabling read mode ($\overline{WR} = 1$) and enabling the memory chip. As a result memory contents appear at the output.

10.45 MEMORY EXPANSION

In many applications the required memory capacity, i.e. the number of words or size of the word cannot be satisfied by a single available memory IC chip. Therefore, several identical ICs can be cascaded to provide an expanded memory of larger capacity. The Possibilities of expansion include:

(i) Expanding word size

(ii) Expanding word capacity

(iii) Expanding word size and capacity.

(i) Expanding word size. If it is required to have memory of word size n and word size of the available memory ICs is N $(n > N)$, then a number of similar ICs can be combined together to get the desired word size. The number of ICs required is an integer next higher to the value n/N. These are connected as:

(i) Connect the corresponding address lines of each IC individually.

(ii) Connect the control inputs (RD, WR and CS) of each memory together. Similarly, connect CS, WR and RD inputs.

As a result the number of data inputs output lines will be equal to the product of the number of chips used and word size of the chip. Figure 10.63 show two 32 × 4 memories being used to create 32 × 8 memory. The corresponding address lines are common so that any address activates each row of same number in each memory and produces an 8 bit word.

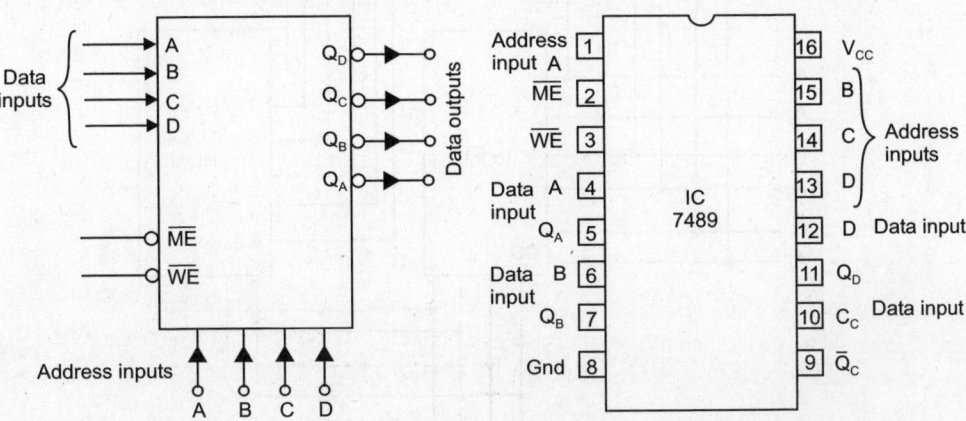

Fig. 10.62: 7489 RAM IC (a) Functional diagram, (b) Pin diagram

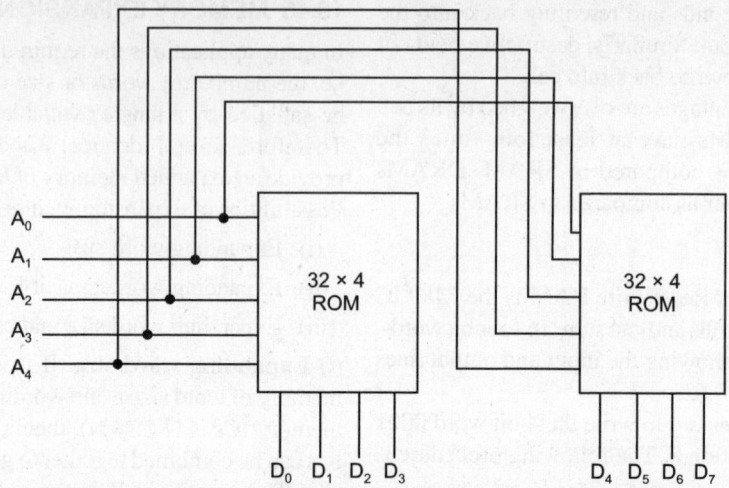

Fig. 10.63: Two 32 × 4 memories connected to make a 32 × 8 memory

(ii) Expanding word capacity. Memory ICs can be combined together to produce a memory with desired number of address locations, Thus to have a memory of capacity *m* words using memory IC with *M* words each. The number of ICs required is an integer higher next to value *m/M*. These ICs are connected as:

(i) Connect the corresponding address lines of each IC individually.

(ii) Connect the control inputs (RD, WR, CS, etc.) of each IC together.

(iii) Use a decoder of proper size and connect each of its outputs to one of the CS terminals of memory IC.

Figure 10.64 shows how two 32 × 4 memories are connected to make a 64 × 4 memory.

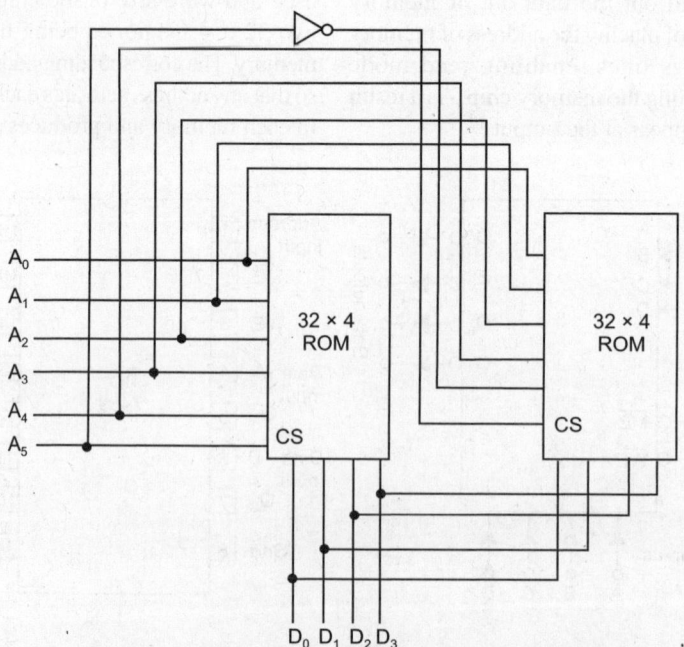

Fig. 10.64: Two 32 × 4 memories combined to form 64 × 4 memory

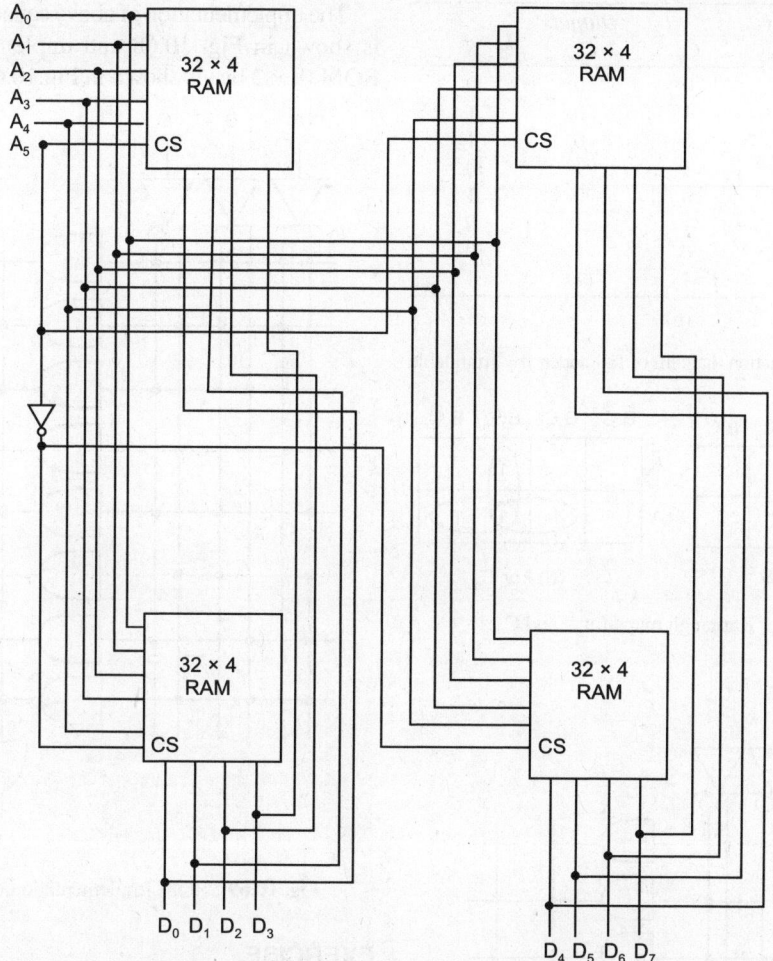

Fig. 10.65: 32×4 memories combined to form 64×8 memory

When chip select is 1, memory gets activated. Then with $A_5 = 1$ the left memory is activated, while with $A = 0$, it will be right memory. Thus the 32 addresses 000000 to 011111 produces output words D_0 to D_3 from right memory and the addresses from 100000 to 111111 from the left memory.

(iii) Expanding word size and capacity. Word size and capacity can be expanded by combining the two previous configurations. Figure 10.65 illustrates how four 32×4 memories are combined to form 64×8 memory.

Example 10.1: Implement full adder using a PLA and ROM.

Solution: The truth table and functional diagram for the full adder are shown in Fig. 10.66. The K-maps for the outputs S and C_{out} are given in Fig. 10.67 and minimized expressions are given by

$$S = \overline{A}\overline{B}C_{in} + \overline{A}B\overline{C}_{in} + A\overline{B}\overline{C}_{in} + ABC_{in}$$
$$C_{out} = AB + AC_{in} + BC_{in}$$

(a)

| | Inputs | | Outputs | |
A	B	C_{in}	C_{out}	S
0	0	0	0	0
0	0	1	0	1
0	1	0	0	1
0	1	1	1	0
1	0	0	0	1
1	0	1	1	0
1	1	0	1	0
1	1	1	1	1

(b)

Fig. 10.66: (a) Function diagram of full adder, (b) Truth table

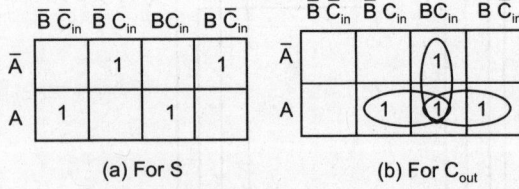

(a) For S   (b) For C_{out}

Fig. 10.67: Karnaugh maps for S and C_{out}

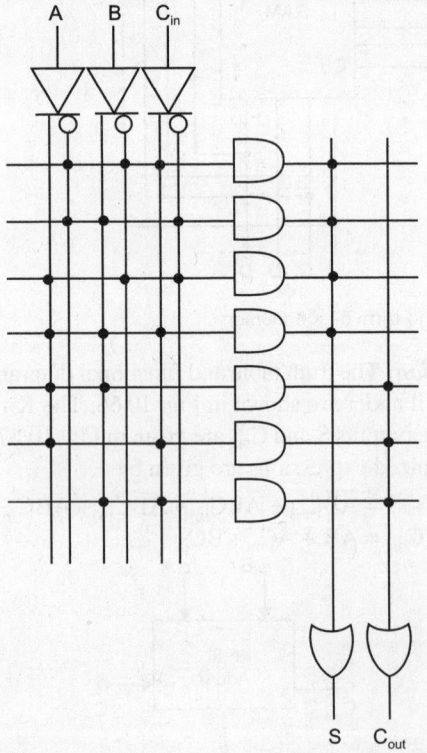

Fig. 10.68: PLA Implementation of full adder

The implementation of above equations using PLA is shown in Fig. 10.68 and implementation using ROM (8×2 bit) is shown in Fig. 10.69.

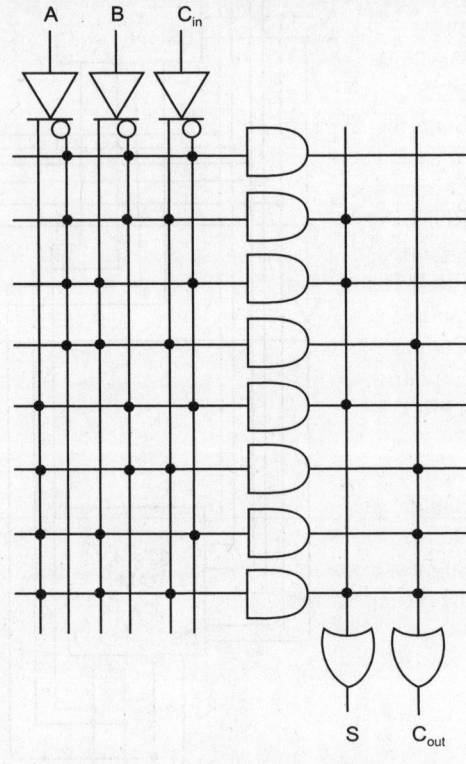

Fig. 10.69: ROM Implementation of full adder

EXERCISE

1. Define a decoder. Show how to decode the 4-bit code 1011 (LSB).

2. Define a demultiplexer. Show how to convert a decoder into a demultiplexer. Indicate how to add a strobe to this system.

3. Define a multiplexer. Draw a logic block diagram of a 4-to-1 line multiplexer.

4. Define an encoder. Indicate a diode matrix encoder to transform a decimal number into a binary code.

5. Define a read-only memory. Show a block diagram of an ROM. What is stored in the memory? What hardware constitutes the memory elements?

6. Write the truth table for converting from a binary to a Gray code. Write the first six lines of the truth table for converting a Gray into a binary code.

7. Show the system for a 4-bit odd-parity checker.

8. Design a BCD to Excess-3 code converter using PLA.

9. Design an excess-3 to BCD code converter using PLA.

10. Design BCD-to-seven segment decoder using PLA.

11. State the full name of the following memory devices and describe very briefly the function of each one:
ROM, PROM, EPROM, EAROM, RAM, PLA.

12. Describe the methods of storing data in ROMs, PROMs and EPROMs. What are their relative merits?

13. Compare ROM, PAL and PLA.

14. Explain the difference between an EPROM and a PROM.

15. Explain why an EPROM is or is not a volatile memory.

16. Explain the term volatile memory.

17. Define the term mask-programmable.

18. Suggest a suitable arrangement for expanding the bit capacity of CCDs.

19. Suggest a suitable arrangement for expanding the word length of CCDs.

20. Explain the working of a BCD-to 7 segment decoder by drawing the truth table and Karnaugh maps.

21. How many Boolean functions of 4 variables can the following ROMs store:
  (i) 256×4 ROM   (ii) 512×8 ROM
  (iii) 1024×4 ROM   (iv) 16×4 ROM

22. What is the minimum number of input and output terminals required by the ROMs of problem 21.

23. Describe the advantages of an EEPROM over an EPROM. How is an EPROM erased?

24. How are data stored in a CCD memory? What are the principal advantages and disadvantages of the CCD memory?

25. How does a PLA differ from a ROM?

26. Design a logic circuit to generate (i) an even parity bit and (ii) an odd parity bit for a 3-bit binary input.

27. Explain the difference between a MUX and a DEMUX.

11

Analog to Digital (A/D) and Digital to Analog (D/A) Converters

11.1 INTRODUCTION

Digital systems are being widely used for control, communication, computers and instrumentation, etc. In many such applications, the signals are not available in the digital form, e.g. an analog voltage to be processed using digital techniques may be a dc voltage or a time varying voltage. The process of converting an analog signal to a digital form involves a sequence of following four processes

 (i) Sampling
 (ii) Holding
 (iii) Quantizing
 (iv) Encoding

These processes are not performed separately. Sampling and holding are done simultaneously using a circuit known as sample and hold circuit. These operations are required for conversion of time varying analog signals and not for dc signals. Quantizing and encoding are done simultaneously using a circuit known as an analog to digital converter (A/D converter or ADC)

The output of the system may be required in the analog form, therefore the digital output may be required to be back in analog form. The process is referred to as digital to analog conversion (D/A converter or DAC).

Some example of A/D and D/A converters are:

1. A digital system can be used to monitor the ambient temperature of an oven and if it exceeds a certain limit, it should reduce the fuel input. Here an A/D converter is required to convert the output of the sensor (which converts temperature to an analog electrical signal) to digital form. If the temperature exceeds the specified limit, digital output produced is to be converted to analog form in order to control the device which reduces the fuel input.

2. A digital voltmeter measures an analog voltage and displays the voltage in numeric form. This uses an A/D converter. The output in this case is not required to be converted back to analog form (i.e D/A converter is not required)

3. A digital communication system is used to transmit massages which are in the form of analog electrical signals. This requires an A/D converter at the transmitting end and a D/A converter at the receiving end.

In microprocessor based process control system, A/D and D/A converters are often used and are referred to as peripherals or I/O devices.

Since D/A converters are used as sub systems in many A/D converters, we will first discuss D/A converters.

11.2 DIGITAL TO ANALOG CONVERSION

The D/A converter can be considered as a decoding device which accepts a digitally coded signal D and an analog reference p as inputs and provides an analog output A related to the input as

$$A = pD$$

D is a digital word of given number of bits and can be represented as

$$D = \left(b_{N-1} + \frac{b_{N-2}}{2^1} + \frac{b_{N-3}}{2^2} + \dots \dots \frac{b_{N-N}}{2^{N-1}} \right)$$

where N = total number of bits. b_1, b_2, b_3.......... etc. are the bit coefficients which are quantized to be either 1 or 0.

$$\therefore \quad A = P \, [b_{N-1} + b_{N-2} 2^{-1} + b_{N-3} 2^{-2} + \dots \dots$$
$$+ b_0 \, 2^{-(N-1)}]$$

There are two types of commonly used D/A converters. These are based on:

1. Weighted-resistor D/A converter, or
2. R-2R ladder D/A converter.

The actual implementation of a D/A system contains four separate parts viz:

(i) A reference voltage corresponding to parameter p.

(ii) Binary switches to simulate the binary coefficients b_0, b_1......b_{N-1}

(iii) A resistive weighting network (1 or 2 above). and

(iv) an output summing means.

11.3 WEIGHTED-RESISTOR D/A CONVERTER

An actual D/A configuration incorporating all four of the above essential components is shown in Fig. 11.1. The blocks S_1, S_2.......S_N are electronic switches which are digitally controlled.

Let us assume an N-bit straight binary input to the resistor network through switches S_1.....S_N. It produces a current I corresponding to logic 1 at the MSB, I/2 corresponding to logic 1 at the next lower bit, $I/2^2$ for logic 1 at the next lower bit and so on and $I/2^{N-1}$ for logic 1 at the LSB position (i.e. switch S_1 connects the R resistor to the reference voltage and when a 0 is present on the MSB line, switch S_1 connects the resistor to the ground line and so on). Thus the switch is a single pole double throw (SPDT) electronic switch. The OP AMP acts as a current to voltage converter. The total current produced will be proportional to the digital input. This current can be converted into a corresponding voltage by the OP AMP.

This D/A converter circuit is referred to as a *weighted resistors D/A converter* since the resistance

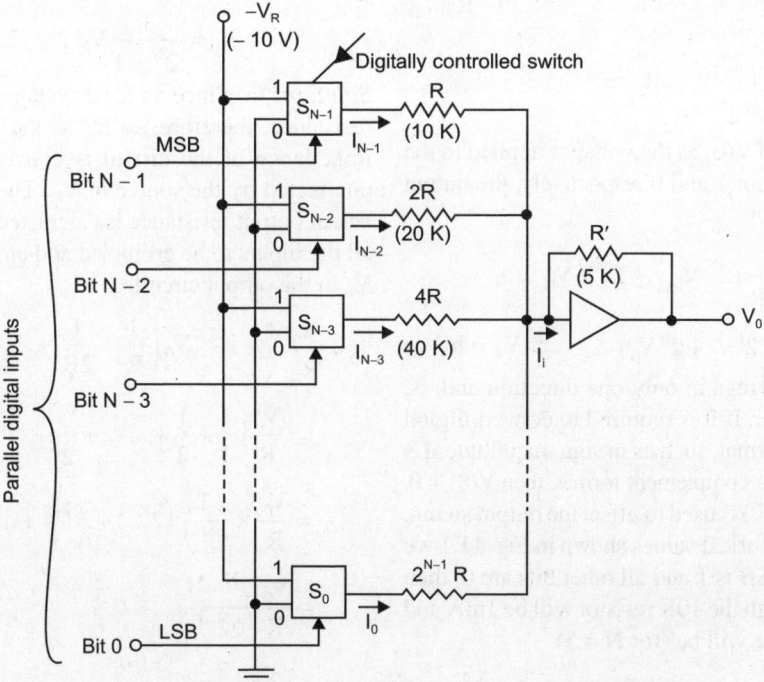

Fig. 11.1: A D/A converter with binary weighted resistors

values are weighted in accordance with the binary weights.

In the circuit of Fig. 11.1 voltage applied to a resistor is V(1) (or −V_R) if the switch connected to it is in position 1 and V(0) (= 0), if it is in position 0. The current I_i is given by

$$I_i - I_{N-1} + I_{N-2} + I_{N-3} + \ldots + I_2 + I_1 + I_0$$

where $I_{N-1} = \dfrac{V_{N-1}}{R}$,

$$\left. \begin{aligned} I_{N-2} &= \frac{V_{N-2}}{2R} \\ &\vdots \\ I_0 &= \frac{V_0}{2^{N-1}R} \end{aligned} \right\} \text{ where } V_n = V(1) \text{ if } b_n = 1$$

$$= V(0) \text{ if } b_n = 0$$

The analog output voltage V_o of an N-bit straight binary D/A converter is related to the digital input by the equation

$$V_0 = K\,(2^{N-1}\,b_{N-1} + 2^{N-2}\,b_{N-2} + \ldots + 2^2 b_2 + 2^1 b_1 + b_0)$$

For the circuit of Fig. 11.1, V(0) = 0 and V(1) = −V_R and

$$V_0 = -(-V_R)\left[\frac{R'}{R}b_{N-1} + \frac{R'}{2R}b_{N-2} + \ldots \frac{R'}{2^{N-1}R}b_0\right]$$

Thus here

$$K = \left(\frac{R'}{2^{N-1}R}\right)V_R$$

Or, with V(1) and V(0) as the voltages applied to the resistor network for 1 and 0 respectively, the output voltage is given by

$$V_0 = \frac{R'}{2^{N-1}R}(2^N V_{n-1} + 2^{N+2} V_{N-2} +$$

$$\ldots + 2^1 V_1 + 2^0 V_0) \qquad \therefore V_n = b_n V_R)$$

The output swings in only one direction and, is, therefore unipolar. If it is required to convert digital data in bipolar format, such as in sign magnitude, 1's complement or 2's complement format, then V(0) ≠ 0. In such a case, V(0) is used to offset the output swing.

Using the numerical values shown in Fig. 11.1 we see that if the MSB is 1 and all other bits are 0, then the current through the 10K resistor will be 1mA and the output voltage will be (for N = 5)

$$V_0 = \frac{5}{16 \times 10} \times 16 \times 10 = 5\text{V}.$$

Similarly, we see that the weight of the LSB becomes

$$V_0 = \frac{5}{16 \times 10} \times 10 = 1 \times \frac{5}{16}\text{V}.$$

If all five bits are 1, the output becomes

$$V_0 = \left(1 + \frac{1}{2} + \frac{1}{4} + \frac{1}{8} + \frac{1}{16}\right) \times 5 = 31 \times \frac{5}{16}$$

Therefore, the analog output V_0 is proportional to the digital input, the proportionality factor is 5/16 for the circuit of Fig. 11.1 (if N = 5).

Example 11.1: Figure 11.2 shows a binary weighted resistor D/A converter (a) shows that the outputs resistance is independent of the digital word and that

$$R_0 = \left(\frac{2^{N-1}}{2^N - 1}\right)R$$

(b) Shows that the analog output voltage for the MSB is

$$V_0 = \left(\frac{2^{N-1}}{2^N - 1}\right)V_R$$

(c) Shows that the analog output voltage for the LSB is

$$V_0 = \frac{1}{2^N - 1}V_R$$

Solution: (a) Since an ideal voltage source has zero resistance, therefore, so far as the effective output impedance of the circuit is concerned, it will be unaffected by the source of V_R, Therefore the equivalent output resistance is calculated by considering all the inputs to be grounded and an effective source V_0 in the output circuit.

$$i_0 = \frac{V_0}{R_0} \text{ or } i_0 = V_0\left(\frac{1}{R} + \frac{1}{2R} + \ldots + \frac{1}{2^{N-1}R}\right)$$

$$= \frac{V_0}{R}\left(\frac{1}{1} + \frac{1}{2} + \ldots + \frac{1}{2^{N-1}}\right)$$

$$= \frac{V_0}{R} \times \frac{1}{2^{N-1}}\left(2^{N-1} + 2^{N-2} + \ldots + 4 + 2 + 1\right)$$

$$= \frac{V_0}{R}\frac{2^N - 1}{2^{N-1}}$$

$$R_0 = \frac{2^{N-1}}{2^N - 1}R$$

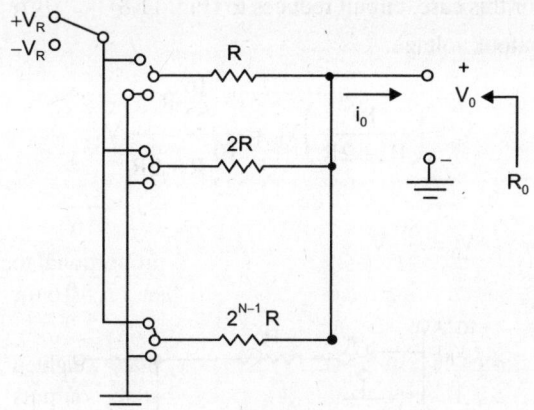

Fig. 11.2

(b) Only MSB is present and all others are zero (i.e. grounded)

$\therefore$ 2R, 4R, 8R.........2^{N-1}R are in parallel. Their equivalent resistance Z' is given by

$$\frac{1}{Z'} = \frac{1}{2R} + \frac{1}{4R} + \ldots + \frac{1}{2^{N-1}R}R$$

$$= \frac{1}{R} + \frac{1}{2R} + \frac{1}{4R} + \ldots + \frac{1}{2^{N-1}R} - \frac{1}{R}$$

or $\quad \dfrac{1}{Z'} = \dfrac{1}{R} \times \dfrac{1}{2^{N-1}} \left(\dfrac{2^{N-1}}{1} + \dfrac{2^{N-1}}{2} \right.$

$$\left. + \ldots 4 + 2 + 1 \right) - \frac{1}{R}$$

$$= \frac{1}{2^{N-1}R} (2^{N-1} + 2^{N-2} + \ldots 4 + 2 + 1) - \frac{1}{R}$$

$$= \frac{1}{2^{N-1}R} \times (2^N - 1) - \frac{1}{R} \qquad \ldots(i)$$

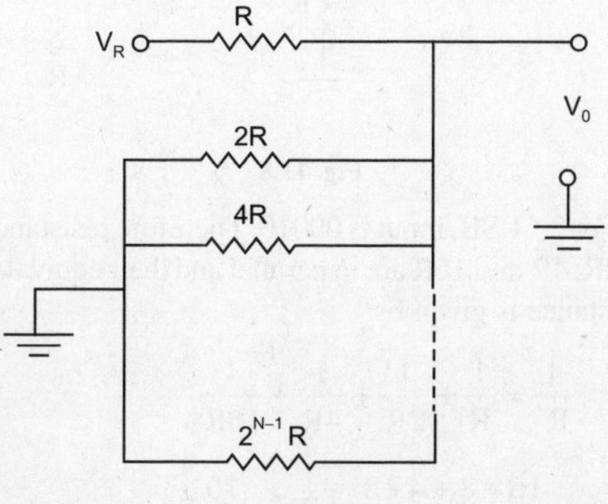

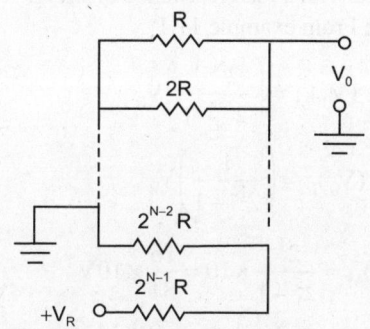

Fig. 11.3

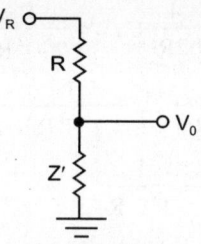

Fig. 11.4

$$V_0 = \frac{Z'}{R + Z'} V_R = \frac{1}{\dfrac{R}{Z'} + 1} V_R$$

$$= \frac{1}{\dfrac{2^N - 1}{2^{N-1}} - 1 + 1} V_R \qquad \text{Using eq (i) for R/Z')}$$

$$V_0 = \left(\frac{2^{N-1}}{2^N - 1} \right) V_R$$

(c) For LSB, R, 2R, 4R.......2^{N-2} R are in parallel and at zero. Their equivalent resistance Z' is given by

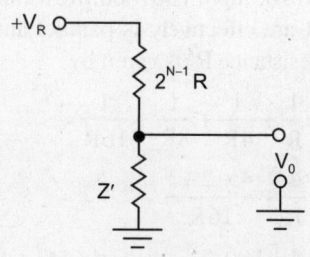

Fig. 11.5

Fig. 11.6

$$\frac{1}{Z'} = \frac{1}{R} + \frac{1}{2R} + ... + \frac{1}{2^{N-2}R}$$

$$= \frac{1}{R} + \frac{1}{2R} + ... + \frac{1}{2^{N-1}R} - \frac{1}{2^{N-1}R}$$

$$= \frac{2^N - 1}{2^{N-1}R} - \frac{1}{2^{N-1}R} \qquad ...(ii)$$

$$V_0 = \left(\frac{Z'}{Z' + 2^{N-1}R}\right)V_R = \frac{V_R}{\dfrac{2^{N-1}R}{Z'} + 1}$$

Using eq (ii) for 2^{N-1} R/Z')

$$= \frac{V_R}{2^N - 1 + 1} \quad \text{or} \quad V_0 = \frac{1}{2^N - 1}V_R$$

Example 11.2: What are the output voltages caused by each bit in a 5 bit weighted resistor D/A converter if input levels are

$$0 = 0V \text{ and } 1 = +10V.$$

Suppose the value of R for the 5-bit divider is 1KW. Determine the current that each input digital voltage source must be capable of supplying. Also determine the full scale output voltage of the ladder when it is terminated with a load resistance of 2KW.

Solution: From example 11.1

for MSB $(V_0)_M = \left(\dfrac{2^{N-1}}{2^N - 1}\right)V_R$

for LSB $(V_0)_L = \left(\dfrac{1}{2^N - 1}\right)V_R$

$$\therefore \quad (V_0)_M = \frac{2^4}{2^5 - 1} \times 10 = \frac{16}{31} \times 10V$$

$$(V_0)_L = \frac{1}{2^5 - 1} \times 10 = \frac{10}{31}V.$$

For 2nd MSB, input is 01000. Resistance R, 4R, 8R and 16R are effectively is parallel and so, their equivalent resistance R' is given by

$$\frac{1}{R'} = \frac{1}{R} + \frac{1}{4R} + \frac{1}{8R} + \frac{1}{16R}$$

$$= \frac{1}{R} + \frac{4 + 2 + 1}{16R}$$

$$= \frac{1}{R}\left[\frac{16 + 7}{16R}\right] = \therefore R' = \frac{16}{23}R$$

for this case, circuit reduces to (Fig. 11.8) output voltage

$$V_0 = \left(\frac{R'}{R' + 2R}\right)V_R = \frac{\dfrac{16}{23}R}{\dfrac{16}{23}R + 2R}V_R.$$

or $\quad V_0 = \dfrac{80}{31}V$

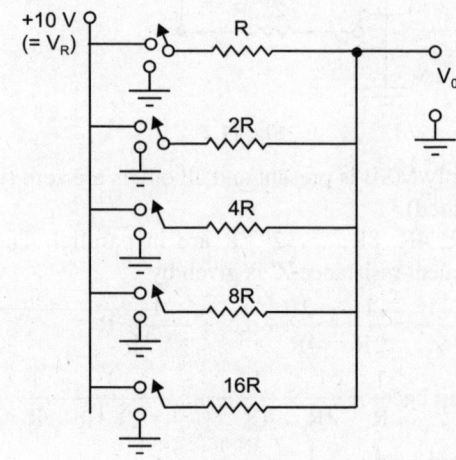

Fig. 11.7

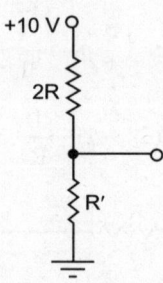

Fig. 11.8

For 2nd LSB, input is 00010. Therefore resistances R, 2R, 4R and 16R are in parallel and their equivalent resistance is given by

$$\frac{1}{R'} = \frac{1}{R} + \frac{1}{2R} + \frac{1}{4R} + \frac{1}{16R}$$

$$= \frac{16 + 8 + 4 + 1}{16R} \Rightarrow R' = \frac{16}{29}R$$

circuit reduces to (Fig. 11.9)
Output voltage

$$V_0 = \left(\frac{R'}{R' + 8R}\right) V_R$$

$$= \frac{\frac{16}{29}R}{\left(\frac{16}{29} + 8\right)R} \times 10$$

$$\therefore \quad V_0 = \frac{20}{31} V.$$

For 3rd LSB, input is 00100. Resistances R, 2R, 8R and 16R are in parallel. Equivalent resistance is given by

$$\frac{1}{R'} = \frac{1}{R} + \frac{1}{2R} + \frac{1}{8R} + \frac{1}{16R}$$

$$= \frac{(16 + 8 + 2 + 1)}{16R} \Rightarrow R' = \frac{16}{27}R$$

circuit becomes as shown in Fig. 11.10

Output voltage $V_0 = \left(\frac{R'}{4R + R'}\right) V_R$

$$= \frac{\frac{16}{27}R \times 10}{\left(4 + \frac{16}{27}\right)R} = \frac{40}{31} V$$

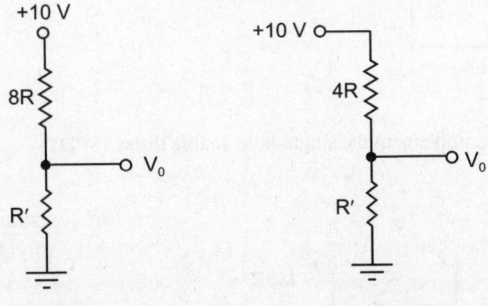

Fig. 11.9   **Fig. 11.10**

Now, current that each input digital voltage source is capable of supplying is given by:

For $\quad LSB = i_0 = \dfrac{-V_0 + V_R}{16000} \qquad (\therefore R = 1\ K\Omega)$

$$= \left(\frac{-10}{31} + 10\right)\frac{1}{16000} A$$

$$= \frac{310 - 10}{31} \times \frac{1}{16000} A = \frac{300}{31 \times 16} mA = \frac{300}{496} mA$$

$$\approx 0.60\ mA.$$

For 2nd LSB, $i_2 = \dfrac{V_R - V_0}{8000}\left(10 - \dfrac{20}{31}\right) \times \dfrac{1}{8} mA$

$$= 1.17\ mA$$

for 3rd LSB $\quad i_3 = \left(10 - \dfrac{40}{31}\right) \times \dfrac{1}{4} mA = 2.17\ mA$

for 2nd MSB $\quad i_4 = \left(-\dfrac{80}{31} + 10\right) \times \dfrac{1}{2} = 3.71\ mA$

for MSB $\quad i_5 = \left(10 - \dfrac{160}{31}\right) = 10 - 5.16 = 4.84\ mA$

Now, output voltage is maximum for maximum input, i.e. when all inputs are at logic 1.

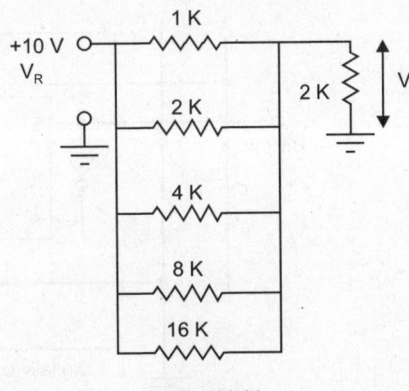

Fig. 11.11

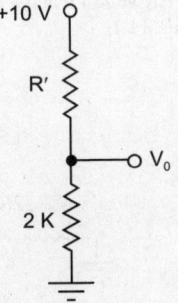

Fig. 11.12: Equivalent circuit of Fig. 11.11

Equivalent resistance of the ladder is given by

$$\frac{1}{R'} = \frac{1}{1} + \frac{1}{2} + \frac{1}{4} + \frac{1}{8} + \frac{1}{16}$$

$$= \frac{16 + 8 + 4 + 2 + 1}{16} \Rightarrow R' = \frac{16}{31} K\Omega$$

$$(V_0)_{max} = \frac{2K}{2 + R'} \times 10$$

$$= \frac{2K}{\left(2 + \frac{16}{31}\right)K} \times 10 = \frac{310}{39} = 7.9 V$$

11.4 IMPLEMENTATION OF THE SWITCHING DEVICE (SPDT SWITCH)

This is shown in Fig. 11.13 Q_1 and Q_2 are p-channel MOS transistors. The S-R FLIP-FLOP is also implemented with MOSFETs and holds the bit on the corresponding bit line the resistance R_1 depends on the bit under consideration for example, for the N-2 bit, $R_1 = 2R$

Digitally controlled, switch works as follows:

Let us assume that logic 1 corresponds to -10V and logic 0 corresponds to 0V(negative logic). A 1 on the bit line sets the FF at Q = 1 and $\overline{Q}$ = 0. Thus, transistor Q_1 is ON, connecting the resistor R_1 to the reference voltage-V_R,while transistor Q_2 remains OFF. Similarly a 0 on the bit line sets the FF at Q = 0 and $\overline{Q}$ = 1. Then Q_1 is OFF and Q_2 is ON. This 0 at the input bit line connects the resistor R_1 to the ground terminal. Thus the FF in conjunction with MOSEETS Q_1 and Q_2 act as SPDT.

11.5 BINARY LADDER

This is shown in Fig. 10.14 for a four bit input.

Output voltage will be a properly weighted sum of the digital inputs. Let us first assume, for the moment that r.h.s. (i.e. output) is open circuited. Then assume that all inputs are at ground. Consequently, the l.h.s. 2R and 2R resistors can be combined to form an equivalent resistor of value $\frac{2R \times 2R}{2R + 2R} = R$ as shown in Fig. 11.15.

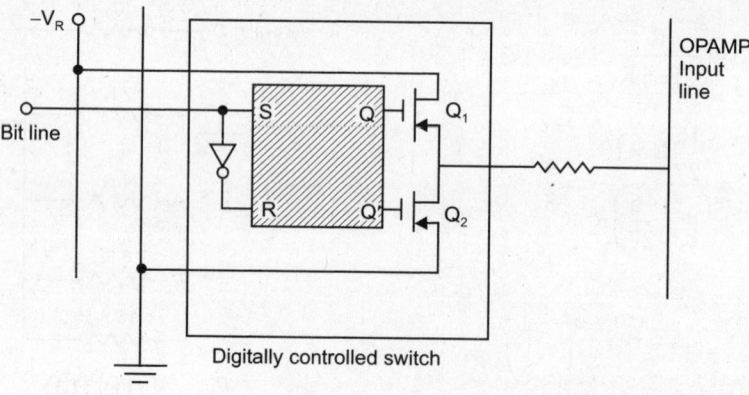

Fig. 11.13: An MOS FF and a pair of MOSFETS Q_1 and Q_2, implement the single-pole double throw (SPDT) switch of Fig. 11.1

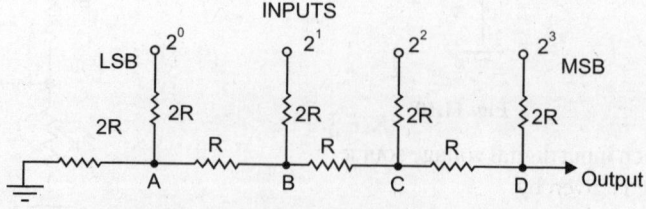

Fig. 11.14: A binary ladder for a 4-bit input

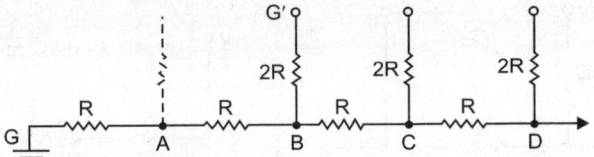

Fig. 11.15

Now, moving to node B, the resistance BG(= R + R) and BG′ (= 2R) can be combined as in Fig. 11.16 and so on.

Thus, the total resistance looking from any node back toward the terminating resistor or out toward the digital input is 2R. This is true regardless of whether the digital inputs are at ground or +V (because the internal impedance of an ideal voltage source is zero ohm and digital inputs are assumed to be ideal voltage sources). We will now discuss use of this ladder to determine output voltages for various inputs.

Example 11.3: <u>Input 1000</u>

Since there are no voltage sources to the left of D, the entire network to the left of this node can be replaced by a resistance 2R. Therefore, output voltage is

$$+V\left(\frac{2R}{2R+2R}\right)=\frac{+V}{2}$$

Example 11.4: <u>Input 0100</u>

Since there are no voltage sources to the left of C, the entire network to the left of this node can be replaced by a resistance of 2R. We now replace the network to the left of C with its Thevinin's equivalent by cutting the circuit on the jagged line (Fig. 11.19).

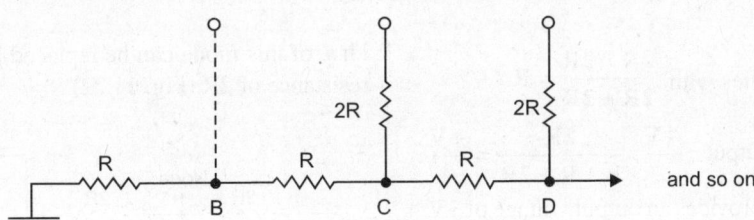

and so on.

Fig. 11.16

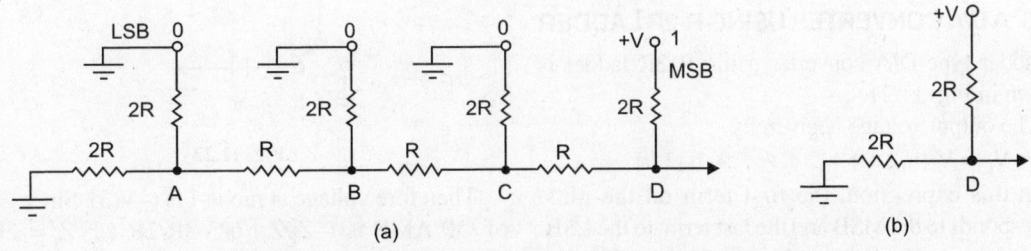

Fig. 11.17: (a) ladder for input 1000 (b) equivalent circuit of (a)

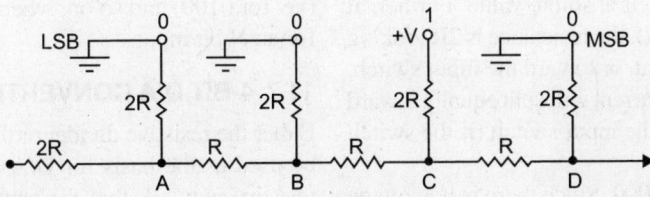

Fig 11.18

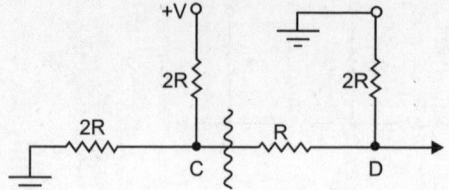

Fig. 11.19

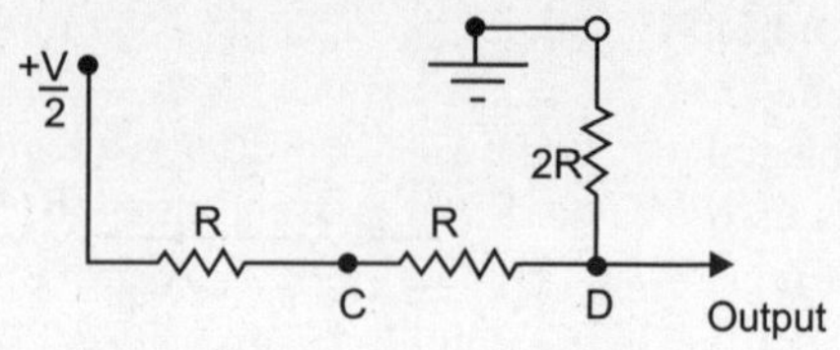

Fig. 11.20: Equivalent circuit of Fig. 11.19

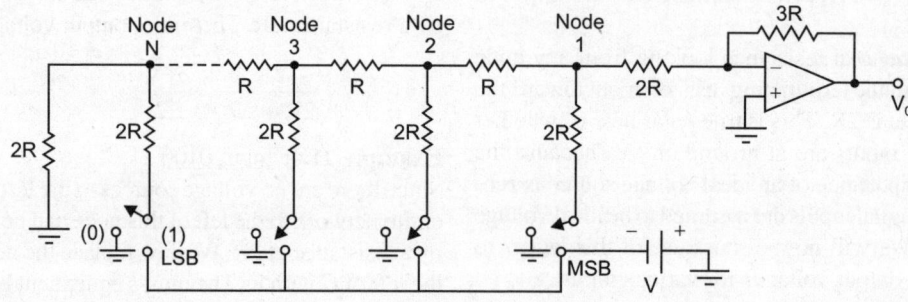

Fig. 11.21

$Vc = + V/2$ in series with $\dfrac{2R \times 2R}{2R + 2R} = R$.

For Fig. 11.20, output $= \dfrac{+V}{2} \cdot \dfrac{2R}{R + R + 2R} = \dfrac{+V}{4}$.

i.e. second MSB provides an output voltage of $+V/4$. Similarly third MSB provides an output voltage of $+V/8$. LSB provides an output voltage of $+V/2^N$.

11.6 A D/A CONVERTER USING R-2R LADDER

A ladder type D/A converter using R-2R ladder is shown in Fig. 11.21.

The output voltage is given by

$$V_0 = V [b_1 2^{-1} + b_2 2^{-2} + \dots + b_N 2^{-N}]$$

In this expression, the first term on the r.h.s. corresponds to the MSB and the last term, to the LSB. It is to be noted here that in Fig. 11.21 the ladder is a current splitting device, therefore ratio of resistances is more critical than their absolute value. Further, at any of the nodes (1 to N), the resistance is 2R looking to the left, or to the right, or toward the input switch. Hence, at every node, current will split equally toward the left and right, and the input switch (if the switch is 'ON', i.e. 1).

Consider an input 1000. Since there is no voltage source to the left of node 1, the entire network to the

l.h.s. of this mode can be replaced by an equivalent resistance of 2R (Fig. 11.22)

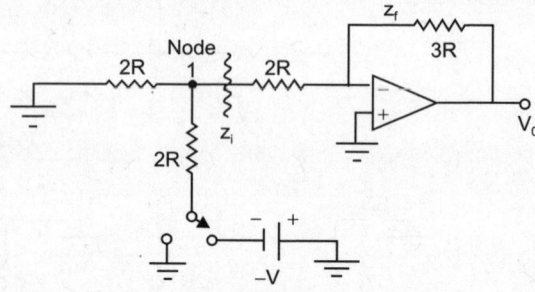

Fig. 11.22

Therefore voltage at mode 1 is $(-V/3)$. Since gain of OP AMP is $(-Z_f/Z_i)$ or $-3R/2R$ ($\because Z_i = 2R$). Therefore $V_0 = (-V/3)(-3R/2R) = V/2$. Similarly, when 2nd MSB is 'ON' (all others OFF), $V_0 = V/4$ (i.e. for 0100) and so on, when LSB is 1, $V_0 = V/2N$ for an N bit input.

11.7 4-BIT D/A CONVERTER

Either the resistive divider or the resistive ladder can be used as the basis for D/A converter. It is in the resistive network that the actual translation form a digital signal to analog voltage takes place. However, there is a need for additional circuitry to complete the design.

An integral part of the D/A converter is a register which is used to store the digital information. The simplest register is formed by use of RS FFs, with one FF per bit. Level amplifiers are also required between the registers and the resistive network. The use of level amplifiers ensures that the digital signals presented to the resistive network are all of the same level and are constant. Then, gates are required on the input of the register so that FFs can be set with the proper information from the digital system. Figure 11.23 shows the block diagram of a complete D/A converter and the complete schematic of a 4-bit D/A converter is shown in Fig. 11.24. Here, the resistive network used is of the ladder type. Each level amplifier has two inputs: one input is +10V from the precision voltage source and the other, from a FLIP-FLOP. The amplifiers work in such a way that when the input from a FF is High, the output of the amplifier is at + 10V, when the input from the FF is Low, the output is 0V.

The four FFs form the register necessary for storing the digital information. The FF on the right represents MSB and the FF on the left represents the LSB. Each

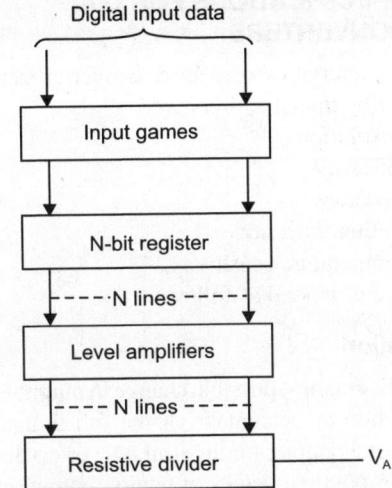

Fig. 11.23: Block diagram of a complete D/A converter

FF requires a positive level at R or S input to reset or set it. With the gating scheme shown, however, the FFs need not be set or reset each time new information is entered. The data are entered into the register each time the strobe (READ IN) pulse occurs. V_A is the analog output voltage.

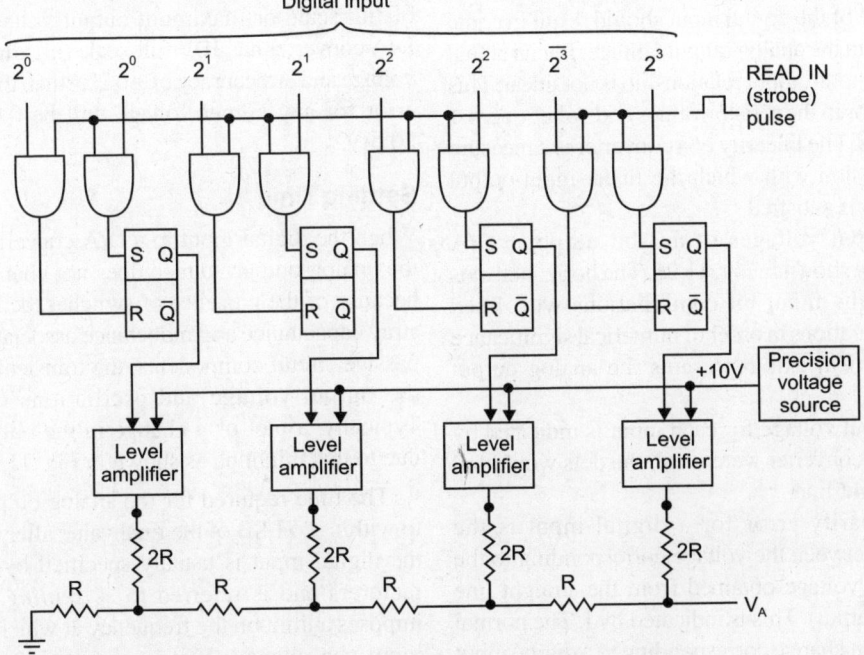

Fig. 11.24: Complete schematic of a 4-bit D/A converter

11.8 SPECIFICATIONS FOR D/A CONVERTERS

The characteristics of a D/A converter generally specified by manufacturers are:

1. Resolution
2. Linearity
3. Accuracy
4. Setting time, and
5. Temperature sensitivity

These are discussed as follows:

Resolution

This is the smallest possible change in output voltage as a traction or percentage of the full scale output range. For example, for an 8-bit converter, there are 2^8 or 256 possible values of analog output voltage, hence the smallest change in the output voltage is 1/255 of the full-scale output range. Its resolution is described as one part in 255 or 0.4 percent. Alternatively, the number of bits accepted at the input can itself be used as the resolution. For example, an 8-bit D/A converter has an 8-bit resolution.

Linearity

In a D/A converter, equal increments in the numerical significance of the digital input should result in equal increments in the analog output voltage. For an actual circuit, the input-output relationship is not linear. This is due to error in the resistor values and voltage across the switches. The linearity of a converter is a measure of the precision with which the linear input-output relationship is satisfied.

The output voltages of a 3-bit unipolar D/A converter are shown in Fig. 11.25. The horizontal axis represents the input bit combinations with fixed interval separations in order of numerical significance and the vertical axis represents the analog output voltage.

The output voltage for each input is indicated by a dot. If the converter were ideal, the dots would fall on the straight line.

The linearity error for a digital input is the difference between the voltage corresponding to the dot and the voltage obtained from the straight line (expected output). This is indicated by Î. The normal analog output change corresponding to a digital input change equivalent to the LSB is indicated by Δ.

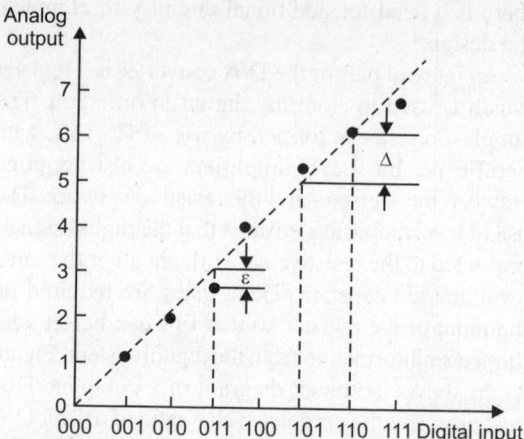

Fig. 11.25: Output versus input of a D/A converter

The linearity of a D/A converter is generally specified by comparing ∈ with Δ′. e.g. the linearity of a commercial D/A converter unit is specified as "less than ±½ LSB". This means that |∈ | < ½ Δ.

Accuracy

The accuracy of a D/A converter is a measure of the difference between the actual output voltage and the expected output voltage. It is specified as a percentage of full-scale or maximum output voltage, e.g. if a D/A converter has 10V full-scale (maximum) output voltage and an accuracy of ± 0.2%, then the maximum error for any output voltage will be 0.002 × 10 = 20 mV.

Settling Time

When the digital input to a D/A converter changes, the analog output voltage does not change abruptly because of the presence of switches, active devices, stray capacitance and inductance associated with the passive circuit components, the transients appear in the output voltage and oscillations may occur. Typically, a plot of a change in the output voltage due to this might be as shown in Fig. 11.26.

The time required for the analog output to settle to within ±½ LSB of the final value after a change in the digital input is usually specified by the manufacturers and is referred to as *settling time*. This imposes a limit on the frequency at which the digital input can change. If it is operated at too high a frequency it may not have a time to settle to the correct

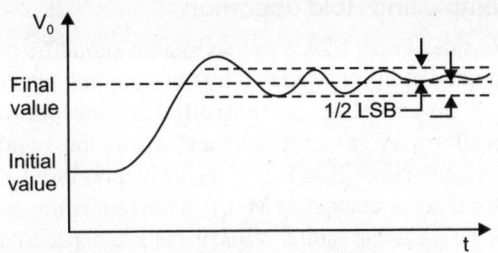

Fig 11.26: Settling time of a D/A converter

output voltage before being switched on to the next digital input.

Example 11.5: A D/A converter has successive binary inputs of 100101, 010111, 001100, 011101. If a level-1 signal switches +5V to the ladder network through the gates of the figure shown and a level 0 signal switches 0V, determine analog sample voltage output for each signal.

Solution: Consider Fig. 11.27. Assuming gain of OP AMP = ∞, X is essentially at ground potential. The input currents are

$$i_1 = \frac{V_R}{R_1}, i_2 = \frac{V_R}{R_2}, \ldots\ldots i_N = i_1 = \frac{V_R}{R_N},$$

in the OP AMP, $i_0 = \sum_{J=1}^{N} i_j = \sum_{J=1}^{N} \frac{V_R}{R_j}$

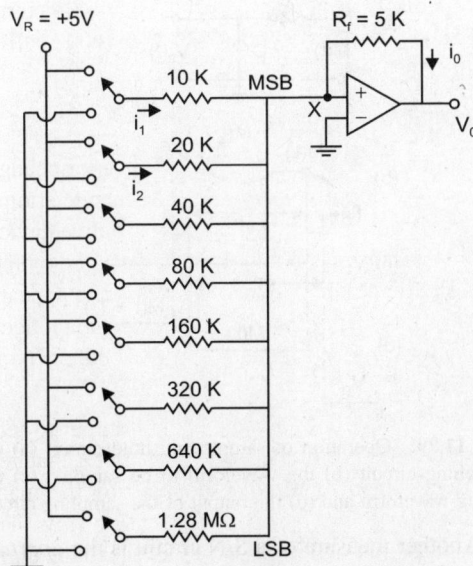

Fig. 11.27

From OP AMP theory, we know that

$$V_0 = -R_f \, i_o$$

$$\therefore \quad -V_0 = R_f \sum_{J=1}^{N} \frac{V_R}{R_j}, R_j = \frac{1}{2^{j-1} R}$$

Here $R_b = 5kW$

$$R = 10KW$$

$$V_R = +5W$$

$$V_0 = \frac{R_f V_R}{R} \left[a_0 + \frac{a_1}{2} + \ldots + \frac{a^{N-1}}{2^N - 1} \right]$$

$$= \frac{5 \times 5}{10} \left[a_0 + \frac{a_1}{2} + \frac{a_2}{2^2} + \frac{a_3}{2^3} + \frac{a_4}{2^4} + \frac{a_5}{2^5} \right] \quad (\therefore N = 6)$$

$$(V_0)_{100101} = 2.5 \left[1 + 0 + 0 + \frac{1}{2^3} + 0 + \frac{1}{2^5} \right]$$

$$= 2.5 \times 37 \times 2^{-5} \text{ V}$$

$$(V_0)_{010111} = 2.5 \left[0 + \frac{1}{2} + 0 + 0 + \frac{1}{2^3} + \frac{1}{2^4} + \frac{1}{2^5} \right]$$

$$= 2.5 \times 23 \times 2^{-5} \text{ V}$$

$$(V_0)_{001100} = 2.5 \left[0 + 0 + \frac{1}{2^2} + \frac{1}{2^3} + 0 + 0 \right]$$

$$= 2.5 \times 12 \times 2^{-5} \text{ V}$$

$$(V_0)_{011101} = 2.5 \left[0 + \frac{1}{2} + \frac{1}{2^2} + \frac{1}{2^3} + 0 + \frac{1}{2^5} \right]$$

$$= 2.5 \times 29 \times 2^{-5} \text{ V}$$

11.9 SAMPLE AND HOLD CIRCUIT

A data-acquisition system receives signals from a number of different sources and transmits these signals in suitable form to a computer. A multiplexer selects each signal in sequence and then the analog information is converted into a constant voltage over the gating-time interval by means of a *sample-and-hold circuit*. The constant output of the sample-and-hold is then converted to a digital signal by means of an A/D converter.

A simple form of *sample-and-hold* circuit is shown in Fig. 11.28a consists of a switch S in series with a capacitor.

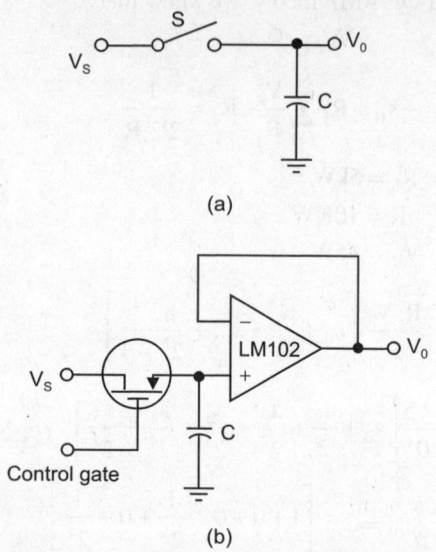

(a)

(b)

Fig. 11.28: Sample and hold circuit (a) Schematic (b) practical

The voltage across the capacitor tracks the input signal during the time T_g (gating time) when a logic control gate closes S, and holds the instantaneous value attained at the end of the interval T_g when the control gate opens S. The switch may be a relay (for very slow wave forms), a sampling diode-bridge gate, a BJT switch or a MOSFET controlled by a gating signal.

The circuit shown in Fig. 11.28b is one of the simplest practical sample-and-hold circuits. A negative pulse at the gate of the p-channel MOSFET turns the switch S 'on', and the holding capacitor C charges with a time constant $R_{on}C$ to the instantaneous value of the input voltage. In the absence of a negative pulse, the switch is turned 'off' and the capacitor is isolated from any load through the LM102 OP AMP. Thus it will hold the voltage impressed upon it. The value of the capacitance should be such that it retains the stored voltage during the HOLD period. The maximum input bias current for the LM 102 is 10nA, so, with a 10 μF capacitor, the drift rate during the HOLD period will be less than 1mV/s.

When the hold capacitor is larger then 0.05 mF, an isolation resistor (~10K) should be included between the capacitor and the + input of the OP AMP. This resistor will protect the amplifier in case the output is short circuited or the power supplies are abruptly shut down while the capacitor is charged.

Sample and Hold operation

Consider Fig. 11.29. Here an analog signal M(t) is sampled at times separated by the sampling interval T_s. The sampling is controlled by the control waveform V_c, which closes and opens the switch. During the time $T_c(= R_{on}C)$, the switch is closed and the capacitor charges to M (t_i). In the remaining time $T_H = T_S - T_C$ the sample value is held on capacitor C. The output waveform is shown in Fig. 11.29d (The hold sample values are indicated by solid levels. The output waveform between hold samples is of no special interest and is indicated by dashed portion of the waveform).

Acquisition time and Aperture time

One way to specify the quality of a sample and hold circuit is the *acquisition time*. This is the time needed to get an accurate sample (typically, to within 0.1 percent) after the switch is closed. Ideally, acquisition time is zero, but in a real sample-and-hold circuit, the charging time constant of the hold capacitor and other factors produce a non-zero acquisition time (Acquisition time is the time it takes for the capacitor to change from one level of holding voltage to the new value of input voltage after the switch has closed).

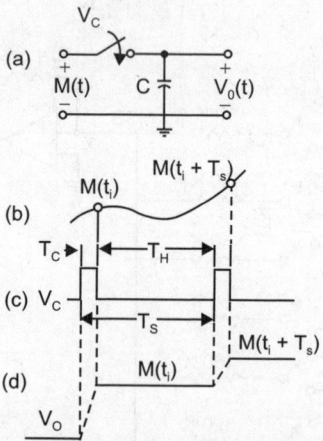

Fig. 11.29: Operation of sample and hold circuit (a) the switching circuit (b) the waveform to be sampled (c) the gating waveform and (d) the output of the sampling circuit

Another measure of a S/N circuit is the *aperture time* which is the delay between the time that the pulse is applied to the switch and the actual time the switch closes (in other words, it is the time required for the switch to open). It is typically less than 100 ns.

Application of a sample and hold circuit

Sample and hold circuit is used for sampling of analog signals which are to be fed to a A/D converter. It is advantageous to have pulsed output from A/D converters. For this, the analog signal has to be sampled at fast rate before conversion to digital form. The sampling gates will provide sampled amplitudes each of which will be converted to a digital code by the A/D converter. The A/D conversion takes place in the time intervals between the samples. Therefore, to utilize more of the time between the samples, a sample and hold circuit is used.

11.10 ANALOG-TO-DIGITAL CONVERTERS

Need for A/D Conversion

Data in a physical system normally appear in electrical analog form, e.g. a temperature difference would be represented by the output of a thermocouple. The need therefore arises for a device that converts the analog information into digital form.

11.10.1 Quantization and Quantization Error

In a D/A converter, the possible number of digital inputs is fixed. For example, in a 3-bit D/A converter there are 8 possible inputs. In contrast, in an A/D converter, the input analog voltage can have any value in a range whereas the digital output can have only 2^N discrete values for an N-bit A/D converter. Thus the whole range of analog voltage is required to be represented in 2^N interval, and each interval then corresponds to a digital output.

Consider an analog voltage in the range 0 to V and a 3-bit digital output for any voltage in this range. Let us divide the whole range in 8 intervals (3-bit output) of size S = V/8. (Fig. 11.30).

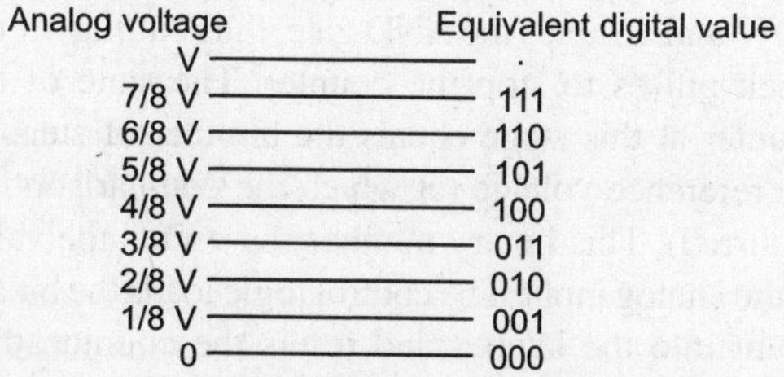

Analog voltage	Equivalent digital value
V	
7/8 V	111
6/8 V	110
5/8 V	101
4/8 V	100
3/8 V	011
2/8 V	010
1/8 V	001
0	000

Fig. 11.30: Quantization process

Each interval is assigned a unique digital value. This process is referred to as *quantization*. From Fig.11.30, we observe that the whole range of voltage in an interval is represented by only one digital value. Thus, there is an error referred to as *quantization error* involved in this process. In the present case, the maximum quantization-error for any analog input voltage in the given range in V/8.

The quantization error can be reduced if we choose the middle six intervals of size S = V/7 and the top and bottom intervals of size S/2 = V/14 as shown in Fig. 11.31.

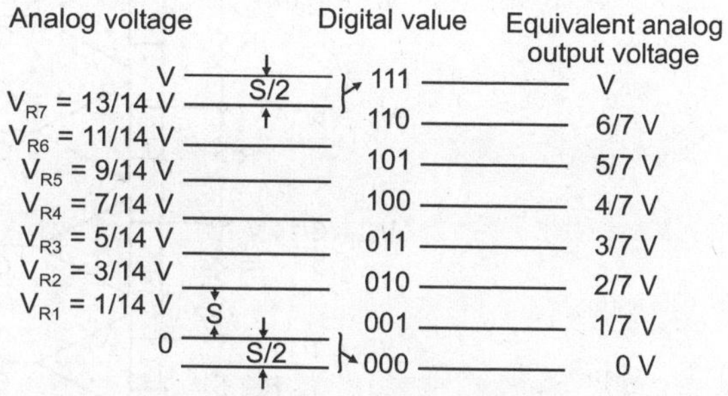

Analog voltage		Digital value	Equivalent analog output voltage
	V	111	V
V_{R7} = 13/14 V	S/2		
V_{R6} = 11/14 V		110	6/7 V
V_{R5} = 9/14 V		101	5/7 V
V_{R4} = 7/14 V		100	4/7 V
V_{R3} = 5/14 V		011	3/7 V
V_{R2} = 3/14 V		010	2/7 V
V_{R1} = 1/14 V	S	001	1/7 V
0	S/2	000	0 V

Fig. 11.31: Quantization with maximum error = S/2

Figure 11.31 shows that the maximum quantization error will be S/2 = V/14, for any analog input voltage in the range 0 to V. We can also express quantization error in terms of LSB, e.g. in the above case, the maximum quantization error is ±½ LSB.

Some of the commonly used A/D converters are discussed in the subsequent sections.

11.11 PARALLEL-COMPARATOR A/D CONVERTER

A 3-bit parallel-comparator A/D converter is shown in Fig. 11.32. V_a is the analog voltage to be converted into digital form. The voltage corresponding to full scale is V, from which the reference voltages V_{R1}, V_{R2}.....are generated using the resistor network. The voltage V_a is compared simultaneously with the reference voltage by using comparators.

A7-bit out put is obtained from the comparators which is stored in latches. This 7-bit digital signal is converted to a 3-bit output by using a decoder circuit. The comparator output and the 3-bit digital output for each interval of the analog voltage are given in Table 11.1.

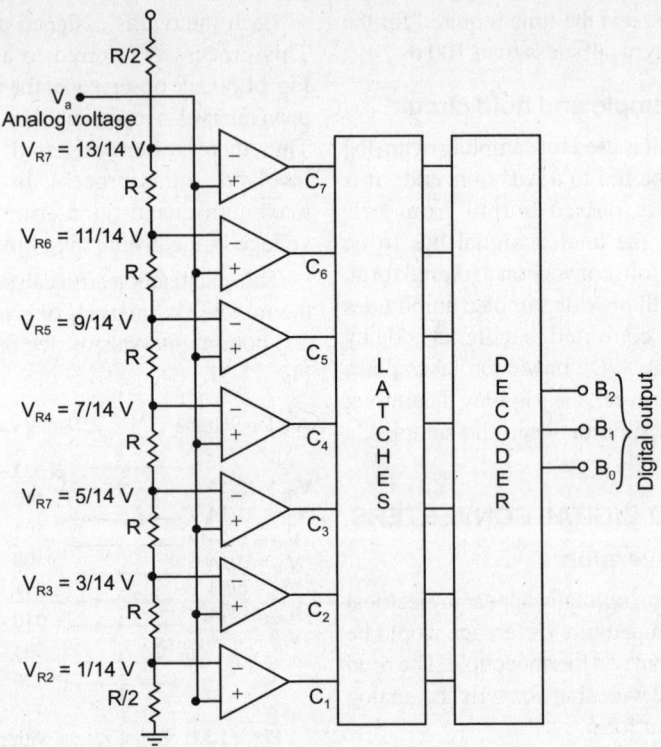

Fig. 11.32: 3-bit parallel comparator A/D converter

Table 11.1: Comparator outputs and digital output of parallel comparator A/D converter

Analog input	Comparator output							Digital output		
Va	C_7	C_6	C_5	C_4	C_3	C_2	C_1	B_2	B_1	B_0
$0 \leq V_a \leq V_{R1}$	0	0	0	0	0	0	0	0	0	0
$V_{R1} \leq V_a \leq V_{R2}$	0	0	0	0	0	0	1	0	0	1
$V_{R2} \leq V_a \leq V_{R3}$	0	0	0	0	0	1	1	0	1	0
$V_{R3} \leq V_a \leq V_{R4}$	0	0	0	0	1	1	1	0	1	1
$V_{R4} \leq V_a \leq V_{R5}$	0	0	0	1	1	1	1	1	0	0
$V_{R5} \leq V_a \leq V_{R6}$	0	0	1	1	1	1	1	1	0	1
$V_{R6} \leq V_a \leq V_{R7}$	0	1	1	1	1	1	1	1	1	0
$V_{R7} \leq V_a \leq V$	1	1	1	1	1	1	1	1	1	1

11.12 A/D CONVERTER USING A STAIRCASE RAMP

This method is also known as the digital ramp or the *counter-method*. It employs a D/A converter and a binary counter to generate the digital value of an analog input. This is illustrated in Fig. 11.33.

Operation: Assume that the counter begins RESET and the output of the D/A converter is zero. Now, an analog voltage is applied to the input. When it exceeds the reference voltage (output of D/A), the comparator switches to a HIGH output state and enables the AND. The clock pulses begin advancing the counter through its binary states, producing a staircase reference voltage from the D/A converter. The counter continues to advance from one binary state to the next, producing successively higher steps in the reference voltage. When the staircase reference voltage reaches the analog input voltage, the comparator output goes LOW and disables the AND gate, thus cutting off the clock pulses to stop the counter. The state of the counter at this stage equals the number of steps in the reference voltage (at which the comparison has occurred). This binary number represents the value of the analog input. The control logic loads the binary count into the latches and resets the counter, thus beginning another count-sequence to sample the input

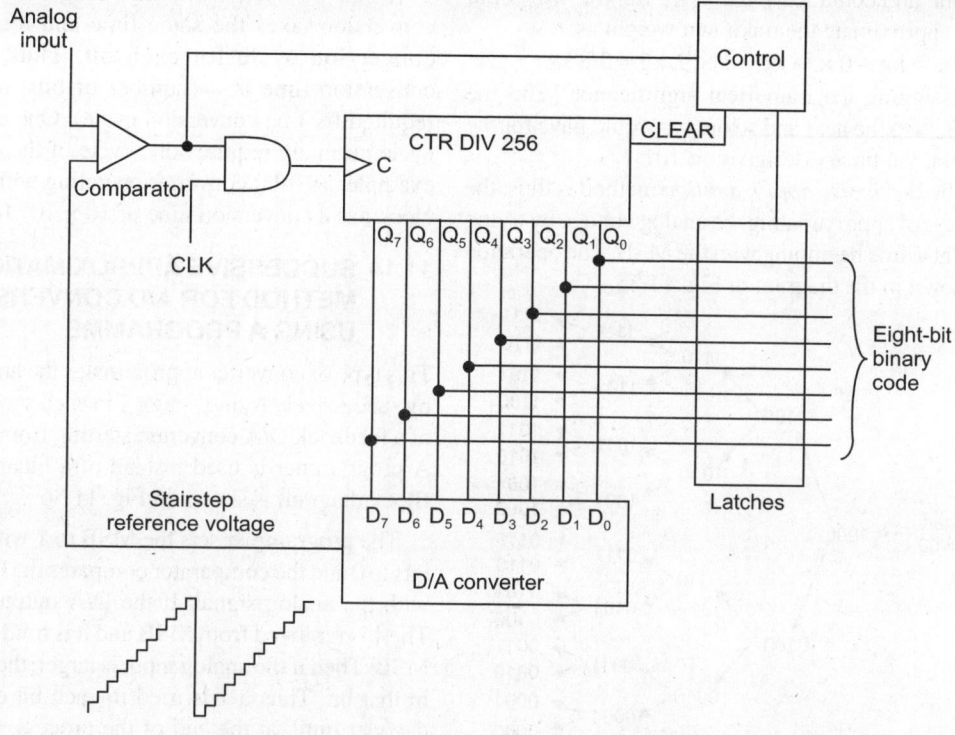

Fig. 11.33: Staircase-ramp A/D converter eight bits

voltage-value. As long as the analog input v_s is greater than v_d, the comparator output is high and therefore, the AND gate is open (1 state) for transmission of clock pulses to the counter. When v_d increases and exceeds v_s, the comparator output changes to low value; the AND gate is disabled. This stops counting at the instant v_s equals v_d. (Fig. 11.34).

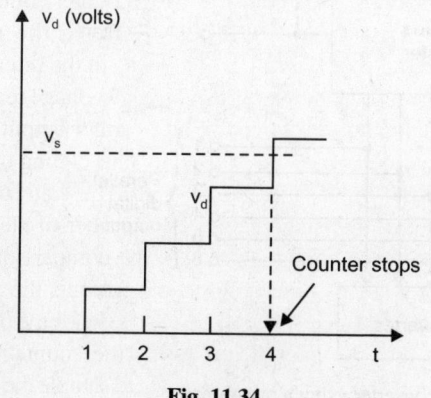

Fig. 11.34

11.13 A/D CONVERSION USING SUCCESSIVE APPROXIMATION METHOD

Principle of the Method: Suppose that we have an object whose weight is unknown but is in the range 0 to 1 kg. Suppose further that a balance is available and a set of known weights of ½, ¼, $^1/_8$ kg, etc. These known weights are to be used in a succession of trials to determine the unknown weight.

Now, with the unknown weight on one side of the balance, we place the ½ kg on the other side. If we find W > ½ kg, we leave the ½ kg weight on the scale and add the ¼ kg weight. If, on the contrary, W < ½ kg, we remove the ½ kg weight and put on the ¼ kg weight. Thus, we continue to try weights successively smaller by a factor of 2. When the last trial weight added tilts the balance toward the side of the known trial weights, we remove the last weight added and try the next smaller weight. Then, if we find that we could leave ½ kg weight, had to remove the ¼ kg weight, and could leave the $^1/_8$ kg weight, we would then approximate the unknown weight as

$1 \times \frac{1}{2}$ kg $+ 0 \times \frac{1}{4}$ kg $+ 1 \times \frac{1}{8}$ kg $= 5/8$ kg

Assigning the numerical significance ½ to the MSB, ¼ to the next and so on, we would have for the weight, the binary designation 101.

The *successive approximation* method is thus, the process of approximating the analog voltage by trying 1 bit at a time beginning with the MSB. The operation is shown in the diagram in Fig. 11.35.

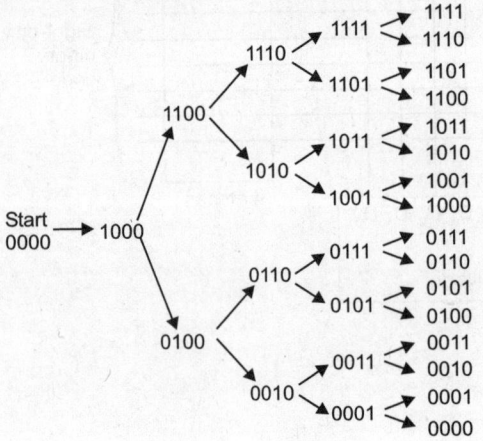

Fig. 11.35: Successive approximation by starting from the MSB

It can be seen from this diagram that each conversion takes the same time and requires one conversion cycle for each bit. Thus the total conversion time is = (number of bits, n) × (time required for one conversion cycle). One conversion cycle normally requires one cycle of the clock. For example, a 10-bit converter operating with a I-MHz clock has a conversion time of 10×10^{-6}J $= 10\mu$s.

11.14 SUCCESSIVE APPROXIMATION METHOD FOR A/D CONVERSION USING A PROGRAMME

This type of converter approximates the analog input by successively trying a digit 1 in each successive bit of a feedback D/A converter, starting from the MSB. A programmer is used instead of a binary counter. Block diagram is shown in Fig. 11.36.

The programmer sets the MSB to 1 with all other bits to 0 and the comparator compares the D/A output with the analog signal. If the D/A output is larger. The 1 is removed from MSB and it is tried in the next MSB. Then if the analog input is larger, the 1 remains in that bit. Thus a 1 is tried in each bit of the D/A decoder until, at the end of the process, all bits are calculated. The decision making logic, concerning whether to leave a trial 1 or replace it with a 0 is

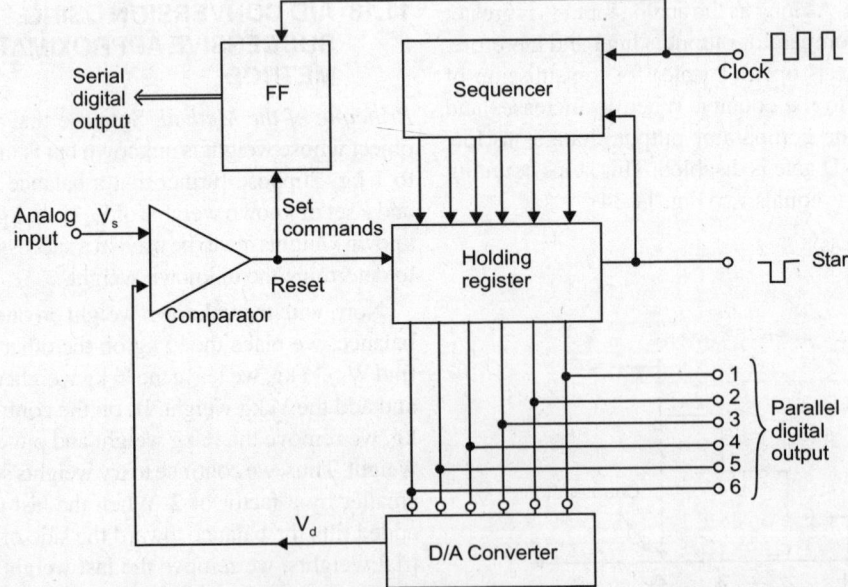

Fig. 11.36: Successive approximation A/D converter using a programmer

performed by the comparator and the sequencer blocks. In this type of a converter, each bit of the digital word is calculated one at a time starting with MSB (*see* also Fig.11.35). The set and reset commands of the comparator in each approximation step indicate the word in a serial form. It can be shown that for N-bit resolution, at most, N successive approximations are needed to reach the final value. The successive approximation method is the most versatile converter because it offers highest conversion rate and also provides parallel output capability.

The 3751 MOS/LSI (Fairchild semiconductor) is a 120bit A/D converter monolithic circuits which makes use of the successive approximation technique. The device is available in a 36 pin dual-in-line package.

11.15 A 3-BIT SUCCESSIVE APPROXIMATION A/D CONVERTER

A diagram of 3-bit successive approximation A/D converter is shown in Fig. 11.37.

This converter is designed to convert an analog waveform into binary code neglecting the sign bit. In this A/D converter we have allowed five equal time intervals to accomplish a single A/D conversion: three of the intervals are used to determine the 3 digital bits, a fourth interval to read the digital output and fifth interval to clear the converter (to make it ready for the next conversion).

The five D-type flip-flops FFA to FFE are connected to form a MOD-5 ring counter. Such a counter provides at its outputs Q_A to Q_E five waveforms, only one of which is at logic level 1 at any time. The level 1 is transferred from A to B to C etc with each successive clock cycle. The three flip-flops FF1, FF2 and FF3 are used to register the digital bits with FF1 for the LSB and FF3 for the MSB.

The conversion cycle begins with $Q_A = 1$ while $Q_B = Q_C = Q_D = Q_E = 0$. Then FF3 is set while FF2 and FF1 are reset. Then $Q_3 = 1$ and $Q_2 = Q_1 = 0$. The input 100 is thus presented to the 3-bit D/A converter,

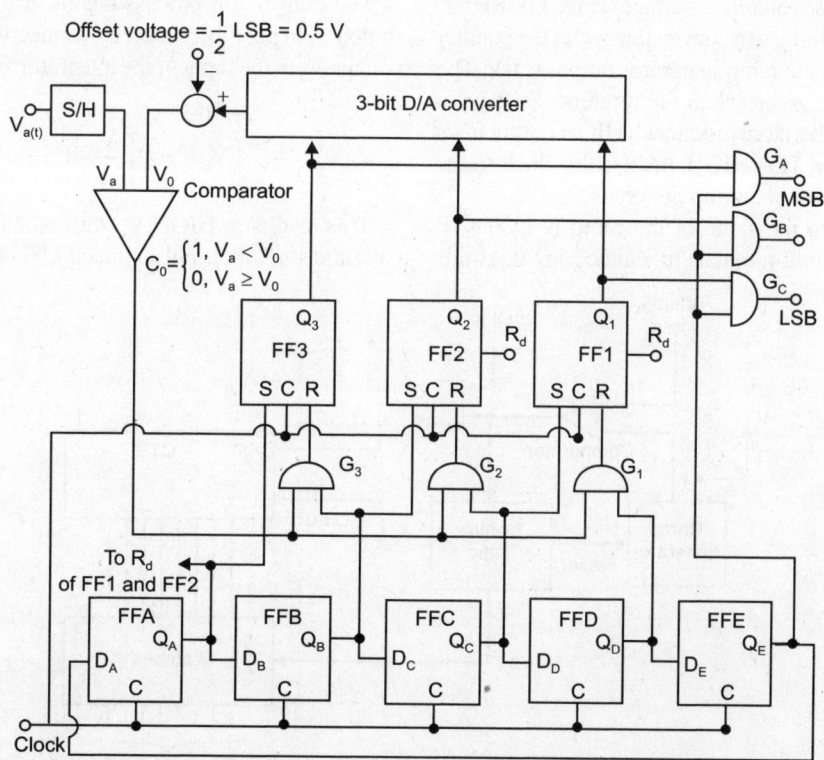

Fig. 11.37: A 3-bit successive approximation A/D converter (5) (After H. Taub and D. Schilling, Digital Integrated Electronics McGraw-Hill, (1977))

which provides a corresponding analog output V_0. The comparator output C_o will be then = 0 or 1 depending on whether $V_a{}^3 V_0$ or $V_a < V_0$. During the next clock interval, $Q_B = 1$ while $Q_A = Q_C = Q_D = Q_E = 0$. With $Q_B = 1$, the AND gate G_3 is enabled and FF3 is reset if $C_0 = 1$ and left in the set state if $C_0 = 0$. Thus, altogether we have tentatively assigned a logic 1 to the most significant position and at the beginning of the second clock interval, this bit remains or is changed to logic 0 depending on the comparison of V_a and V_0.

During succeeding clock intervals, the trial is repeated for the bits in the next two places the interval when $Q_E = 1$ is an interval when no comparisons are made and we can read the digital output. Thus Q_E is used to strobe the output gates G_A, G_B and G_C.

11.16 SINGLE SLOPE A/D CONVERTER

This type of converter does not require a D/A converter. It uses a linear ramp generator to produce a constant-slope reference voltage. (Fig. 11.38)

At the beginning of a conversion cycles the counter is RESET and the ramp generator output is 0V. The analog input is greater than the reference voltage at this point and therefore produces a HIGH output from the comparator. This HIGH enables the clock to the counter and starts the ramp generator.

Assume that the slope of the ramp is 1V/ms. It will increase until it equals the analog input, at this

point, the ramp is RESET and the binary or BCD count is stored in the latches by the control logic. Let us assume that the analog input is two volts at the point of comparison. This means that the ramp is also 2V and has been running for 2ms. Since the comparator output has been HIGH for 2ms, 200 clock pulses have been allowed to pass through the gate to the counter (if clock frequency = 100 KHz). At the point of comparison, the counter is in the binary state representing decimal 200. With proper scaling and decoding, this binary number can be displayed 2.00 V. This type of conversion is used in digital voltmeters.

11.17 DUAL-SLOPE A/D CONVERTER

The block diagram of a dual-slop A/D converter is shown in Fig. 11.39. It has four major blocks

1. an integrator
2. a comparator
3. a binary counter, and
4. a switch driver.

The conversion process begins at t = 0 with the switch S_1 in position 0 thereby connecting the analog voltage V_a to the input of the integrator. The integrator output

$$v_0 = -\frac{1}{\tau}\int_0^t V_a dt = -\left(\frac{V_a}{\tau}\right)t$$

This results in HIGH V_c, thus enabling the AND gate and the clock pulses reach CK of the counter

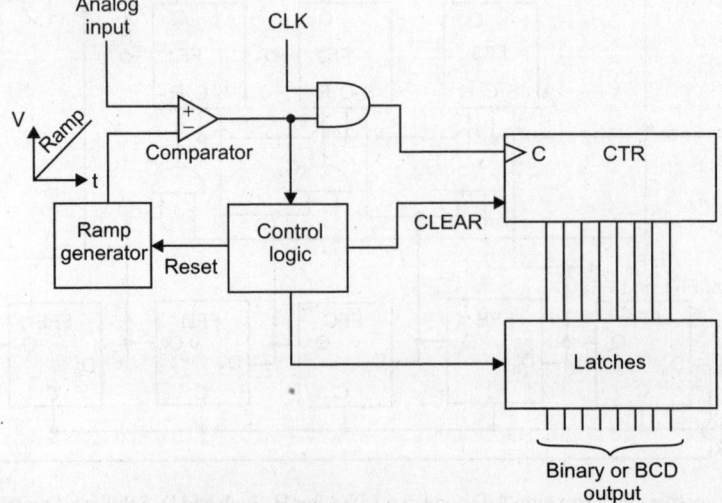

Fig. 11.38: Single slope A/D converter

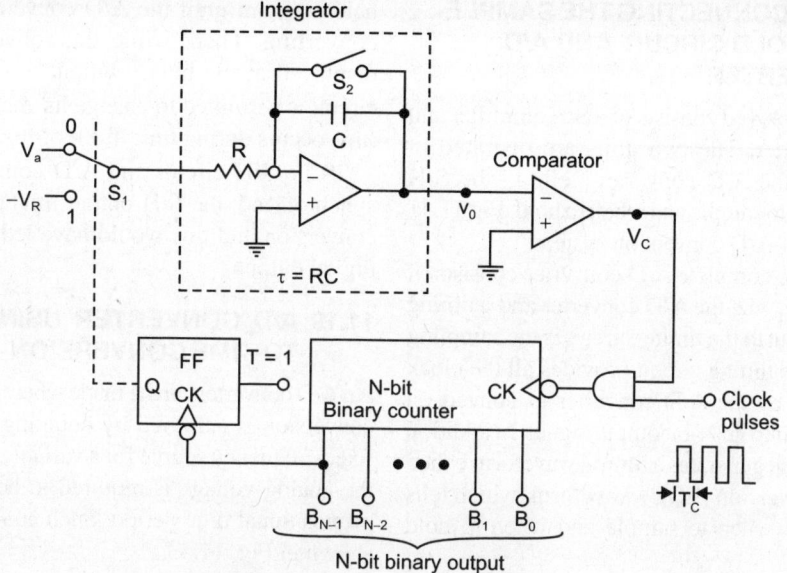

Fig 11.39: Dual-slop A/D converter

which was initially clear. The counter counts from 00...00 to 111....11 when 2^N-1 clock pulses are applied. At the next clock pulse 2^N, the counter is cleared and Q becomes 1. This controls the state of S_1 which now moves to position 1 at T_1, thereby connecting $-V_R$ to the input of the integrator now starts to move in the positive direction. The counter continues to count until $v_0 < 0$. As soon as v_0 goes positive at T_2, V_c goes LOW disabling the AND gate. The counter will stop counting in the absence of the clock pulses. The waveforms of voltages v_0 and V_c are shown in Fig. 11.40. The time T_1 is given by

$$T_1 = 2^N T_c$$

where T_c is the time period of the clock pulses. When the switch S_1 is in position 1, the output voltage of the integrator is given by

$$v_0 = -\frac{V_a}{\tau} T_1 + \frac{V_R}{\tau}(\tau - T_1)$$

$v_0 = 0$ at $t = T_2$

Therefore

$$T_2 - T_1 = \frac{V_a}{V_R} T_1 = \frac{V_a}{V_R} 2^N T_c.$$

Let the count recorded in the counter be n at T_2. Therefore

$$T_2 - T_1 = nT_c = \frac{V_a}{V_R} 2^N T_c.$$

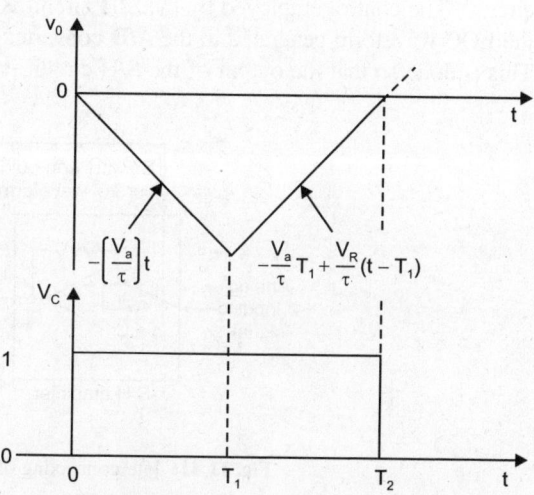

Fig. 11.40: Waveforms of dual-slope A/D converter

which gives

$$n = \frac{V_a}{V_R} \cdot 2^N$$

This shows that the output of the counter is proportional to the analog voltage V_a. The count recorded in the counter is numerically equal to analog voltage V_a of $V_R = 2^N$. (This type of converter is used in digital voltmeters)

11.18 INTERCONNECTING THE SAMPLE AND HOLD CIRCUIT AND A/D CONVERTER

The system employed consists of a S/H amplifier and an A/D converter. The two units are operated in synchronism, the A/D converter 'telling' the S/H amplifier when to sample and when to hold. Fig 11.41 shows the S/H - A/D conversion system.

Note that the complete A/D converter consists of two subsystems, viz. the A/D converter and a timing circuit. The input to the timing circuit is the sampling pulse train. The timing circuit provides all the clock pulses needed by the A/D converter to convert an analog sample into an N-bit output signal. In addition the timing circuit generates a timing waveform called the end-of-conversion (EOC) waveform, which tells the S/H circuit when to sample and when to hold (Fig. 11.42).

It is to be noted here that the sampling pulse train is applied to the A/D converter (and not to the S/H circuit). The control employed by the S/H circuit is the EOC waveform generated in the A/D converter. This is done so that the output of the S/H circuit is held constant until the A/D converter has finished converting. Then, while the converted output is displayed at the binary output terminals, the S/H circuit is permitted to change its analog output level (this occurs during time T_c).

If the S/H circuit and A/D converter were not synchronized, the S/H output might change during conversion and this would have led to an incorrect digital output.

11.19 A/D CONVERTER USING VOLTAGE-TO-TIME CONVERSION

An A/D converter can be made where analog to digital conversion is achieved by counting the cycles of a fixed-frequency source for a variable period. For this, the analog voltage is required to be converted to a propositional time period. Such an A/D converter is shown in Fig. 11.43.

A negative reference voltage-V_R is applied to an integrator, whose output is connected to the inverting input terminal of the comparator. The analog voltage V_a is applied at the non-inverting input terminal of the comparator. The output of the comparator Vc is

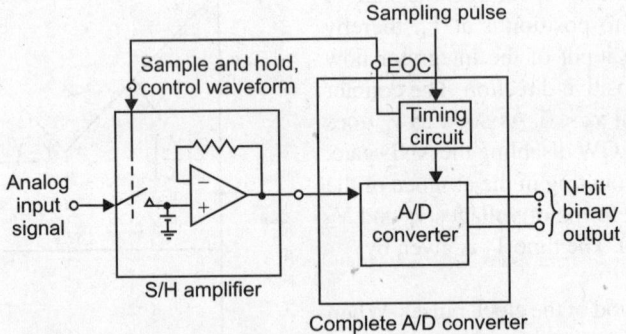

Fig. 11.41: Interconnecting of the S/H circuit and A/D converter

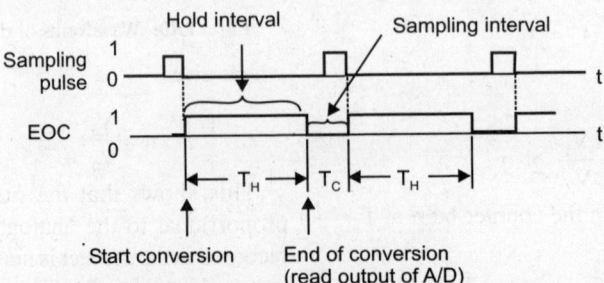

Fig. 11.42: Waveforms for Fig. 11.41

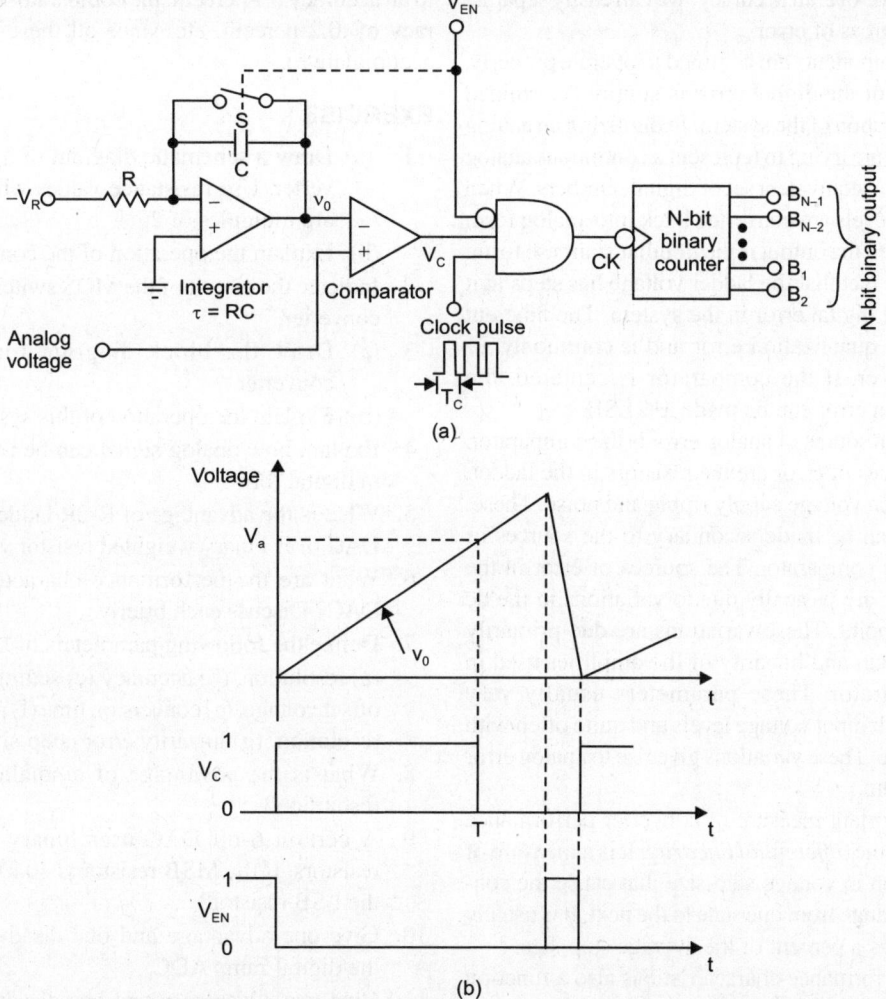

Fig. 11.43: An A/D converter using voltage-to-time conversion (a) schematic circuit, (b) waveforms

at logical level 1 as long as the output of the integrator v_0 is less than V_a. When v_0 crosses V_a at $t = T$, V_c goes LOW. The AND gate is enabled when V_{EN} is LOW and switch S remains open. When V_{EN} goes HIGH, the switch S is closed, thereby discharging the capacitor. Also, the AND gate is disabled. The various waveforms are shown in Fig.11.43b. When the AND gate is enabled, the clock pulses will reach the clock input terminal CK of the counter. The output of the counter is the digital output corresponding to V_a.

The time T is given by $T = \dfrac{\tau}{V_R} \cdot V_a$

which shows that T is proportional to V_a. The counter reading is n at $t = T$, then

$$n = T = \frac{f_c \tau}{V_R} V_a$$

where f_c is the clock frequency. The count n is proportional to V_a.

11.20 A/D ACCURACY AND RESOLUTION

Since the A/D converter is a closed loop system, involving both analog and digital systems, the overall accuracy must include errors from both, and to determine the overall accuracy; we can easily separate the two sources of error.

If all components are assumed to operate properly, the source of the digital error is simply determined by the resolution of the system. In digitizing an analog voltage, we are trying to represent a continuous analog voltage by an equivalent set of digital numbers. When the digital levels are converted back into analog form by the ladder, the output is the familiar staircase form. The simple fact that the ladder voltage has steps in it leads to the digital error in the system. The inherent error is the quantization error and is commonly ±1 bit. However, if the comparator is centered, the quantization error can be made ±½ LSB.

The main source of analog error is the comparator. Other sources of error are the resistors in the ladder, the reference voltage supply ripple and noise. These, however, can be made secondary to the sources of error in the comparator. The sources of error in the comparator are basically due to variations in the dc switching point. These variations are due primarily to offset, gain and linearity of the amplifier used in the comparator. These parameters usually vary slightly with input voltage levels and quite often with temperature. These variations give rise to analog error in the system.

An important measure of converter performance is given by the *differential linearity*. It is a measure of the variation in voltage step-size that cause the converter to change from one state to the next. It is usually expressed as a percent of the average step-size.

This performance characteristic is also a function of the conversion method and is best for the converters having counters usually have better differential linearity than do the successive approximation type converters.

The next logical question is what should be the relative order of magnitudes of the analog and digital errors? Then, in general, it is considered good practice to construct converters having analog and digital errors of approximately the same magnitudes. For example, an 8-bit converter may have a quantization error of $\dfrac{1}{256} \cong 0.4$ percent. It would then be reasonable to construct this converter to an accuracy of 0.5 percent in an effort to achieve an overall efficiency of 1.0 percent. (This might mean to construcd ladder

to an accuracy 0.1 percent, the comparator to an accuracy of ,0.2 percent, etc. since all these errors are accumulative).

EXERCISE

1. (a) Draw a schematic diagram of a D/A converter. Use resistance values whose ratios are multiples of 2.
   (b) Explain the operation of the converter.
2. Indicate the circuit of the MOS switch in a D/A converter.
3. (a) Draw the block diagram for an A/D converter.
   (b) Explain the operation of this system.
4. Explain how analog signal can be represented in digital form.
5. What is the advantage of R-2R ladder network DAC over binary weighted resistor version.
6. What are the performance characteristics of DAC? Discuss each briefly.
7. Define the following parameters of DACs:
   (a) resolution, (b) accuracy (c) setting time, (d) offset voltage, (e) conversion time (f) percentage resolution, (g) linearity error (step-size)
8. What is the advantage of a smaller (timer) resolution?
9. A certain 6-bit DAC uses binary weighted resistors. If the MSB resistor is 40 kW, what is the LSB resistor?
10. Give one advantage and one disadvantage of the digital-ramp ADC.
11. Give two advantages and one disadvantage of the dual-slope ADC.
12. List the advantages and disadvantages of single ramp, dual ramp and successive approximation ADCs.
13. What are the performance characteristics of ADCs? Explain each briefly.
14. Explain following terms as applicable to ADC (i) Resolution, (ii) Quantization (iii) Conversion rate.
15. An 8-bit ADC has a full scale input range of 10V. Find the resolution and the quantization error.
16. What is the function of the sample and hold circuit?

17. Why multiplexing is used in ADCA?

18. What is the binary equivalent weight of each bit in a 6-bit resistive divider?

19. Draw the schematic for a 6-bit resistive divider.

20. Assume that the divider in problem 19 has +10V full scale output, and find the following:

    (a) The change in output voltage due to a change in the LSB.

    (b) The output voltage for an input of 110110.

21. How many bits are required in a binary ladder to achieve a resolution of 1mV of full scale is +5V?

22. Which is the fastest ADC and why?

23. With the help of neat diagrams, explain the working of the following DACs and ADCs:

    (a) R-2R ladder network type DAC

    (b) Weighted-resistor type DAC.

    (c) Dual-slope ADC

    (d) Successive-approximation type ADC

12

The Microprocessors

12.1 INTRODUCTION

The most important technological invention of recent times is the microprocessor (computer on a chip).

In 1971, the engineers of the Intel Corporation of America were able to develop a micro-programmable computer on chip using integrated circuits. This device (Intel 4004) consisted of about 2300 transistors on a chip which was fabricated using silicon-gate p-channel MOS technology. It was later named microprocessor.

Since the introduction of Intel 4004, a 4-bit microprocessor (in November 1971), a large number of other microprocessors have been developed, which have found application in a large variety of products such as laboratory instruments, calculators, process-control systems, aircraft flight control systems, computers, etc. Table 12.1 lists some of the important landmarks in the evolution of microprocessors.

The developments in the field of microprocessors have resulted in tremendous growth in the power of microprocessors. The data-bus width and the memory size, which can be used with the most popular Intel family of microprocessors, is given in Table 12.2.

The clock-frequency has increased from about 3 MHz for 8085 Am processor to 66 MHz for 80486 m processor and the number of instructions which can be executed in one second has gone up from ~0.5 MIPS (million instructions per second) for 8085 A to 54 MIPS for 80486. The most recent Intel µP (Pentium Pro) has a clock frequency of 200 MHz.

More and more powerful and faster microprocessors are expected to be available in the near future.

Table 12.1: Evolution of microprocessor

Micro-processor name	Manu-facturer	Distinction
4004	INTEL	The first microprocessor (1971)
8008	INTEL	First 8-bit microprocessor (1972)
8080A	INTEL	First n-channel, second generation microprocessor (1974)
6800	Motorola	First + 5 V-only micro-processor (1974)
PACE	National Semi-Conductor	First 16-bit microprocessor (1974)
1802	RCA	First CMOS microprocessor (1974)
8048	INTEL	First 8-bit single-chip micro-computer (1976)
8088	INTEL	First 8-bit processor with 16-bit internal architecture (1979)
2920	INTEL	First analog-signal processor (1979)
80386	INTEL	First 32-bit microprocessor (1982)
Pentium	INTEL	First 64-bit microprocessor (1993)

Table 12.2: Popular Intel microprocessors

Microprocessor	Data bus width	Memory size
8085A	8	64K
8086	16	1M
8088	8	1M
80186	16	1M
80188	8	1M
80286	16	16M
80386 SX	16	16M
80386 SL	16	32M
80386 DX	32	4G
80486 SX	32	4G
80486 DX	32	4G
Pentium	64	4G

Some of the popular 8-bit and 16-bit microprocessors with their features are given in Table 12.3. The design and operation of microprocessor are based on the digital circuits which we studied in the previous chapters.

which would act as the "brain" of a computer. These circuits were called the *central processing unit* (CPU). The CPU could perform the basic arithmetic operations such as addition, subtraction, logic operations such as ANDing and ORing and control operations. Thus it could process data.

A CPU cannot be used alone. There are other components which are needed to make a computer. We need, for example, a memory—a place where data can be stored until the CPU needs it. We also need a way for the CPU to communicate with us for CPU to perform calculations and provide an answer. Then, we need an output device. Figure 12.1 illustrates what a simple microprocessor looks like.

12.3 DEFINITION OF A MICROPROCESSOR

What exactly is a microprocessor?
As the name implies, it must be small (micro) and it must be able to process data (processor). A

Table 12.3: Popular 8-and 16-bit microprocessors with their features

Features/μP	8080A	8085	MC6800	Z80	8748	8086	MC68000	Z8000
Year	1974	1976	1974	1976	1977	1978	1979	1979
Process technology	NMOS	NMOS	NMOS	NMOS	NMOS	HMOS	NMOS	NMOS
Data bus width (bits)	8	8	8	8	8	16	16	16
Address bus width (bits)	16	16	16	16	12	20	24	16/23
Memory space (bytes)	64K	64K	64K	64K	4K	1M	16M	64K/8M
Power-supply voltages	+12v, +5V	−5V, +5V	+5V	+5V	+5V	+5V	+5V	+5V
Packaging	40-pin DIP	40-pin DIP	40-pin DIP	40-pin DIP	40-pin DIP covered with transparent quartz lid	40-pin DIP	64-pin DIP	40/48-pin DIP
Number of basic instructions	78	80	72	158		97	61	110+
Instruction time (μs)	2	1	1	1	96	0.5	0.5	0.75
Resident program memory (bytes)	None	None	None	None	2.5 1K ROM 64 RAM	None	None	None

12.2 COMPUTER HARDWARE

The digital circuits which we have studied in the previous chapters are the building blocks of a computer. In the early days of computers, digital circuits were made by using vacuum tubes and later were built using transistors. Circuits were designed

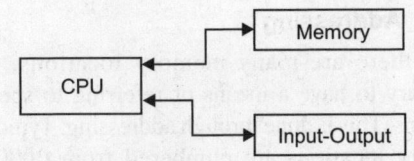

Fig. 12.1: A simplified view of a microprocessor system

microprocessor is a CPU which is constructed on a single silicon chip. A CPU is an electronic circuit which can interpret and execute instructions and control input and output. Figure 12.2 shows block diagram of a complete computer and peripherals.

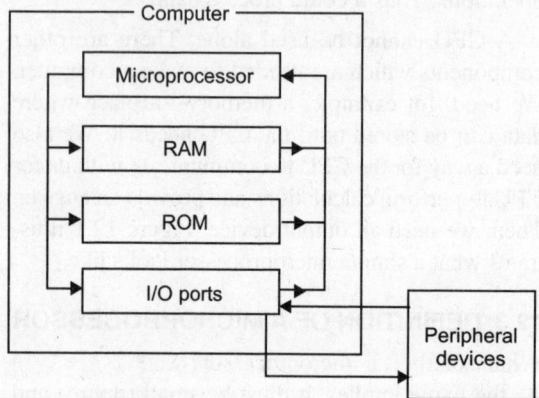

Fig. 12.2: Block diagram of a complete computer and peripherals. (Arrows indicate data flow)

12.4 COMPUTER ARCHITECTURE

Let us review some terms related with computer architecture.

12.4.1 Memory

Memory is needed so that data and instructions could be stored somewhere.

Data and instructions which can be lost after power is removed are stored in RAM. Data and instructions which must never be lost are stored in ROM. Remember that ROM is a type of memory which cannot have its contents changed once the ROM chip is manufactured. PROM and EPROM are used in the same way as ROM but can be programmed after being manufactured (PROM) or even programmed more than once (EPROM). PROM and EPROM differ from RAM in that they require special equipment to program them.

12.4.2 Addressing

Since there are many memory locations, it is necessary to have a means of referring to specific locations. This is done through addressing. Typically, memory locations are numbered from 0000 (in hexadecimal numbering) to the highest location used

by that particular computer. This sequential number which is assigned to each location is its address. (Fig. 12.3)

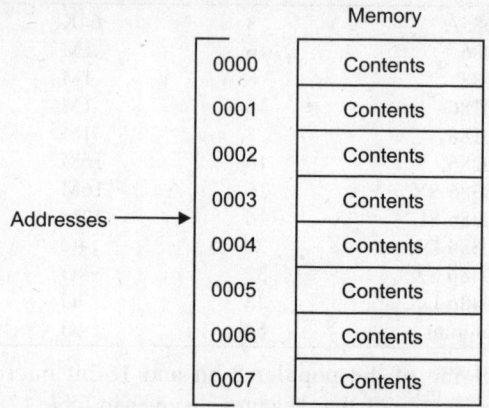

Fig 12.3: Memory addressing

Each memory location has an address and contents. The address is necessary to specify which memory location to *read* information from or *write* information into. The contents of a memory location is the information itself.

12.4.3 Address Bus

Most microprocessors can store information and instructions in a wider range of memory locations. Usually, the memory locations are in a memory chip rather than in the microprocessor. The microprocessor needs a way to tell the memory chip which memory location it wants to put data into or take data from. It does this through the *address bus*. (Fig. 12.4)

The address bus is a communication link between the microprocessor and the memory chips. It is simply a group of electrical paths which are connected to RAM, ROM and the I/O chips. Through this bus, the microprocessor can specify the address of any memory location in any chip or device. Note that information travels on the address bus in only one direction, from the microprocessor to memory and I/O.

12.4.4 Data Bus

Once the microprocessor has specified which memory location or device it wants to put data into or take data from, it then needs a set of electrical paths for the information to travel on. This set of paths is called the *data bus*. It is this set of electrical paths

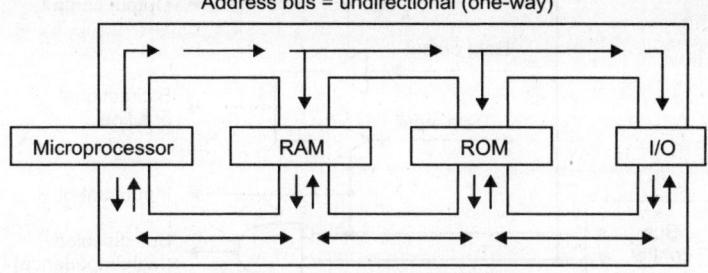

Fig. 12.4: Data bus and address bus

that allows data to flow from one chip to the next. Note that information on the data bus travels both to and from the microprocessor, memory and I/O devices. 8-bit microprocessors have a data bus that is 8-bits wide; 16 bit microprocessor has a data bus 16 bits wide and so on.

12.5 BUS BUFFER

The bus system in a microprocessor is commonly implemented by means of bus buffers constructed with three state gates (1, 0 and high impedance state). The high impedance state behaves as if the output is disabled or floating which means that it cannot affect or be affected by an external signal at the terminal. The graphic symbol of a three state gate is shown in Fig. 12.5 there is a normal input and a control input.

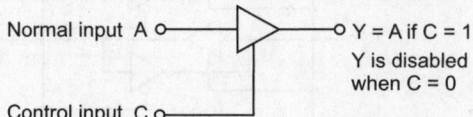

Fig. 12.5: Symbol of a three state buffer gate

When C = 1, the gate behaves as any conventional buffer* with the output equal to the normal input. When C = 0, the output is disabled due to gate going to high impedance state regardless of the value of normal input. This feature is not available in other gates. Because of this feature, a large number of three state gate outputs can be connected with wires to form a common bus line without endangering loading effects. However, no more than one gate should be in the active state at any given time.

Three-state logic diagrams

Figure 12.6 shows three-state logic diagrams.

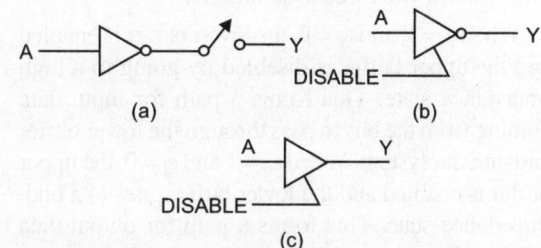

Fig. 12.6: Three-state logic diagrams (a) Equivalent circuit of an inverter (b) Logic symbol of inverter (c) Logic symbol of buffer

Figure 12.6a is an equivalent circuit for a three-state inverter. When DISABLE is low, the switch is closed and the circuit acts as an ordinary inverter. When DISABLE is high, the switch is open and the Y output is floating or disconnected. Figure 12.6b shows the logic symbol for a three state inverter,

(A low DISABLE results in normal inverter action, but a high DISABLE floats the Y output). By modifying the design, we can produce a three state buffer, whose logic symbol is shown in Fig. 12.6c. When DISABLE is low, the circuit acts as a non-inverting buffer and Y = A. When DISABLE is high, the output floats.

12.6 BIDIRECTIONAL BUS BUFFER

A bidirectional bus can be constructed with bus buffers to control the direction of information flow. One line of a bidirectional bus is shown in Fig. 12.7.

---

* A buffer is a device that isolates two other devices, typically a buffer has a high input impedance and a low output impedance.

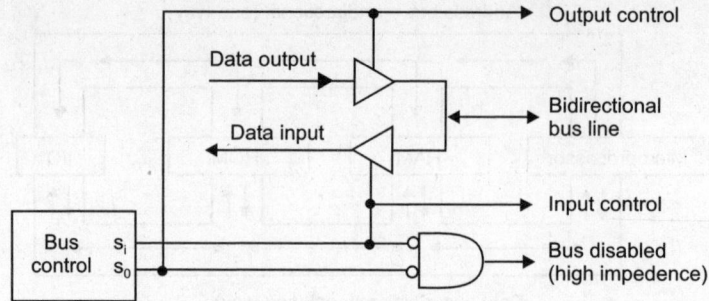

Fig. 12.7: Bidirectional bus buffer

The bus control has two selection lines, s_1 for input transfer and s_0 for output transfer. These selection lines control two three state buffers.

When $s_i = 1$, and $s_0 = 0$, the lower buffer is enabled and the upper buffer is disabled by going to a high impedance state. This forms a path for input data coming from the bus to pass through the lower buffer and into the system. When $s_0 = 1$ and $s_i = 0$, the upper buffer is enabled and the lower buffer goes to a high impedance state. This forms a path for output data coming from the system to pass through the upper gate and out to the bus line. The bus line can be disabled by making s_i and s_0 both 0. This makes both buffers into a high impedance state, preventing any input or output transfer of information through the bus line. This condition is required when an external source is using the common bus line to communicate with some other component.

12.7 AN-8-BIT BIDIRECTIONAL DATA BUS

An ideal µP needs a large number of pins (n) for input and (m) for output. A real µP cannot afford to have a large number of pins (usually m = n) and it is possible to use the same pins for inputs and outputs by using bidirectional pins. The number of such pins is referred to as the *data path width*, and collectively these pins are referred to as *data bus*. The data bus in a µP is bidirectional and is represented as shown in Fig. 12.8 (The word *bus* is used for a group of lines).

The number of lines forming the bus is written by its side. The number of lines in the data bus shown in Fig. 12.8 is 8. The number is also known as *bus width*. The organization of bidirectional bus is shown in Fig. 12.9 which is self explanatory.

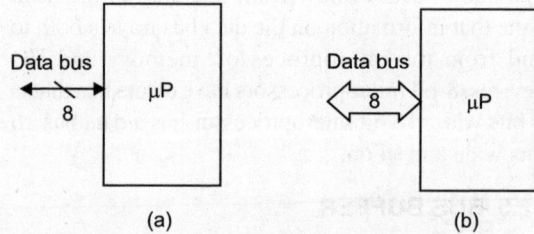

Fig. 12.8: Representation of bidirectional bus

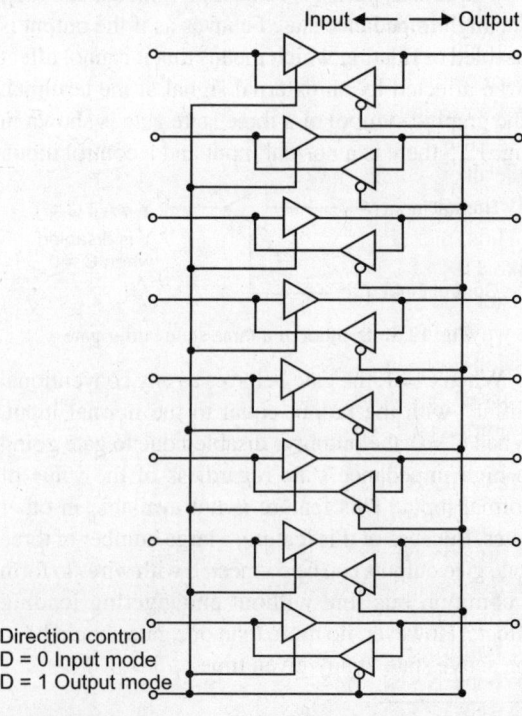

Fig. 12.9: Organization of an 8-bit bidirectional bus

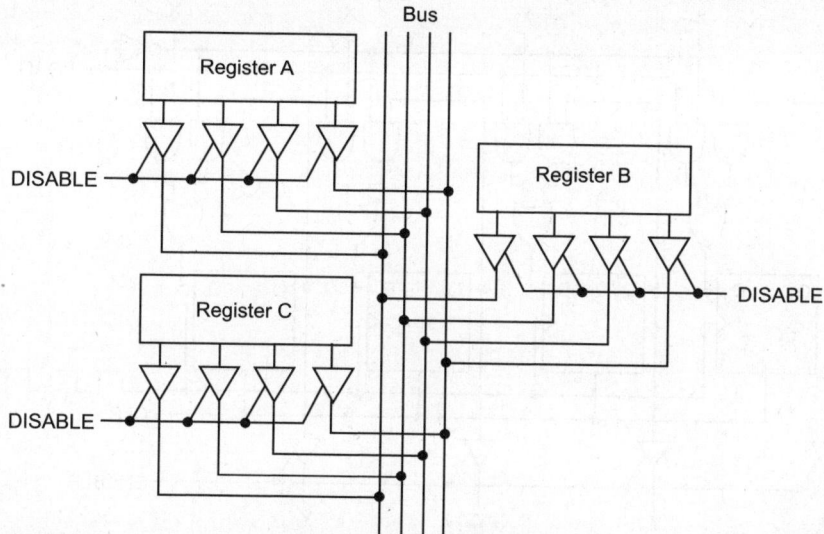

Fig. 12.10: Three-state bus control

12.8 THREE-STATE BUS CONTROL FOR REGISTERS

Figure 12.10 shows some registers (A, B and C) connected to a common bus. The three state buffers control the flow of binary data between these registers.

Suppose, we want the contents of register A to appear on the bus, all we have to do is make DISABLE low for register A but high for registers B and C. Then all the three-state switches on register A are closed while all other three state switches are open. As a result, only the contents of register A appear on the bus.

Thus, the idea in any bus-organized system is to make DISABLE high for all registers except the register whose contents are to appear on the bus. In this way many registers can time-share the same transmission-path. Not only it does reduce the wiring-cost but also it has simplified the architecture and design of computers.

12.8.1 Bus Organization of Registers in a Computer

Bus organized registers employ three state switches which convert two-state output of a register to a three state output. For instance, Fig. 12.11 shows a three state buffer register.

When ENABLE is low, the Y outputs float. When ENABLE is high, the Y outputs equal Q outputs:

$$Y = Q$$

We already know how the rest of the circuit works. (it is a controlled buffer register). When LOAD is low, the contents of the register are unchanged. When LOAD is high, the next positive clock edge loads X_3 $X_2 X_1 X_0$ into the register.

Figure 12.12 shows four registers A, B, C and D connected to a bus. (A bus is a group of wires that transmit a binary word, here wires W_3, W_2, W_1 and W_0 are a bus) i.e. a common transmission path between the three state registers.

The input data bits for register A come from the W-bus; at the same time, the three-state output of register A connects back to the W-bus. Similarly, the other registers B, C and D have their inputs and outputs connected to the W-bus. In Fig. 12.12 all control signals are in uncomplemented form: therefore the registers have active high inputs. Consequently, a load input (L_A to L_D) must be high to set up for loading and an enable signal (E_A to E_D) must be high to connect an output to the bus.

12.8.2 Word Transfer between Registers

To begin with, the same clock signal drives all registers but nothing will happen until one applies high control inputs, i.e. as long as all LOAD and ENABLE inputs are low, the registers are isolated from the bus.

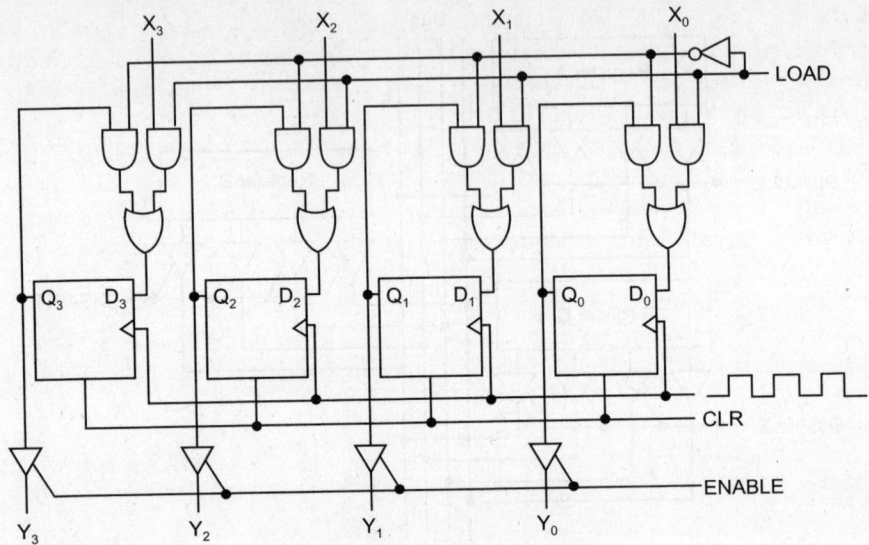

Fig. 12.11: Three state buffer register

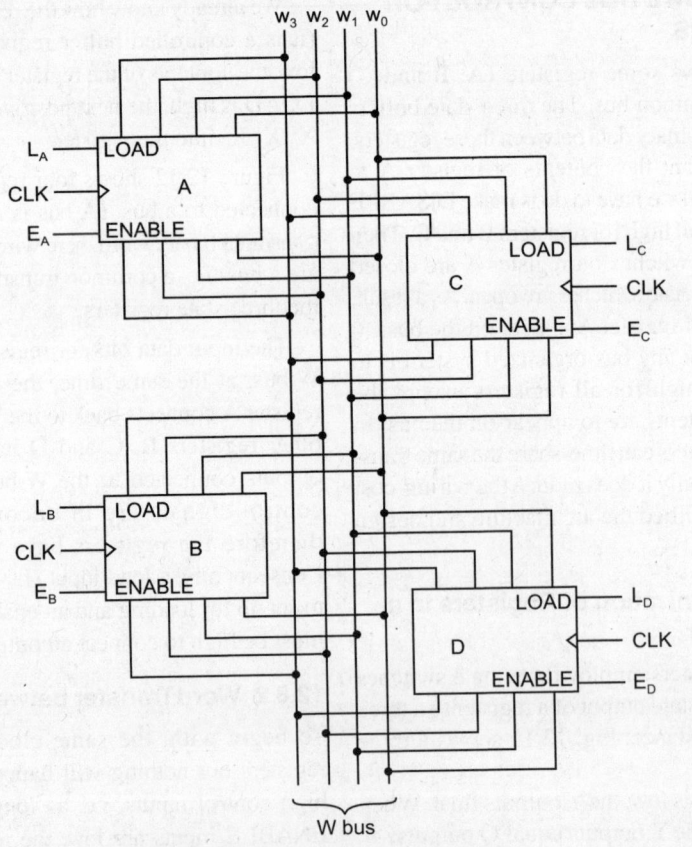

Fig. 12.12: Registers connected to a bus W

To transfer a word from one register to another, make the appropriate control inputs high. For instance, here is how to transfer the contents of register A to the register D. Make E_A and L_D high, then the contents of register A appear on the bus and register D is set up for loading. When the next positive clock edge arrives, word A is stored in register D.

Similarly, to transfer word C into register B, make E_C and L_B high. The high E_C closes the three state switches of register C, placing word C on the bus. The high L_B sets up register B for loading. When the next positive clock edge arrives, word C is stored in register B.

The bus organization simplifier the wiring and operation of computers. Figure 12.13 shows an

stored). The address can be specified in the normal binary form or in a more compact form using hexadecimal notation. Figure 12.14 depicts the interfacing of a memory with μP.

Some of the latest μPs, such as Intel 8085 A and 8086 use the same bus for transmitting addresses as well as for transferring data. This means that a bus is used for dual purpose: it is used as an address bus when an address is to be sent and as a data bus when data transfer is to take place. For example, in Intel 8085 AμP, the lower byte of the address is sent over the address/data (AD) bus with pins marked as AD_0-AD_7 and the higher byte of the address is sent over a dedicated (i.e. exclusively for address) 8-bit address

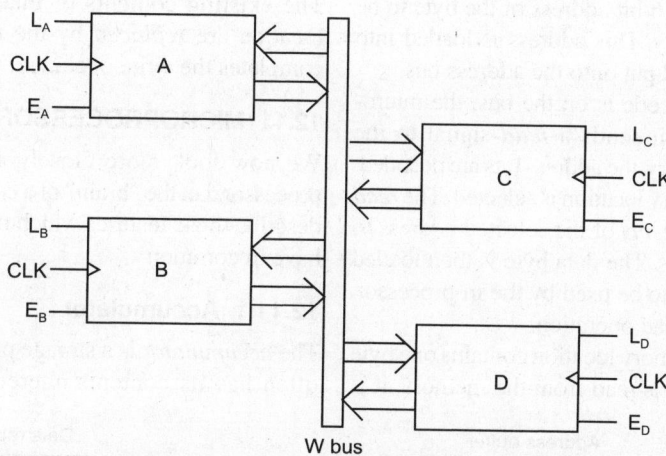

Fig. 12.13: Simplified diagram of Fig. 12.12

abbreviated form of Fig. 12.12. The solid arrows represent words going into and out of registers and the solid bar represents the W bus.

12.9 INTERFACING OF MEMORY WITH MICROPROCESSOR (μP)

A μP needs memory to store information and fetch information from it when required. For this purpose, both the μP and the memory need a set of lines called the *address bus*. The size of the memory which can be addressed with P lines in the address bus is 2^P referred to as *address space* or *memory space*. In many real μPs, the address bus width is 16 which can address upto $2^{16} = 65536 (= 64$ K) memory locations. (The information is stored in the form of bytes and at every memory location, one byte is

bus with pins marked A_8-A_{15}. This type of operation, where a single bus is used for two different functions,

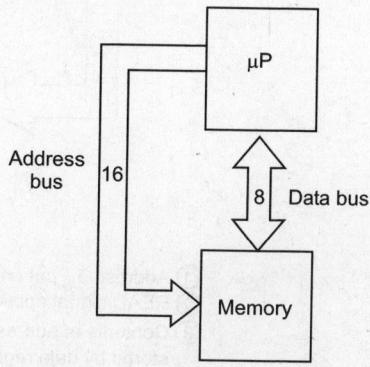

Fig. 12.14: Interfacing of memory with μP

is known as *multiplexing*, which results in saving of pins on the IC chip.

12.10 READ AND WRITE OPERATIONS

In the following discussion on read and write operations, a memory capacity of 65,536 bits is assumed because it represents a common memory size used in many microprocessors. (This memory is designated as 64 K).

12.10.1 Read Operation

To transfer a byte of data from the memory to the microprocessor, a *read* operation must be performed as illustrated in Fig. 12.15. To begin, a *program counter* contains the 16-bit address of the byte to be read from the memory. This address is loaded into the *address buffer* and put onto the address bus.

Once the address code is on the bus, the microprocessor control unit sends a *read* signal to the memory. At the memory, the address bits are decoded, and the desired memory location is selected. The *read* signal causes the *contents* of the selected address to be put on the data bus. The data byte is then loaded into the data register to be used by the m-processor. This completes the read operation.

Note that each memory location contains one byte of data. When a byte is read from the memory, it is not destroyed but remains in the memory. This process of "copying" the contents of a memory location without destroying the contents is called *nondestructive readout*.

12.10.2 Write Operation

In order to transfer a byte of data from the µP to the memory, a write operation is required as depicted in Fig. 12.16.

The memory is addressed in the same way as during a read operation. A data byte being held in the data register is put onto the data-bus and the µP sends the memory a *write* signal. This causes the byte on the data bus to be stored at the selected location in the memory as specified by the 16-bit address code. The existing contents of that particular memory location are *replaced* by the new data byte. This completes the write operation.

12.11 MICROPROCESSOR ARCHITECTURE

We now look more closely at the actual microprocessor, i.e. the "brain" of a computer. We will now describe those features which most microprocessors have in common.

12.11.1 Accumulator

The *accumulator* is a storage place or register which often has its contents altered in some way. The

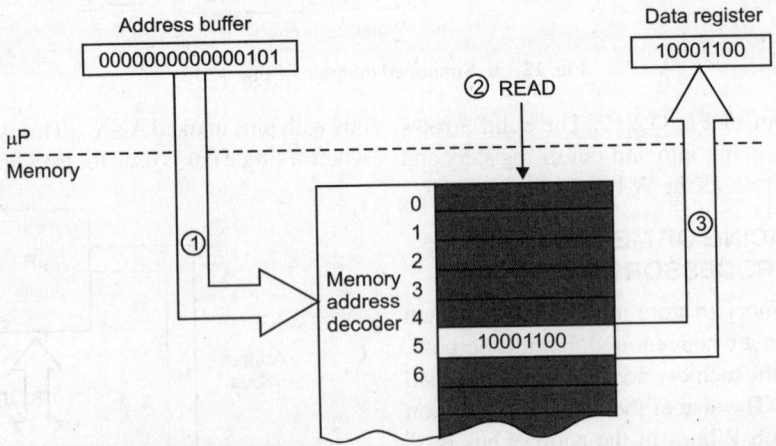

① Address 5_{10} put on address bus.
② READ signal applied.
③ Contents of address 5_{10} in memory put on data bus and stored by data register.

Fig 12.15: Read operation in a typical microprocessor

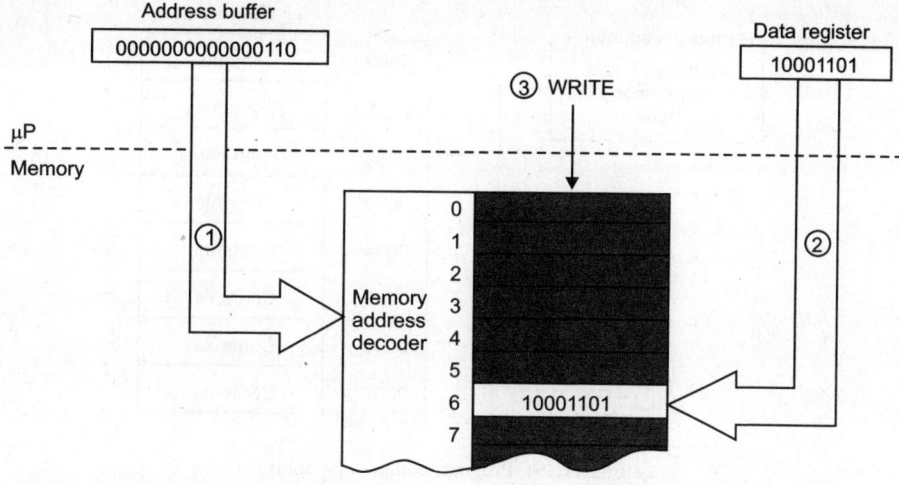

Address buffer
0000000000000110

Data register
10001101

③ WRITE

μP
Memory

Memory
address
decoder

0
1
2
3
4
5
6   10001101
7

①
②

① Address put on address bus.
② Data put on data bus.
③ WRITE signal causes data to be stored in address 6_{16}.

Fig. 12.16: Write operation in a typical μP

microprocessor can take the contents of the accumulator and the data coming in, perform some operation on the two and place the result back in the accumulator (Fig. 12.17). For example, the microprocessor might find the 1's complement of the contents of the accumulator and place the result in the accumulator in place of the original number.

There may be one or more than one accumulator in a microprocessor.

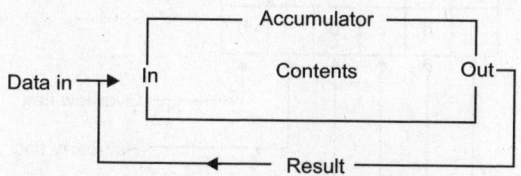

Accumulator

Data in → In   Contents   Out

Result

Fig. 12.17: Accumulator operation

12.11.2 General Purpose Registers

General purpose registers are similar to the accumulator with the difference that operations involving two pieces of data are usually not performed in them with the result (unlike accumulator) not going back into the register itself. The microprocessor can often alter the contents of the register. Figure 12.18 shows the operation of a general purpose register.

One may wonder why a microprocessor needs general purpose registers although it has RAM to

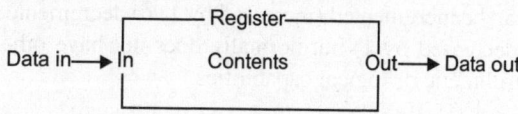

Register

Data in → In   Contents   Out → Data out

Fig. 12.18: General purpose register operation

temporarily store information. The answer is speed. Data in registers can be accessed and moved more quickly than data in RAM.

12.11.3 Program Counter (Instruction Pointer)

We already know that instructions are stored in memory. Consider the fact that there can be tens of thousands or even millions of memory locations, it is obvious that the microprocessor must keep track of the location from which it will be getting its next instruction. This is the job of the *program counter*. It is a very special register whose only job is to keep track of the location of the next instruction which the microprocessor will use. Figure 12.19 illustrates this operation.

The program counter "points" to the address of the next instruction to be retrieved and used by the microprocessor. The act of "getting" an instruction

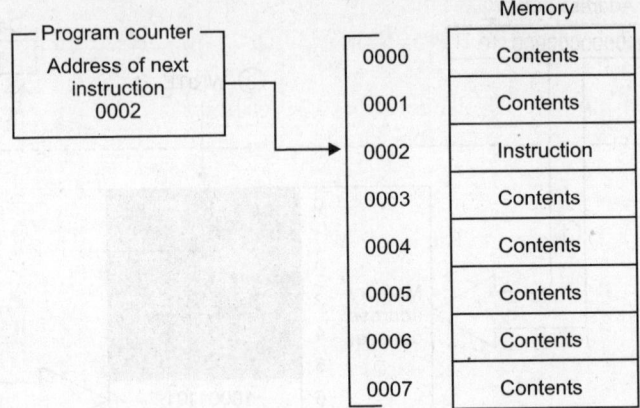

Fig. 12.19: Program-counter operation

is usually referred to as *fetching* the instruction. The period of time needed for this is called the *fetch cycle*.

12.11.4 Index Registers

An *index-register* is used to help locate a data. The index register is normally used as an aid in accessing data in tables stored in memory. The index register (s) can be incremented (increased by 1) or decremented (decreased by 1) but normally does not have other arithmetic or logical capabilities.

12.11.5 Status Register

The status register (also called the *condition code register* or *flag register*), is a special register which keeps track of certain facts about the outcome of arithmetic, logical and other operations.

This register makes it possible for the micro-processor to be able to test for certain conditions and then to perform alternate functions based on those conditions. This is done through the use of *flags*. The status register is divided into individual bits which have their own unique functions. Each bit is called a *flag*. Each flag keeps track of or "flags us" concerning certain conditions. Figure 12.20 shows a model of a typical status register.

When referring to flags, the following logic is used. If some condition has come to be or is true, the flag uses a 1 to say" yes, this is true or has happened". If that condition has not occurred, the flag uses a 0 to say "No, this is not true or has not happened" (causing

a flag to become 1 is called *setting* a flag. Causing a flag to become 0 is called *clearing* a flag)

Some of the commonly used flags are:

Carry: If a carry is generated from MSB as a result of certain operation, the carry flag is set 1 otherwise it is reset 0.

Zero: If the result of an operation is zero, the zero flag is set, otherwise it is reset.

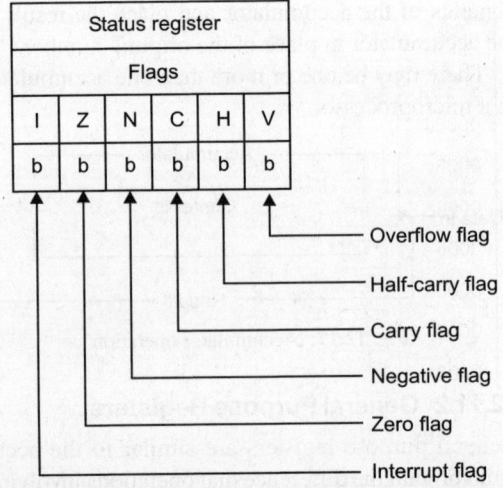

Fig 12.20: Model of a typical status register. b's represent bits

Sign: If the result of an operation produces 1 as MSB in the accumulator, the sign flag is set, otherwise it is reset.

Parity: If the result of an operation makes the parity of the bits in the accumulator even, the parity flag is set, otherwise it is reset.

Auxiliary carry (half-carry): If an operation produces carry out from the lower order four-bits, the auxiliary carry flag is set, otherwise it is reset. It is used for BCD arithmetic.

Overflow: Subtraction is performed using 2's complement representation of numbers. If the result of an operation produces overflow then the overflow flag is set, otherwise, it is reset.

12.11.6 Stack and Stack Pointer

The stack is a special place in memory. The stack is most often used to store certain critical pieces of data during subroutines and interrupts. The structure of the stack is a *first-in-last-out* (FILO) type of structure. Unlike main memory, where one can access any data item in any order, the stack is designated so that we can access only the top of the stack. If one wants to place the data in the stack, at must go on top. If one wishes to remove data from stack, it must be on top before it can be removed.

Structure of a stack is shown in Fig. 12.21. Let us see how this situation has come into being.

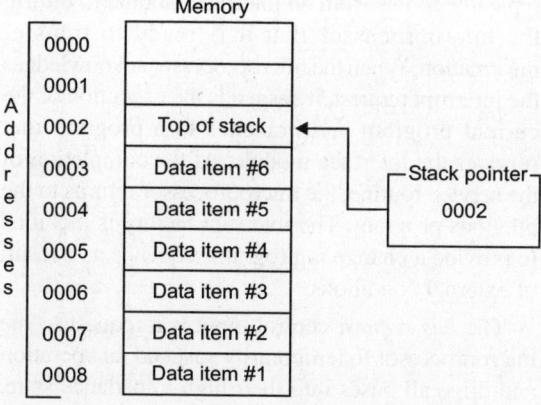

Fig. 12.21: Typical stack and stack pointer

To do that, refer to Fig. 12.22. Data item #1 is the first item we wish to place on the stack. At this time, the stack pointer is "pointing" to memory location 0008. Therefore data item #1 will be placed in the stack at that memory location. The act of putting a piece of data in the stack is called *pushing* data onto the stack.

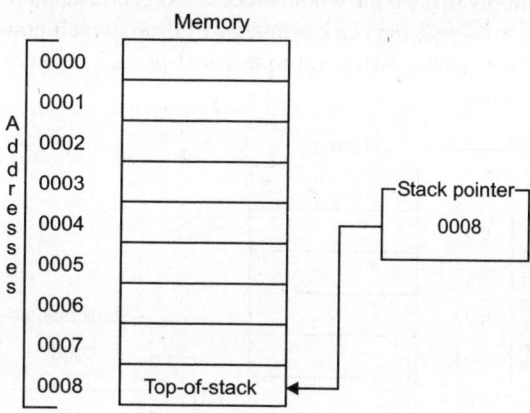

Fig. 12.22: Typical stack and stack pointer

It is as though the data is being pushed in from the top. Now look at Fig. 12.23. We have pushed data item #1 onto the stack and the stack pointer has been decremented or decreased by one, which means that it is now pointing to memory location 0007.

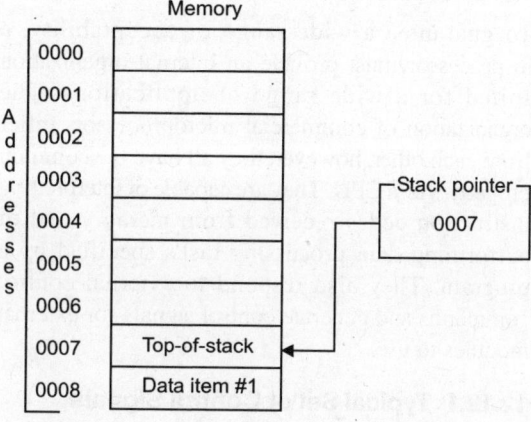

Fig. 12.23: Typical stack and stack pointer

Location 0007 is the top-of-stack now. Now let us push data item #2 onto the stack. The stack will appear as in Fig. 12.24. When data item #2 was *pushed* onto the stack, it went into the location the stack pointer was pointing to which was 0007. The stack pointer was then decremented to 0006. This process will be repeated until it appears as in Fig. 12.21.

At some point, we will need this data in the stack, so we will remove it from the top of the stack. This is called *popping* or *pulling* the data from the stack. We simply reverse the whole process. As each data item is removed, the stack pointer will drop, which now means that it will point to the next-greater memory address.

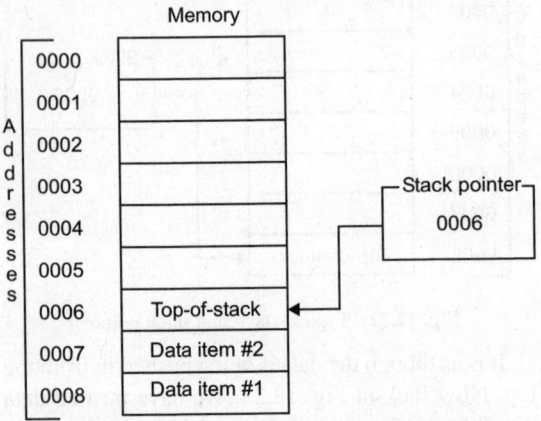

Fig. 12.24: Typical stack and stack register

12.12 MICROPROCESSOR ORGANIZATION

To guarantee a wide range of acceptability, a m-processor must provide an internal organization suited for a wide range of applications. The organization of commercial microprocessors differ from each other, however, they all have the common property, viz. a CPU. They are capable of interpretting instruction codes received from memory and of performing data processing tasks specified by a program. They also respond to external control commands and generate control signals for external modules to use.

12.12.1 Typical Set of Control Signals

Proper operation of a microprocessor requires that certain control and timing signals be provided to accomplish specific functions and in addition, other control lines be monitored to determine the state of the microprocessor. A typical set of control lines available in most microprocessor is shown in Fig. 12.25. Also shown are the data bus, address bus and input power supply.

The *clock* input is used by the microprocessor to generate multiphase clock pulses that provide timing and control for internal functions. (Sometimes, some microprocessors and external clock generator to supply the clock pulses). The clock pulses are used by external modules to synchronize their operations with the operations of the microprocessor.

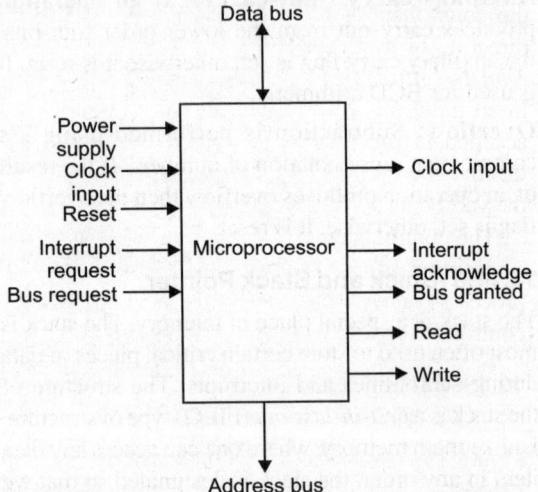

Fig. 12.25: Control signals in a microprocessor

The *reset* input is used to reset and start the microprocessor after power is turned on or any time the user wants to start the process from the beginning.

The *interrupt* request into the microprocessor typically comes from an interface module to inform the microprocessor that it is ready to transfer information. When the microprocessor acknowledges the interrupt request, it suspends the execution of the current program and branches to a program that services the interface module. At the completion of the service routine, the microprocessor returns to the previous program. The interrupt facility is included to provide a change in program sequence as a result of external conditions.

The *bus-request* control input is a request to the microprocessor to temporarily suspend its operation and drive all buses into their high-impedance state. When the request is acknowledged, the microprocessor responds by enabling the *bus granted* control output line. Thus when an external device wishes to transfer information directly to memory, it requests that the microprocessor relinquish control of the common buses. Once the buses are disabled, the device that originated the request takes control of the address and data buses to conduct memory transfers without processor intervention. (This feature is called *direct memory access*).

The read and write are control lines that inform the component selected by the address bus of the direction of transfer expected in the data bus. The *read* line informs the selected unit that the data bus is in an input mode and that the microprocessor will accept data from the data bus. The *write* line indicates that the microprocessor is in an output mode and that the valid data are available on the data bus.

The control signals given in Fig. 12.25 constitute a minimum set of control functions for a microprocessor and microprocessors may have additional control features for special functions.

12.13 INTERNAL ORGANIZATION OF A TYPICAL MICROPROCESSOR

Figure 12.26 shows the block diagram of a central processor unit (CPU) enclosed within a microprocessor chip. (This is similar to the 8085 microprocessor except that the F and G registers are called H and L in 8085).

Externally it provides a bidirectional data bus, an address bus and a number of control lines. Here, we have shown only the control lines associated with the bus transfer. The data bus is designated by the symbol DBUS and consists of eight lines. The information contained in the eight lines is called a *byte* (i.e. an 8-bit word). The address bus, designated by the symbol ABUS consists of 16 lines to specify $2^{16} = 64$ K (K = 1024) possible addresses. Thus the microprocessor is capable of communicating with a memory unit of 64 K bytes.

Internally, the microprocessor has six processor registers labeled B through G, an accumulator register designated as A and a temporary register T.

These registers are 8-bits wide and can accommodate a byte. The ALU operates on the data stored in A and T and the result of the operation is transferred to A or through an internal bus, to any one of the other six processor registers.

The status register holds the status bits of an operation (such as end carry from the ALU, the sign bit value and a zero result indication). The operation

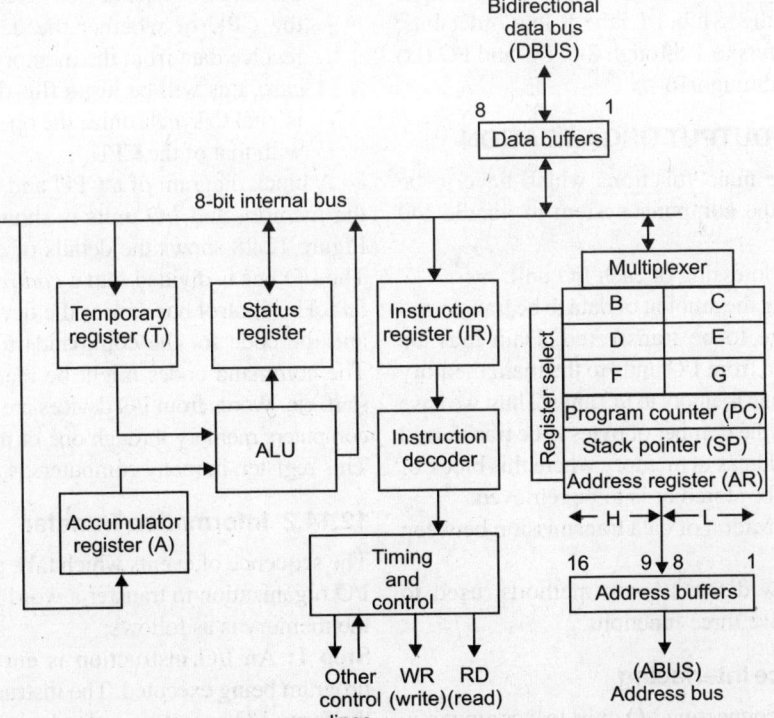

Fig. 12.26: Block diagram of a typical microprocessor

code of an instruction is transferred to the instruction register (IR), where it is decoded to determine the sequence of micro operations needed to execute the instruction. The timing and control supervise all internal operations in the CPU and the external control line in the microprocessor.

The address buffers receive information from three sources: the program counter (PC), the stack pointer (SP) and the address register (AR).

PC maintains the memory address of the current program instruction and is incremented after every instruction fetch. AR in used for temporary storage of addresses that are read from memory. SP is used in conjunction with a memory stack.

The address bus can also receive address information from a pair of processor registers. Three pairs can be formed to provide a 16-bit address. They are labeled with the combined register symbols BC, DE and FG. Each processor register contains 8 bits and when combined with the one adjacent to it, forms a register pair of 16 bits. It is sometimes convenient to partition the three 16 bit registers PC, SP and AR into two parts. The symbol H designates the 8 high-order bits and the symbol L, the 8 low-order bits. Thus PC (L) refers to 1 through 8 of PC and PC (H) refers to bits 9 through 16.

12.14 INPUT-OUTPUT ORGANIZATION

There are three main functions which have to be performed by the computer system to handle I/O units. These are:

(i) Unique addressing of each I/O unit.

(ii) Specifying the amount of data to be transferred and where to be transferred. Data may be transferred from I/O units to the main memory or from main memory to I/O units. Thus, we have to specify the number of bytes to be transferred and the address in memory where this block of data is to be stored or is to be retrieved.

(iii) Synchronization of data transmission between units.

We will now discuss some methods, used to perform the above three functions.

12.14.1 Device Interfacing

One method of connecting I/O units to a computer is to connect an I/O bus to the CPU and connect all devices to this I/O bus. Each I/O device has associated with it a device controller which controls the operation of the device.

The device controller sends to a device interface unit status signals and data and receives data and commands from it to control the device operation. The interface unit should have the following features.

(i) A device address decoder. Each device has a unique address. When the CPU wants to send data to a device or receive data from it then it places the device's address on the I/O bus. This address is decoded by the address decoder in the interface unit.

(ii) A data register in which the data to be moved to or from the memory is stored. In some devices such as a teletype, this may be a one byte register. In case of faster peripheral devices (such as magnetic disks) it would be a one word register and there will be another register to store the number of words to be moved in or out of the device.

(iii) A status register which is used to specify where the data from the device is ready to be read by the CPU or whether the device is ready to receive data from the memory. In the simplest case, this will be just a flip-flop. This register is vital to synchronize the operation of devices with that of the CPU.

A block diagram of a CPU and its connection to the memory and I/O units is shown in Fig. 12.27. Figure 12.28 shows the details of an interface unit. The I/O bus is divided into a *control bus* and a *data bus*. The control bus carries the device address code and the code for the commands to be carried out. The command codes might be read, write, rewind, start, etc. Words from I/O devices are transferred to the computers memory through one of the CPU registers. This register, in many computers is the accumulator.

12.14.2 Information Transfer

The sequence of events which take place in a simple I/O organization to transfer a word from a device to the memory is as follows:

Step 1: An I/O instruction is encountered in the program being executed. The instructions may be of the form: I/O operation code, device address, device command.

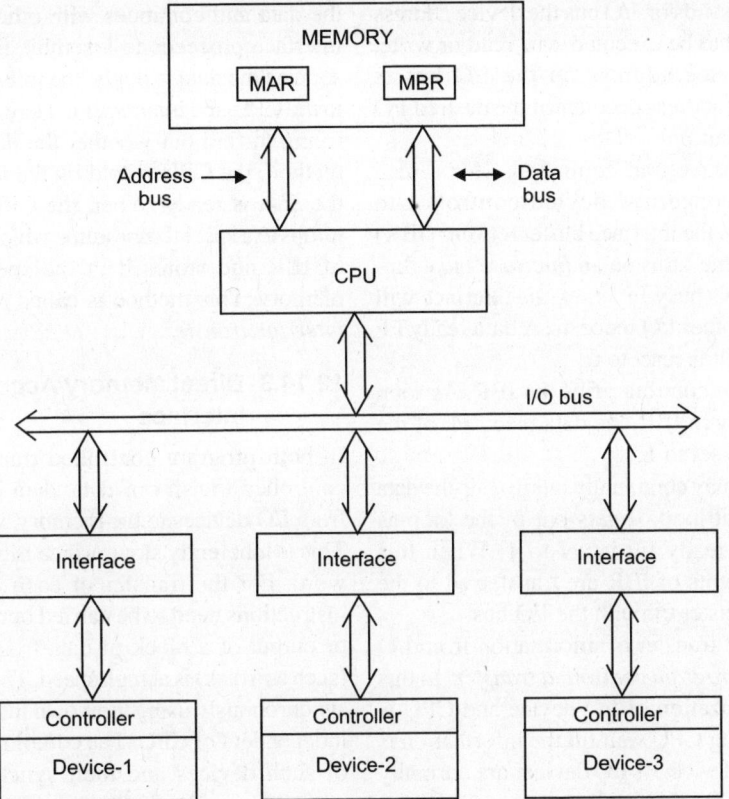

Fig. 12.27: Communication between CPU and peripheral devices

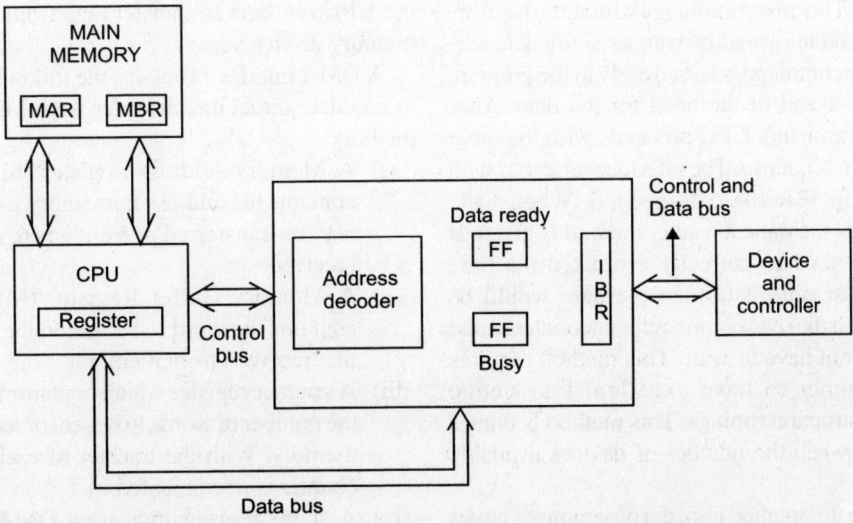

Fig. 12.28: Details of CPU-peripheral communication

Step 2: The CPU sends on I/O bus the device address and the command to be executed, viz. read or write.
Step 3: The device address on the I/O bus is recognized by the address decoder of the desired I/O device interface unit.
Step 4: The interface unit commands (for a read command) the concerned device controller to assemble a word in the interface buffer register (IBR) and at the same time turns on an *interface busy* flip-flop. As long as this busy FF is set, the interface will not entertain any other I/O requests. A data ready FF in the interface unit is reset to 0.
Step 5: The device controller fills the IBR. As soon as the data is ready in IBR, the data ready FF of the device interface is set to 1.
Step 6: The CPU may continually interrogate the data ready FF in a wait loop. It gets out of the loop as soon as the data ready FF is set to 1. When this happens, the contents of IBR are transferred to the specified CPU register through the I/O bus.

This method of transfer of information from I/O to CPU is called *program controlled transfer*. In this method, synchronization of I/O device and CPU is achieved by making CPU wait till the information is assembled by the device. As I/O devices are normally much slower (~ 1000 times) compared to CPU, this method of synchronization is not desirable as it wastes CPU time. There are two ways in which the CPU waiting time can be reduced.

Method 1: The programmer estimates the time required to read information from a specified device. A start reader command is issued early in the program several steps ahead of the need for the data. After giving this command, CPU proceeds with the other instructions in sequence. The interface proceeds with assembling in IBR the data needed. When CPU actually needs the data, a read command is given. If the programmer had correctly estimated the time needed to assemble data then the data would be available when the read command is encountered and CPU would not have to wait. This method requires the programmer to have excellent knowledge regarding instruction timings. This method becomes complicated when the number of devices available is large.

Method 2: In this method also, the programmer issues a start reader command well ahead of the need for

the data and continues with other instructions. The interface proceeds to assemble the data in IBR. As soon as the data is ready, the interface sends a signal to the CPU and *interrupts* it. Here, instead of the CPU trying to find out whether the data is ready, in this method, the CPU is told by the interrupt signal that the data is ready. When the CPU is interrupted, it jumps to a special subroutine which reads the contents of IBR and stores it in the specified location in memory. This method is called *program controlled interrupt transfer*.

12.14.3 Direct Memory Access (DMA) Interface

In both program controlled transfer and interrupt controlled transfer of data, data is transferred to or from I/O devices to the memory, viz. a CPU register. This is inherently slow as data is transferred word by word. For the transfer of each data word several instructions need to be carried out by the CPU. Input or output of a block of data from a fast peripheral (such as a disk) is at high speed. These devices operate synchronously using their own internal clock and are independent of CPU. The combination of high speed of such devices and their synchronous operation makes it impractical to use program controlled transfer on interrupt controlled transfer via CPU.

Another method which uses a Direct Memory Access (DMA) interface unit eliminates the need to use CPU registers to transfer data from I/O units to memory or vice versa.

A DMA interface contains the following registers to facilitate direct data transfer from I/O devices to memory:

(i) A Memory Address Register (MAR) which contains the address in memory to which data is to be transferred or from where data is to be received.

(ii) A Memory Buffer Register (MBR) which contains the word to be sent to the memory or that received from memory.

(iii) A counter register which contains the count of the number of words to be sent or received from memory. With the transfer of each word, the counter is decremented by I.

(iv) A status register indicating DMA busy/free, data ready and other status information.

(v) A register containing the I/O command to be carried out. This command is received from the CPU and stored in DMA to enable the DMA interface to carry out the command independent of CPU.

The configuration of the DMA interface and its connection to the CPU through buses is shown in Fig. 12.29. The DMA interface functions as follows:

Step 1: When an I/O instruction is encountered by the CPU, it sends through the control bus the device address and the I/O command to the DMA interface.

The address in memory where a word is to be stored or from where a word is to be received is also sent to DMA by the CPU. If a group of words are to be stored or retrieved, then the address of the first word in the group and the count of the number of words is sent by CPU to the DMA interface. All the information sent by the CPU to the DMA interface is stored locally in the appropriate registers by DMA.

Step 2: After sending the above information to the DMA interface the CPU continues with the next instruction in the program CPU resources are then not required.

Step 3: The DMA uses the device address to select the appropriate device and starts assembling the word in IBR. The DMA busy FF is set. The DMA cannot be accessed by CPU till this busy FF is reset.

Step 4: As soon as IBR is full, it is transferred to MBR of the memory. The address is transferred to the MAR of the memory. (The memory has two independent parts, one connected to the CPU through a pair of buses and the other to the DMA). This allows complete independence of CPU data transfer and I/O data transfer. A signal is sent to CPU by DMA indicating that the data is ready for transfer.

Step 5: In the case of dual ported memory, the DMA interface can transfer information to the memory as long as CPU is not accessing the same part of memory. In other words, if CPU is performing an operation not involving memory then DMA stores or retrieves a word from memory and CPU continues with its work. After the word is transferred, the word counter in DMA is decremented. The DMA then assembles the next word in IBR and performs steps 4 and 5 until the word count is zero.

The above steps are shown as a flow chart in Fig. 12.30.

12.15 MICROPROCESSOR OPERATION

Basically a microprocessor performs the following two operations.

   I. FETCH an instruction from memory, and

   II. EXECUTE the instruction.

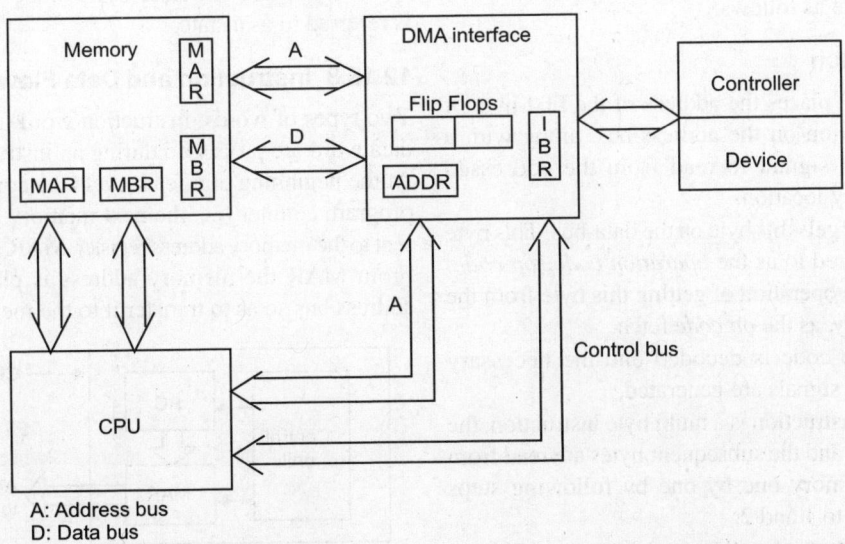

Fig. 12.29: Flow chart explaining DMA interface

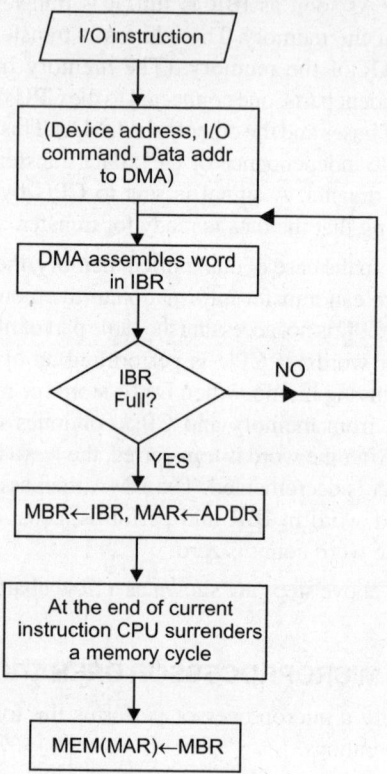

Fig. 12.30: The DMA interface

The various steps involved in performing these operations are as follows:

12.15.1 Fetch

1. The μP places the address of the first byte of instruction on the address-bus, along with a control signal, to read from the addressed memory location.
2. The μP gets this byte on the data-bus. This byte is referred to as the *operation code (op code)* and the operation of getting this byte from the memory, as the *op code* fetch.
3. The op-code is decoded and the necessary control signals are generated.
4. If the instruction is a multi byte instruction, the second and the subsequent bytes are read from the memory one by one by following steps similar to 1 and 2.

The operation of getting instruction bytes from the memory is known as *instruction fetch*.

12.15.2 Execute

After the μP gets the complete instruction (all the bytes of the instruction), it performs the operation specified by the instruction. This is referred to as *execution*.

Thus a μP fetches instructions from a memory, decode and execute them, i.e. perform certain arithmetic and logical operations, accepts data from input devices and sends the results to output devices.

Machine Cycle and State

The necessary steps which are carried out to access a memory or I/O device, constitute a machine cycle. In other words, necessary steps which are carried out to perform a fetch, read or write operation, constitute a *machine cycle*. An instruction cycle consists of a number of machine cycles. In one machine cycle only one operation such as opcode fetch, memory read. memory write, I/O read or I/O write is performed. The first machine cycle of an instruction cycle is an opcode fetch cycle. The single-byte instructions are executed in only one machine cycle. Two-byte and 3-byte instructions need more machine cycles as additional machine cycles are required for reading/ writing data from/into the memory or I/O devices.

A state (or T-state) is one subdivision of an operation performed in one clock period. These subdivisions are internal states synchronized with the system clock. So one clock cycle of the system clock is referred to as a state.

12.15.3 Instruction and Data Flow

Two types of words: instruction word (opcode) and data word are processed during an instruction cycle. At the beginning of a fetch cycle the contents of the program counter (i.e. the next memory address) are sent to the memory address register. MAR, (Fig. 12.31). From MAR the memory address is placed on the address bus so as to transfer it to the memory.

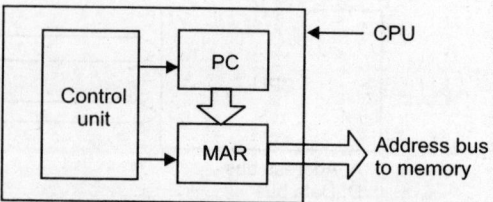

Fig. 12.31: Address transfer to memory

A read control is sent by the CPU to the memory. Having received the address, the memory places the opcode on the data bus. Then the opcode is received in the data register DR (Fig. 12.32). From DR the opcode is sent to the instruction-register, IR. Thereafter the opcode is decoded by the decoder circuitry and then executed.

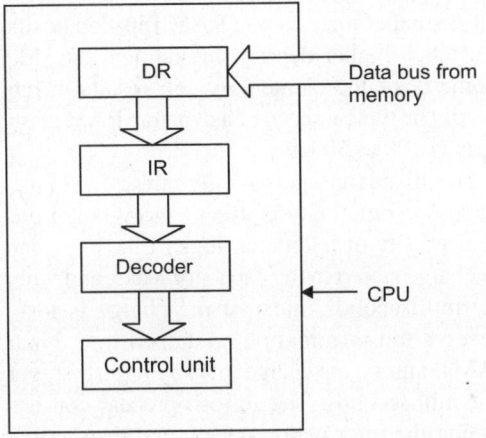

Fig. 12.32: Flow of Instruction word (Opcode)

Any data (or address) read from the memory is also received in data register. From the data register the data are transferred to the accumulator or any other general purpose register depending on the instruction, as shown in Fig. 12.33 If any data are sent to the memory from the CPU that data also flow through DR.

12.16 THE MEMORY

Two main types of memories used in microprocessors systems are RAM and ROM. RAM is a read/write memory. Its contents can be written to, or read from. Reading from a RAM is non-destructive, i.e. it will not destroy the information. The main disadvantage of an LSI RAM in the current state of the technology is that it is *volatile*, whenever the power disappears, the contents of the RAM will be lost. If power goes off then before the system can resume operation, it is necessary to reload the control program from another permanent storage medium such as a diskette or a casette.

The second type of memory is the ROM or read only memory. Once the contents of this memory have been defined by a manufacturing process, they can no longer be altered. The contents can be read but no new contents may be written. A ROM, therefore, is used to store essential programs. It is nonvolatile.

Thus, in an industrial control environment, programs are generally stored in ROM as the programs are seldom changed and should not be reloaded every time power is turned off. The same applies to the programs of single chip microcomputers which are normally used as dedicated control devices and manufactured in quantity.

In a professional or business environment, many different programs are being executed all the time and one program may fill all the available memory. Thus, programs are stored in RAM, which is easily changeable. A small ROM is also required for the

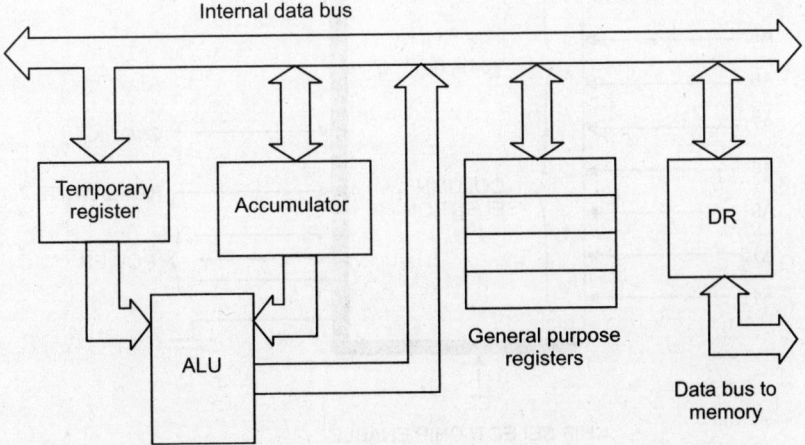

Fig. 12.33: Flow of data word

monitor program which maintains communication with the system via a keyboard and loads other programs from disk or tape into RAM. A typical memory configuration for an industrial control system is 4K ROM and 1K RAM. A typical memory configuration for a business system is 2K ROM and 62 K RAM (1K = 1024). Here K in this context refers to the number of bytes.

12.16.1 Random Access Memory

RAM stands for random access memory, and refers to the fact that any of its contents may be accessed at any time. This is contrast to a serial memory (such as a magnetic tape) where access to the stored data is only in a fixed order. Actually both ROM and RAM are random-access memories, but the terms random-access traditionally refers to read/write memories and is, therefore, only used with RAM. Two technologies are used for RAM memories: static and dynamic.

Static Versus Dynamic RAM

A *static RAM* stores a bit of information within a flip-flop. It is asynchronous and does not require a clock. The contents of a static RAM remain stable forever as long as power is available.

A *dynamic RAM* stores a bit of information as a change. A dynamic RAM uses the gate-substrate capacitance of an MOS transistor as an elementary memory-cell. An illustration of a typical dynamic RAM is depicted in Fig. 12.34. The obvious advantage of a dynamic RAM is that this elementary cell is smaller than a static RAM flip-flop, resulting in a much higher density. In addition, the simpler geometry of the elementary cell results in higher speed. The typical speed of a dynamic RAM memory today is 100 to 500 ns.

The disadvantage of a dynamic RAM is the increased complexity of the memory board due to the necessity of additional logic. Like any change, the charge stored in the capacitor leaks, and within a few milliseconds, most of the charge is lost. To preserve the information contained in a dynamic RAM memory, the charge must by refreshed every 1 or 2 milliseconds. The refresh process consists of reading the information out of, and then writing it back into the memory, thus restoring a full charge.

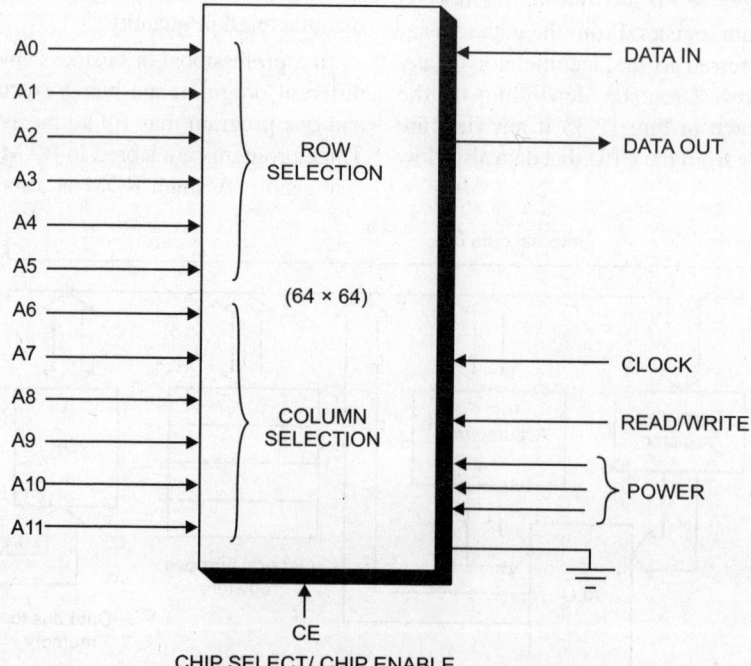

Fig. 12.34: A typical 4K dynamic RAM

To save time, the refresh process reads out a complete row or column at a time. In particular, a typical 4K dynamic RAM (Fig. 12.34), may have 64 columns × 64 rows. Thus, only 64 operations will be needed for the complete refresh of this memory.

12.16.2 Read Only Memory

Recall that a read only memory is a memory whose contents, once written, can only be read. Writing data in a ROM is generally called *programming* the ROM, since a program is what is usually written into it. However, programming here means that the specified bit patterns have been written into the memory. Because a read only memory is intrinsically non-volatile it is nearly always used to store *control programs*.

Four main types of ROMs are used—the pure ROM, the PROM, the EPROM and the EAPOM (electrically alterable ROM).

A ROM is a mask programmed read only memory that can only be produced by the manufacturer. The bit patterns corresponding to the desired contents must be supplied by the user in the standard format. The 0s and 1s are implemented on the memory by either establishing or not establishing connections between rows and columns.

The last manufacturing step or a ROM chip is the metallization steps by which these interconnections are established. Once the customer supplies the bit pattern, the manufacturer can realize a mask for the metallization step and perform this last step. Thus a ROM is said to be *mask-programmed*. Because of cost considerations, the manufacturer normally requires that about 1000 ROMs must be produced at one time. Further, a delay of from three to six weeks is usually required for production. Figure 12.35 shows the internal structure of a 16 K static ROM (the 8316A). It is organized in 2048 words of 8 bits. Its access time is 850 ns maximum. Masked ROMs have many advantages: high bit density, non-volatility and the lowest cost of any type of memory in large quantities.

12.16.3 Cache Memory

Cache is a very high speed small capacity memory introduced as a buffer between the CPU and the main memory. Parts of the current program (and data) are copied from the main memory into the high speed

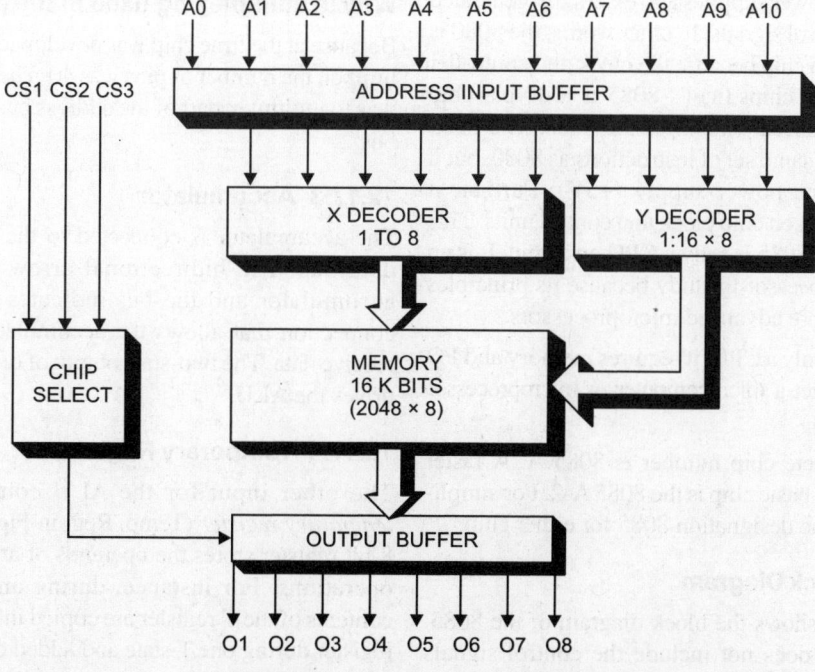

Fig. 12.35: A 16K static ROM (8316A)

cache. When the CPU refers to an instruction (or data), it is fetched from the cache if it is already there. This is called a *hit*. If the required address is not in cache, it is read from the main memory. When the probability of hit is very high, the CPU will be reading an instruction (or data) most of the time from the cache.

The hardware design is such that the existence of cache is not "visible" to a programmer. Typically a cache memory would be ~5 times faster than the main memory and about one hundredth its capacity. The CPU cycle time in a computer is normally much faster than the memory cycle time. Thus every time information is to be read from the memory for use by CPU, this speed mismatch would make the CPU wait. Introduction of cache is an attempt to minimize this waiting time.

12.17 THE 8085 (AN INTRODUCTION)

Intel Corporation introduced the 8080 in 1973. This 8-bit microprocessor set off the microcomputer explosion now taking place throughout the world. Although it was the most popular microprocessor of the early 70's it had several disadvantages, such as needing two power supplies and externally generated clock and control signals. In other words, the 8080 is not a CPU on a chip because the clock and controller are on separate chips. Intel's 8085 is a 40-pin chip that is an enhanced version of the 8080. The 8085 has almost the same set of instructions as 8080, but it needs only one power supply (+5V). Further, it includes its own on chip clock and control units. This means that the 8085 is truly a CPU on a chip. It is an ideal microprocessor to study because its principles are used in more advanced microprocessors.

Since it is only a CPU, it requires memory and I/O chips also to get a microcomputer or microprocessor based systems.

The complete chip number is 8085 A. A faster version of this basic chip is the 8085 A-2. For simplicity, we use the designation 8085 for either chip.

12.17.1 Block Diagram

Figure 12.36 shows the block diagram of the 8085. The drawing does not include the control signals driving each register. Since three state registers need load and enable signals to communicate properly along common bus, therefore, even though they are not shown, control signals drive all the interval registers in Fig. 12.36.

12.17.2 Address, Data and Control Buses in 8085

Near the top of the diagram is an 8-bit internal data bus. This carries instructions and data between the CPU registers.

The external buses are the ones we have to connect to other chips like memory, I/O and so on. Near the bottom left, is the external control bus (RD, WR, ALE......). On the bottom right are the external *address* and *address-data* buses.

The upper 8 address bits are on a separate bus always used for address-bits, this upper section of the address bus is designated A_{15}-A_8. The lower 8 bits are multiplexed. This means that the eight lower bus lines are used for address-bits during some T states and for data-bits during other T states. This is why the bus is labelled address-data bus, designated AD_7-AD_0.

Why is multiplexing used in the 8085?

Because at the time chip was developed, the practical limit on the number of pins was 40. The only solution was to multiplex part of the address bus with the data bus.

12.17.3 Accumulator

The accumulator is connected to the 8-bit internal data bus. The bidirectional arrow between the accumulator and the bus indicates a three-state connection that allows the accumulator to send or receive data. The two-state output of the accumulator drives the ALU.

12.17.4 Temporary Register

The other input for the ALU comes from the *temporary register* (Temp. Reg. in Fig. 12.36). This 8-bit register stores the operands of arithmetic logic operations. For instance, during an ADD C the contents of the C register are copied in the temporary register during one T state and added during another T state.

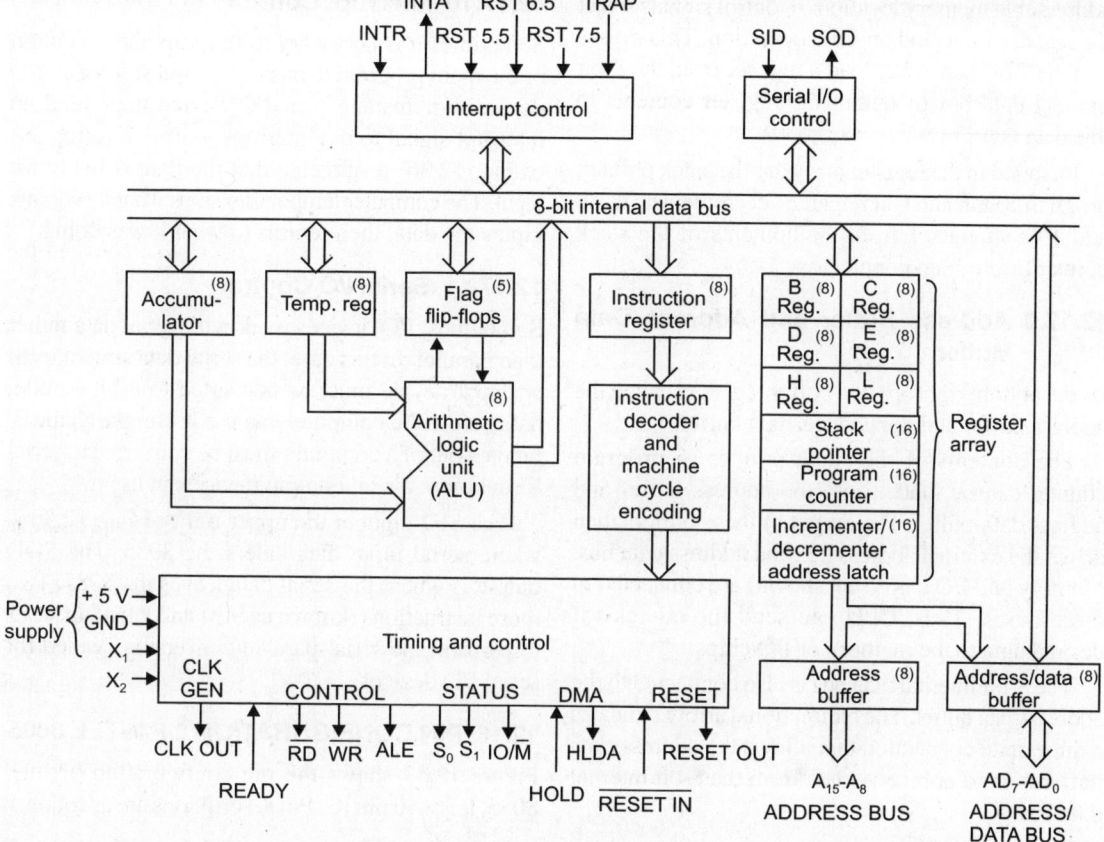

Fig. 12.36: The 8085 block diagram

12.17.5 ALU and Flags

The ALU carries out the arithmetic and logic operations. As shown, the contents of the accumulator and the temporary register are the inputs to the ALU. The ALU result then is stored back in the accumulator.

12.17.6 Instruction Register and Decoder

During the fetch cycle, the op code of an instruction is stored in the instruction register. This op code then drives the instruction decoder and machine cycle encoder.

12.17.7 Timing and Control

The timing and control section includes an oscillator and a controlled sequencer. The oscillator generates the two-phase clock signals (CLK and CLK) that synchronize all registers. The controller sequencer

also produces the control signals needed for internal and external control.

The controller-sequencer is micro-programmed; it has a ROM that stores all the micro-routines needed for executing the instructions. After each instruction is fetched and stored in the instruction register, the op code is decoded to get the starting address of the desired micro-routine.

As each microinstruction is read out of the control ROM, control signals are sent to the internal and external data buses. The effect is to move data between registers, to perform arithmetic-logic operations, to input or output data, etc.

12.17.8 CPU Registers

Notice the array of CPU registers (B, C, D, E, H, L). This register array is like a small on-chip RAM with addressable memory locations. Control signals select the register for a read or write operation. This means that the CPU can either load a register from the 8-bit internal data bus or output the register contents to this data bus.

Included in the register array are the stack pointer, program counter and incrementer-decrementer (it can add 1 or subtract 1 from the contents of the stack pointer or program counter).

12.17.9 Address Buffer and Address Data Buffer

At the bottom right are two buffer registers called the address buffer and the address data buffer.

The contents of the stack pointer or program counter can be loaded into the address buffer and address data-buffer. The output of these buffers then drives the external address bus and address data bus. Memory and I/O chips (not shown) are connected to these buses. Thus, CPU can send the address of desired data to the memory or I/O chips.

The 8-bit internal data bus is also connected to the address-data buffer. The bidirectional arrow indicates a three-state connection that allows the address-data buffer to send or receive data from the 8-bit internal data bus.

12.17.10 Interrupt Control

Sometimes it is necessary to interrupt the execution of the main program to answer a request from an I/O device. For instance, an I/O device may send an interrupt signal to the interrupt control unit (top left of Fig. 12.36) to indicate that the data is ready for input. The computer temporarily stops what it is doing, inputs the data, then returns to what it was doing.

12.17.11 Serial I/O Control

Sometimes, I/O devices work with serial data rather than parallel. In this case, the serial data stream from an input device must be converted to 8-bit parallel data before the computer can use it. Similarly, the 8-bit data out of a computer must be converted to serial form before a serial output device can use it.

The SID input at the upper right of Fig. 12.36 is where serial input data enters the 8085. The SOD output is where the serial data leaves the 8085. Two more instructions (known as SIM and RIM) allow us to perform the serial-parallel conversion needed for serial I/O devices.

12.18 PIN CONFIGURATION OF INTEL 8085

Figure 12.37 shows the pin configuration of Intel 8085. It is a 40 pin IC. Pin descriptions are as follows:

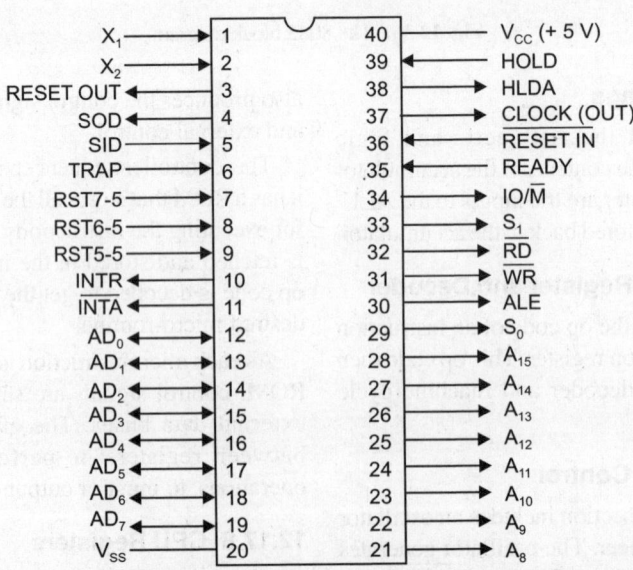

Fig. 12.37: Pin configuration of Intel 8085

A_8-A_{15} **(output):** These are address lines. They form the address bus. They carry 8 MSBs of the memory address or the 8 bits of I/O address.

AD_0-AD_7 **(input/output):** These are address/data lines. They form time multiplexed address/data bus, i.e. they serve dual purpose. They carry 8 LSBs of the memory address or I/O address during the first clock cycle of a machine cycle. Again they are used' to carry data during the second and third clock cycles.

ALE (output): It is an address latch enable signal. It goes high during the first clock cycle of a machine cycle and enables 8 LSBs of the address to get latched into the on-chip latch of peripherals.

IO/$\overline{M}$ (output): It is a status signal to indicate whether the address sent by the microprocessor is for a memory or an I/O device. When it is high the address on the address bus is for an 1/0 device. When it is low the address on the address bus is for the memory.

S_o and S_1 (output): These are status signals issued by the microprocessor to identify the various types of operations given in Table 12.4

Table 12.4: Status Codes for Intel 8085

S_1	S_o	Operations
0	0	HALT
0	1	WRITE
1	0	READ
1	1	FETCH

$\overline{RD}$ **(output):** It is a control signal sent by the microprocessor to control READ operation. The selected memory or I/O device is read when $\overline{RD}$ is low.

$\overline{WR}$ **(output):** It is a control signal issued by the microprocessor to control write operation. The data on the data bus are written into the selected memory or I/O device when $\overline{WR}$ is low.

HOLD (input): When an external device wants to use the address and data bus. it sends HOLD signal to the microprocessor. Having received the HOLD signal the microprocessor completes its current instruction at hand, and then relinquishes the control of buses to allow the external device to use them.

HLDA (output): It is a HOLD acknowledgement signal. It is sent by the microprocessor to the external device to indicate that the HOLD request has been received. On the completion of data transfer the external device removes the HOLD request. The HLDA goes low after the removal of HOLD request. The CPU takes over the control of buses half cycle after HLDA goes low.

INTR (input): It is an interrupt signal of the lowest priority.

INTA (output): It is an interrupt acknowledgement signal. It is issued by the microprocessor after INTR is received.

RST 5.5, 6.5, 7.5 and TRAP (input): These are interrupts. The TRAP is a nonmaskable interrupt and has the highest priority. Others are maskable interrupts. The order of priority is TRAP, RST 7.5, RST 6.5 and RST 5.5.

RESET IN (input): When this signal is applied the CPU is brought to reset condition. The contents of the program counter becomes zero.

RSET OUT (output): This signal indicates that the CPU is being reset.

READY (input): It is an input signal to the microprocessor. It is sent by a peripheral device to indicate whether it is ready to transfer data or not. The microprocessor examines READY signal before data are transferred. If READY is high it shows that peripherals are ready to transfer data. If READY is low the microprocessor waits till READY becomes high. The status of READY is examined in the second clock cycle of the machine cycle.

X_1 and X_2 (input): An external crystal oscillator is connected to these terminals to supply clock for the microprocessor. The crystal oscillator drives an internal circuitry which is within the microprocessor to produce a suitable clock for the operation of the microprocessor.

CLK (output): The clock is also required by some other ICs of the computer. Hence CLK is a clock output from the microprocessor, which can be utilized for the operation of other ICs.

SID (input): It is an input line for serial data. The data received from this are loaded into the 7th bit of the accumulator when RIM instruction is executed.

SOD (output): It is an output line for serial data. The 7th bit of the accumulator is sent through this line when SIM instruction is executed.

12.19 THE 8086 MICROPROCESSOR (AN INTRODUCTION)

The 16-bit 8086 microprocessor was introduced by Intel in 1978. It was fabricated using silicon-gate H-MOS process containing 29000 transistors on a chip of about 225 mils square. The high performance MOS technology resulted in a 5MHz clock rate, making it faster than any microprocessor available at that time.

The 8086 μP has the attributes of both 8-and 16 bit μPs. It has capabilities not available in 8-bit microprocessors such as 16-bit arithmetic, signed 8- and 16 bit arithmetic including multiply and divide, efficient interruptible byte-string operations and improved bit manipulation. It also includes operations such as reentrant code, position independent code and dynamically relocatable programs. The 8086 μP can directly address upto 1 mega byte (MB) of memory and can support multiple processor configurations.

12.19.1 The 8086 Architecture

Figure 12.38 shows internal block diagram of the 8086 microprocessor. It is divided into two units, the *bus interface unit* (BIU) and the *execution unit* (EU). Since the operation in this microprocessor is divided into two units, the speed of operation increases.

The Bus Interface Unit: The bus interface unit handles all transfers of data and addresses on the buses for fetching instructions from the memory, read and write operations for the memory and the I/O devices. It has a queue of 6 bytes for holding prefetched instruction bytes operating in the FIFO configuration. Due to this facility, the BIU can fetch instruction bytes

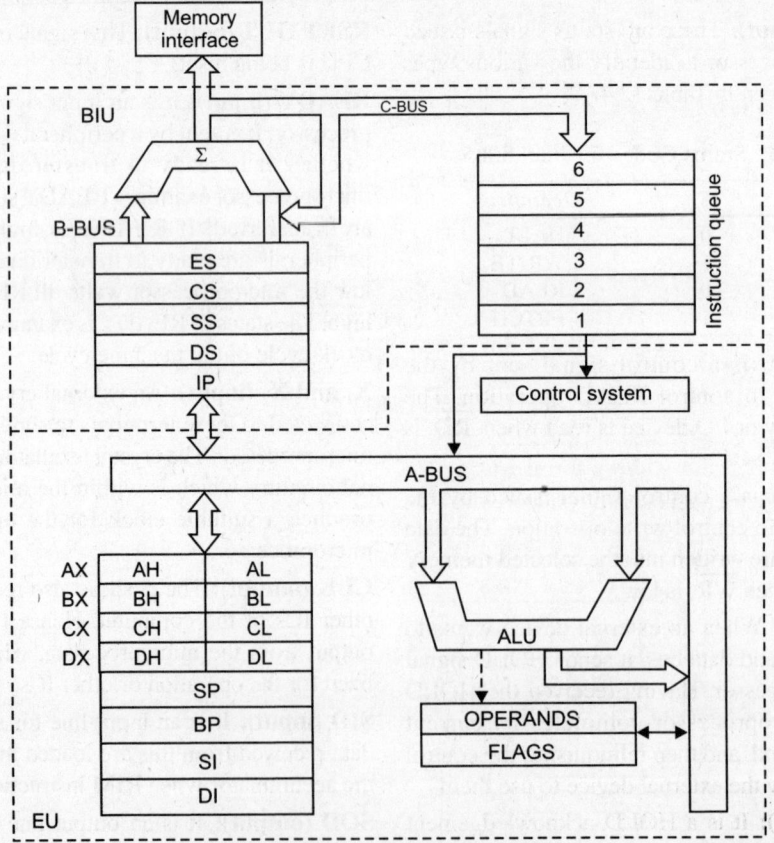

Fig. 12.38: Internal block diagram of the Intel 8086 μP

while the EU is decoding or executing an instruction, which does not require use of the buses.

The 16-bit data is time multiplexed with the sixteen bits of address for both memory and I/O (AD_{15}-AD_0).

The width of the address bus for memory referencing is 20-bit, consisting of A_{19}-A_{16} and AD_{15}-AD_{07} which can address $2^{20} = 1048576 = 1$ MB of memory. The memory is organized in the form of 64 KB segments. The method of addressing memory is as follows.

The BIU contains four 16-bit segment registers: The code segment (CS) register, the stack segment (SS) register, the extra segment (ES) register and the data segment (DS) register. These segment registers are loaded with the most significant 16 bits of the starting addresses of the corresponding memory segments which the 8086 is working with at a particular time. The BIU inserts zeros for the lowest four bits of the 20-bit starting address for a segment.

If the first byte of the word is at an even address, 8086 reads the entire word in one operation. On the other hand, if the first byte of the word is at an odd address, 8086 will read the first byte in one operation and the second byte, in another operation. For an even starting address of the word, the data byte with even address is transferred on the D_7-D_0 bus (lower byte) while odd-addressed data byte is transferred on the D_{15}-D_8 bus lines (higher byte).

Instruction-pointer (IP) register is a 16-bit register which holds the address of the next byte of the instruction code within the code segment of the memory. The value contained in the IP register is added to the 20-bit base address formed using CS register and it generates 20-bit physical address of the code-byte.

The Execution Unit: The execution unit (EU) decodes the instructions fetched from the memory and executes the instructions. It tells the BIU about the location from where to fetch instructions or data at any time. It contains control circuitry, 16-bit ALU, general purpose and index registers, flags, etc.

It contains eight 8-bit general purpose registers, which can also be used in pairs to form four 16-but registers. The 16-bit registers are labeled as AX (AH-AL), BX (BH-BL), CX (CH-CL) and DX (DH-DL) registers. The AX register is called the accumulator.

The stack-segment register (SS) contains the 16-bit (upper) part of the starting address of the stack and the stack-pointer register (SP) contains the 16 bit offset value. The physical address in the stack segment is obtained by shifting the contents of the SS register towards left by four bit positions, adding four zeros in the least significant four bit positions and adding the 16-bit contents of the SP in this 20-bit number.

The base pointer (BP), source index (SI) and destination index (DI) registers are used to hold the 16-bit offset of a data word in data segment (DS) register or extra segment (ES) register. Various addressing modes use these registers in different ways.

The 8086 µP contains nine flags. Six of them are conditional flags which indicate some condition produced by the execution of an instruction and the remaining three are used to control certain operations of the processor. The conditional flags are:

1. Carry flag   : CF
2. Parity flag   : PF
3. Auxiliary carry flag   : AF
4. Zero flag   : ZF
5. Sign flag   : SF
6. Overflow flag   : OF

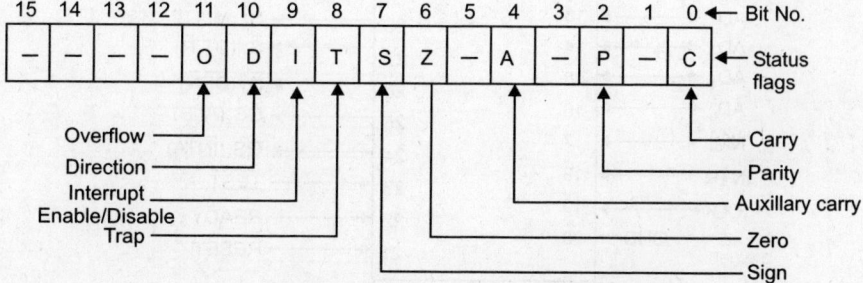

Fig. 12.39: Status flags of Intel 8086

The control flags are

1. Trap flag (TF): used for single stepping through a program
2. Interrupt flag (IF): used to allow/prohibit the interruption of a program
3. Direction flag (DF): used with string instructions.

12.20 PIN CONFIGURATION OF 8086

Figure 12.40 shows the pin diagram of 8086. The signals shown by the side of pins 24 to 31 are for the maximum mode of operation. The signals shown in the brackets against the pins 24 to 31 are for the minimum mode of operation. In Fig. 12.40 the direction shown for signals at pin 30 and 31 are for $\overline{RQ}/GT_0$ and $\overline{RQ}/GT_1$ in the maximum mode of operation. In the minimum mode of operation the direction of HOLD will be inward and that for HLDA outward. The 8288, a bus controller is designed to be used with 8086 in the maximum mode of operation. It generates control signals for memory and I/O devices such as memory read, memory write, I/O read, I/O write, interrupt acknowledge, etc.

12.20.1 Operating Principle

The internal architecture of 8086 is partitioned logically into two processing units: a bus interface unil (BIU) and an execution unit (EU). The memory and I/O devices connected to the 8086 are handled by BIU. Its function is to transfer instruction and data between the processor and peripherals. The function of the execution unit (EU) is to decode and execute instructions.

The BIU fetches opcodes from the memory, and maintains a 6-byte queue for the opcodes. Besides opcode fetching and maintaining a queue its other functions are address relocation, operand fetching and storing, and result storing. When EU desires I/O or memory access, it makes bus access request to BIU. If BIU is not currently busy, it acknowledges the request. The EU sends un-relocated operand address to BIU. The EU receives operand through BIU.

The ED receives the opcode from the queue, decodes and executes it. The queue is a first-in-first-out (FIFO) buffer. The queue is always filled-up by the BIU when at least two bytes of the queue fall

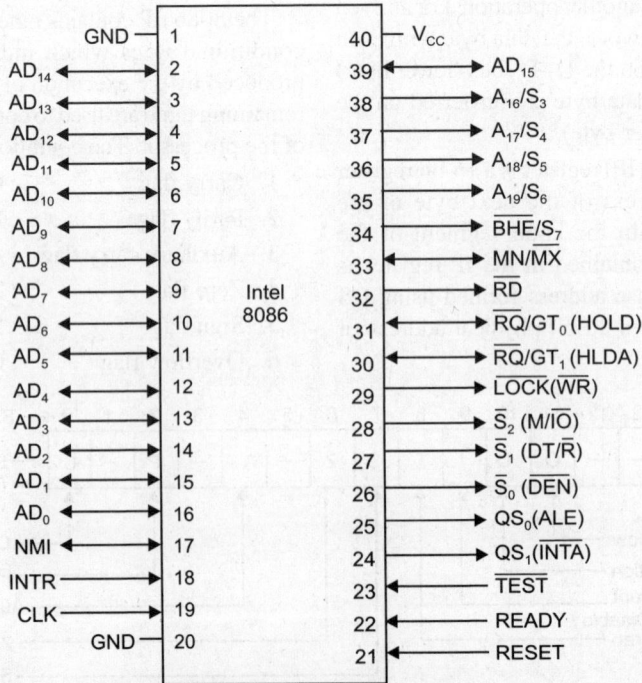

Fig. 12.40: Pin diagram of 8086

vacant. The operation of BIU and EU are independent of each other, but they can interact if required. While EU is executing instructions, the BIU fetches opcodes from the memory. The overlapping of execution of instructions and fetching of instructions makes 8086 a faster processor. Fetching of next instruction while the current instruction is being executed is called *pipelining*.

12.21 NEED FOR PROGRAMMING LANGUAGES

As is evident from the way microprocessors operate and information is stored and transferred between various subsystems, it understands only the binary signals (0 & 1) Therefore, the program is to be written in binary form which is referred to as the *machine language*. Machine language alone is directly executable by a microprocessor for programming in machine language; the programmer needs a table of binary instruction codes. Also he/she has to keep track of the contents of memory locations, registers and flags; of course, he/she should be thoroughly familiar with the details of the microprocessor. However, in machine-language program, it is difficult to determine whether a particular byte corresponds to an instruction, a data or an address. So working with binary numbers is not convenient and we are liable to commit mistakes.

The machine-language program listing can be simplified somewhat by converting the binary codes to octal or hexadecimal format. However, it also does not improve the understandability of the program. Further, the program has to be converted lack to the machine language by some means before it is loaded into memory.

A better alternative to machine language is *assembly language*. In this, names are assigned to instructions, registers, data and memory locations. These names are usually the abbreviations of the names or descriptions of the instructions (mnemonic), addresses or data and are used to aid the designers memory.

The assembly-language program is much easier to write, modify, understand and interpret than the corresponding machine-language program.

12.22 HIGH LEVEL LANGUAGES

Machine language and assembly language are referred to as *low-level languages* because only very simple instructions exist. Over the course of time, people working with computers felt it would be helpful to create languages that were more like English so that it would be easier to communicate with the computer and as a result, were called *high-level languages*, e.g. commands like *print, run, do, next* and *end*. The microprocessor does not understand these English words, but the language changes (enables to interpret or compile) them into machine language before sending them to the microprocessor.

Many high-level languages have been created over the years. FORTRAN (*formula translation*) is a language that handles high-level mathematics very well and is designed for scientists and engineers.

COBOL, which stands for *common business-oriented language*, is tailored to the needs of business.

BASIC, which stands for beginner's all-purpose symbolic instruction code, was designed to be easy for non professional programmers to learn and use.

Pascal, named for the French mathematician Blaise Pascal, is designed to encourage the programmer to adhere to what are considered "correct" programming practices.

There are some languages that are somewhat "in between" the high-level and low-level languages, most notably are C and FORTH.

EXERCISE

1. Define microprocessor in brief.
2. Describe the architecture of microprocessor with diagram.
3. What are the basic operations in a microprocessor?
4. Describe 8-bit microprocessor 8085 and give its block diagram
5. Describe 16-bit microprocessor 8086 with its functional diagram.
6. Explain buses and their types.
7. Describe 16-bit microprocessor 8086 and explain its internal architecture.
8. Discuss the function of a CPU. What are the main sections of a CPU? Discuss the function of each section.

9. What is the purpose of providing registers in a CPU? Describe various registers which are usually provided in a microprocessor.

10. Explain the requirements of a program counter, a stack pointer and status flags in a microprocessor.

11. Explain instruction cycles, machine cycles and states.

12. What are fetch cycles and execute cycles?

13. What status flags are normally provided in a microprocessor?

14. Discuss the function of an index register, memory address register (MAR) and memory buffer register (MBR).

References

1. Anand Kumar A. *Fundamentals of Digital Circuits*, Prentice Hall of India, New Delhi, 2001.
2. Bartee Thomas C. *Digital Computer Fundamentals*, Tata McGraw-Hill, New Delhi, 1985.
3. Floyd Thomas L. *Digital Fundamentals,* Universal Book, New Delhi, 2000.
4. Jain R.P. *Modern Digital Electronics*, Tata McGraw-Hill, New Delhi, 1997.
5. Malvino A.P. and Brown. *Digital Computer Electronics,* Tata McGraw-Hill, New Delhi, 1995.
6. Malvino Albert P. and Leach Donald P. *Digital Principles and Applications*, McGraw-Hill, Singapore, 1981.
7. Millman Jacob and Halkias Christos C. McGraw-Hill, Kogakusha, 1985.
8. Morris Mano, M. *Digital Logic and Computer Design*, Prentice Hall of India, New Delhi, 2001.
9. Morris Robert L. and Miller John R. (Eds.) *Designing TTL Integrated Circuits*, McGraw-Hill, Singapore, 1971.
10. Rajaraman V. and Radhakrishnan T. *An Introduction to Digital Computer Design*, Prentice Hall of India, New Delhi, 1988.
11. Ram B. *Computer Fundamentals*, New Age International (P) Ltd., New Delhi, 1999.
12. Sinha Pradeep K. and Sinha Priti. *Computer Fundamentals*, BPB, New Delhi, 2005.
13. Taub H. and Schilling D. *Digital Integrated Electronics*, McGraw-Hill, Singapore, 2001.
14. Tocci Ronald J. *Digital Systems, Principles and Applications*, Prentice Hall of India, New Delhi, 2000.
15. Zaks Rodnay. *An Introduction to Microprocessor, from Chips to Systems*, BPB, New Delhi, 1985.

Index